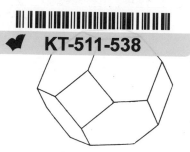

Physical Properties of Semiconductors

Charles M. Wolfe
Department of Electrical Engineering
Washington University

Nick Holonyak, Jr.
Department of Electrical Engineering
University of Illinois—Urbana

Gregory E. Stillman
Department of Electrical Engineering
University of Illinois—Urbana

Prentice-Hall International, Inc.

ISBN: 0-13-669995-2

 © 1989 by Prentice-Hall, Inc.
A Division of Simon & Schuster
Englewood Cliffs, NJ 07632

Printed in the United States of America

10 9 8 7 6 5 4 3 2 1

ISBN 0-13-669995-2

Prentice-Hall International (UK) Limited, *London*
Prentice-Hall of Australia Pty. Limited, *Sydney*
Prentice-Hall Canada Inc., *Toronto*
Prentice-Hall Hispanoamericana, S.A., *Mexico*
Prentice-Hall of India Private Limited, *New Delhi*
Prentice-Hall of Japan, Inc., *Tokyo*
Simon & Schuster Asia Pte. Ltd., *Singapore*
Editora Prentice-Hall do Brasil, Ltda., *Rio de Janeiro*
Prentice-Hall, Inc., *Englewood Cliffs, New Jersey*

Contents

3   Crystal Imperfections   71

4   Equilibrium Distributions   112

5   Transport Properties   139

6   Scattering Processes   175

7   Optical Properties   203

8   Excess Carriers   236

9   Heterostructures   275

10   Surface Structures   312

Appendices

A   Semiconductor Properties (300 K)   339

B   Fermi–Dirac Integrals   343

List of Symbols

a	cubic unit cell dimension (lattice constant), atomic spacing	e	electron heat content
$\mathbf{a}$	acceleration, dislocation axis	e_{pz}	piezoelectric constant
		e^*	Born effective charge
$\mathbf{a}_i$	direct lattice primitive vectors	e_c^*	Callen effective charge
		$\mathcal{E}$	one electron energy
$\mathbf{A}$	vector potential	$\mathcal{E}'$	total electron energy, total energy of system
$\mathbf{b}$	Burgers vector		
$\mathbf{b}_i$	reciprocal lattice primitive vectors	$\mathcal{E}_a$	acceptor energy
		$\mathcal{E}_c$	conduction band minimum energy
$\mathbf{B}$	magnetic flux density		
c	velocity of light (2.998×10^8 m/s)	$\mathcal{E}_d$	donor energy
		$\mathcal{E}_f$	Fermi energy
c_{ij}	elastic constants	$\mathcal{E}_g$	forbidden energy gap ($\mathcal{E}_c - \mathcal{E}_v$)
C	capacitance per unit area		
$\mathbf{d}$	basis vectors for atoms in unit cell	$\mathcal{E}_i$	impurity energy
		$\mathcal{E}_v$	valence band maximum energy
D_n	electron diffusion constant		
D_p	hole diffusion constant	f	nonequilibrium Fermi–Dirac distribution function
$\mathbf{D}$	electric flux density	f_0	equilibrium Fermi–Dirac distribution function

$\mathbf{F}$	force	N_a	total acceptor atom concentration
$F_j(n)$	Fermi–Dirac integral of order j	N_c	effective conduction band density of states
$\mathscr{F}$	electrothermal field	N_d	total donor atom concentration
h	Planck's constant (6.623×10^{-34} J-s)	N_s	surface states per unit area
$\hbar$	$h/2\pi$	N_v	effective valence band density of states
h, k, l	Miller indices	η	refractive index
$\mathbf{H}$	Hamiltonian, magnetic field	p	valence band hole concentration
$\mathbf{J}$	electric current density	$\mathbf{p}$	electron momentum
k	Boltzmann's constant (1.38×10^{-23} J/K)	$\mathbf{P}$	crystal momentum, polarization vector
$\mathbf{k}$	electron wavevector	$P(\mathscr{E})$	probability
$\mathbf{K}$	reciprocal lattice vectors	P_E	Ettinghausen coefficient
L_n	electron diffusion length	P	Thomson heat
L_p	hole diffusion length	$\mathscr{P}$	thermoelectric power
m	free electron mass (9.11×10^{-31} kg)	q	electronic charge, absolute value (1.602×10^{-19} C)
m^*	general effective mass		
m_c	conductivity effective mass	$\mathbf{q}$	photon wavevector
m_{de}, m_{dh}	density-of-states effective mass	$\mathbf{q}_s$	phonon wavevector
m_e	electron effective mass	Q	total charge
m_h	hole effective mass	Q_N	Nernst coefficient
m_h	heavy-hole effective mass	$\mathbf{r}$	direct space vector
m_l	effective mass longitudinal to axis of revolution	r_H	Hall factor
m_l	light-hole effective mass	R	reflectivity, resistance
m_t	effective mass transverse to axis of revolution	$\mathbf{R}$	Bravais or direct lattice vector
M	multiplication factors	R_H	Hall constant
n	conduction band electron concentration	$\mathbf{s}$	electron spin
n_i	intrinsic concentration	S	entropy
n'	total number of electrons	$\mathbf{S}$	Poynting vector ($\mathbf{E} \times \mathbf{H}$)
N	number of primitive unit cells in a crystal	S_{RL}	Righi–Leduc coefficient
		T	temperature, kinetic energy

$\mathcal{T}$	Thomson coefficient	$\lambda_{e,n,p}$	extrinsic Debye length
$\mathbf{u}$	displacement	λ_i	intrinsic Debye length
U	potential energy	μ	chemical potential energy
$\mathbf{v}$	velocity	μ_c	conductivity mobility
V	volume of crystal, voltage	μ_H	Hall mobility
V_0	built-in potential	μ_n	electron conductivity mobility
W	space charge width		
$\mathbf{W}$	heat flow density	μ_0	permeability of free space $(4\pi \times 10^{-7}$ H/m)
$\hat{x}, \hat{y}, \hat{z}$	Cartesian unit vectors		
Z	charge state of an impurity or defect	μ_p	hole conductivity mobility
		Π	Peltier coefficient
α	absorption coefficient	ρ	charge density, resistivity, mass density
$\alpha(x)$	electron impact ionization coefficient		
		σ	scattering cross section, conductivity
$\beta(x)$	hole impact ionization coefficient		
$\gamma(x)$	$\alpha(x) - \beta(x)$	τ_d	dielectric relaxation time
Δ	spin-orbit splitting energy	τ_e	energy relaxation time
$\Delta\mathcal{E}_a$	acceptor ionization energy $(\mathcal{E}_a - \mathcal{E}_v)$	τ_m	momentum relaxation time
$\Delta\mathcal{E}_d$	donor ionization energy $(\mathcal{E}_c - \mathcal{E}_d)$	τ_n	excess electron lifetime, recombination lifetime
ϵ	permittivity ($\epsilon = \epsilon_r\epsilon_0$)	τ_p	excess hole lifetime, recombination lifetime
ϵ_0	permittivity of free space $(8.85 \times 10^{-12}$ F/m)	ϕ_B	potential barrier
$\epsilon_r, \epsilon(\infty)$	dielectric constant	Φ	work function
ζ	electrochemical potential energy ($\mu + q\psi$)	χ	electron affinity
		ψ	electrostatic potential
ζ_n	electron quasi-Fermi energy	$\psi_k(\mathbf{r})$	electron wavefunction
ζ_p	hole quasi-Fermi energy	ω	photon frequency, radial frequency
η	$(\mathcal{E}_f - \mathcal{E}_c)/kT$ for conduction band, $(\mathcal{E}_v - \mathcal{E}_f)/kT$ for valence band	ω_s	phonon frequency
		Ω	volume of primitive unit cell in direct lattice
κ	thermal conductivity	Ω_k	volume of reciprocal lattice for every allowed value of wavevector k
λ	de Broglie wavelength, characteristic length, free space wavelength		
		Ω_K	volume of primitive unit cell in reciprocal lattice

Preface

This book was developed as an intermediate-level textbook on semiconductor physics for electrical engineering students. As such, it should be suitable for beginning graduate students, advanced seniors, and others with a basic knowledge of the subject. At the University of Illinois and Washington University the material has been used as the basis for one- and two-semester graduate courses, respectively, which are prerequisites for other advanced courses in optoelectronic or high-speed devices. Generally, the students in these courses have a good background in quantum and electromagnetic theory and have been exposed to semiconductors at the level of Ben G. Streetman, *Solid State Electronic Devices*, 2nd ed. (Englewood Cliffs, N.J.: Prentice-Hall, 1980). Some knowledge of thermodynamics is also useful.

We anticipate that some readers of this textbook may find the treatment to be rather uneven, with some subjects in less rigorous form and others in greater detail than they would prefer. Other readers may find no reference to their favorite topics. Some of these problems, real or perceived, may be due to inattention on our part or may simply reflect our preconceived notions as to which subjects are most important for engineering students. Other unevenness may be for pedagogical reasons. For example, we have found that students are almost invariably confused by the following concepts: (1) What is a hole? (2) What are the meanings of the terms *Fermi level, chemical potential, electrochemical potential*, and *quasi-Fermi level*? (3) How do you treat excess carrier distributions in and out of thermal equilibrium? (4) How do you construct energy band diagrams for nonuniform materials? (5) What are the driving forces in heterostructures? (6) What happens to the majority carriers when minority carriers are injected? These are some of the topics we have tended to dwell on, perhaps at the expense of others. Clearly, the advantage of and reason for writing a book is to present the perspective of the authors.

Of course, persons other than the authors have made important contributions to the development of this text. Professors Gary A. Davis and Karl Hess have tested the contents in their courses and have made many useful suggestions. Discussions with Professors Marcel W. Muller and Daniel L. Rode helped to clarify many concepts. A number of our graduate students have helped with the problem solutions, and we wish to take this opportunity to thank them. (A solution manual is available.) The manuscript in its many forms was typed by Mrs. Janet W. Gittelman and we thank her for her persistence and patience.

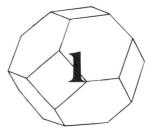

Crystal
Properties

The elements and their compounds which aggregate into the solid state can be classified as amorphous, polycrystalline, or single-crystalline materials. The distinction among these three classes of solids depends on the arrangement of atoms in the material. When the atoms in the material are arranged in a regular manner with a three-dimensional periodicity that extends throughout a given volume of the solid, the material is considered to be a single crystal. In polycrystalline materials the periodic arrangement of atoms is interrupted randomly along two-dimensional sections that can intersect, dividing a given volume of solid into a number of smaller single-crystalline regions or grains. The size of these grains can be as small as several atomic spacings. If, however, there is no periodicity in the arrangement of atoms (the periodicity is of the same size as the atomic spacings), the material is classified as amorphous. The difference among these three classes of solids is shown schematically for a two-dimensional arrangement of atoms in Fig. 1.1.

Although semiconducting properties are observed in all three classes of solids, we will restrict our attention to semiconducting materials in single-crystalline form. There are important theoretical and practical reasons for doing this. Theoretically, when we consider that the spacing between nearest-neighbor atoms in a solid is typically several angstroms ($1\text{Å} = 10^{-8}$ cm), we find that there are 10^{22} to 10^{23} atoms per cubic centimeter. If this enormous number of atoms were arranged randomly in the material, it would be very difficult to construct a useful physical theory of semiconductor behavior. In single crystals, however, the theoretical problems are reduced to

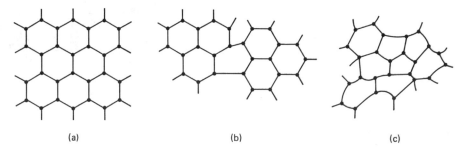

Figure 1.1 Schematic of the difference between (a) a single-crystalline, (b) a polycrystalline, and (c) an amorphous material.

manageable size and we find that many of the important properties of solids are actually determined by the periodicity of the atoms. Practically, the use of single crystals greatly simplifies a number of the processing steps (etching, diffusion, etc.) used in the fabrication of semiconductor devices and permits the high device yields that are characteristic of modern integrated-circuit technology. Also, charge carriers in single crystals exhibit properties that are very useful in device operations. Thus, most useful semiconductor devices are fabricated with single-crystalline material.

In this chapter we consider in detail the crystal structures of the most important semiconductors. The approach we take is to assume that the material is perfectly periodic with no deviations from its periodicity. This, of course, is an idealization since even a perfect single crystal must have surfaces, and some of the most useful physical properties of semiconductors are obtained by introducing defects into the crystal structure (doping). It is, therefore, worthwhile to examine the assumption of perfect periodicity. Considering the surface atoms, if the material has 10^{23} atoms in a centimeter cube, only about 1 atom in 10^8 is on the surface. In many applications, intentionally added impurities produce the dominant deviation from perfect periodicity. Typically, this doping would result in at most 1 impurity atom in 10^3. Thus, in most instances, it is reasonable initially to treat the material analytically as a perfect crystalline structure and later to introduce small perturbations to account for deviations from periodicity. We consider such perturbations in Chapter 3.

1.1 SIMPLE LATTICES

Although no semiconductors crystallize into simple lattices, they form the basis for understanding the more complicated semiconductor structures. We will use them to illustrate some of the more important concepts involved in forming a mathematical description of the crystal lattice (see Fig. 1.2).

A concept most useful in specifying the underlying geometry of a crys-

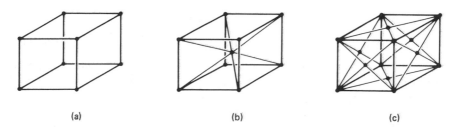

Figure 1.2 (a) Simple cubic, (b) body-centered cubic, and (c) face-centered cubic structures.

tal structure is the Bravais lattice. A *Bravais lattice* is the infinite matrix of points which, together with the atoms or molecules situated at the points, form the crystal structure. It has the property that the arrangement of lattice sites around any given lattice site is the same as that around any other site. Mathematically, a Bravais lattice consists of all points generated by the vectors

$$\mathbf{R} = \sum n_i\mathbf{a}_i, \qquad i = 1, 2, 3 \tag{1.1}$$

where the $\mathbf{a}_i$ are noncoplanar vectors and the n_i take on all integer values. The $\mathbf{a}_i$, which generate the Bravais lattice, are known as primitive vectors.

In the simple cubic structure, which has an atom at each corner of a cube of dimension a, the Bravais lattice can be determined by three mutually orthogonal vectors, each of amplitude a. As indicated in Fig. 1.3, these vectors are

$$\mathbf{a}_1 = a\hat{x}, \qquad \mathbf{a}_2 = a\hat{y}, \qquad \mathbf{a}_3 = a\hat{z} \tag{1.2}$$

where $\hat{x}$, $\hat{y}$, and $\hat{z}$ are Cartesian unit vectors. This set of vectors demonstrates the basic symmetry of the structure, and it is easy to see that the entire Bravais lattice can be constructed with these vectors and (1.1). This set of primitive vectors is not unique, however, in defining the simple cubic Bravais

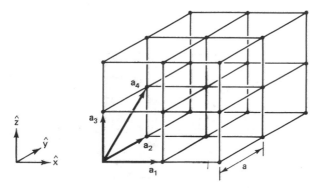

Figure 1.3 Simple cubic Bravais lattice with two sets of primitive vectors.

lattice. For example, the set of vectors

$$\mathbf{a}_1 = a\hat{x}, \qquad \mathbf{a}_2 = a\hat{y}, \qquad \mathbf{a}_4 = a(\hat{y} + \hat{z}) \qquad (1.3)$$

can also be used to construct the lattice as well as an infinite number of other sets. Since it is generally desirable to use primitive vectors which illustrate the symmetry of the structure, the set of vectors defined by (1.2) is preferred for the simple cubic Bravais lattice.

The body-centered cubic (bcc) structure has an atom at each corner of a cube of dimension a and one at the point determined by the intersection of the cubic body diagonals. The Bravais lattice for this structure is shown in Fig. 1.4 with the most symmetric set of primitive vectors. These are given by

$$\mathbf{a}_1 = (a/2)(-\hat{x} + \hat{y} + \hat{z}), \qquad \mathbf{a}_2 = (a/2)(\hat{x} - \hat{y} + \hat{z}), \qquad \mathbf{a}_3 = (a/2)(\hat{x} + \hat{y} - \hat{z})$$

$$(1.4)$$

From the figure it can be seen that the body-centered cubic lattice can also be regarded as two interpenetrating simple cubic lattices, each with cube dimension a.

The face-centered cubic (fcc) Bravais lattice shown in Fig. 1.5 is the most important lattice for semiconductor crystal structures. It consists of lattice sites at the cube corners, with one at each point determined by the intersection of the cubic face diagonals. The most symmetric set of primitive vectors is

$$\mathbf{a}_1 = (a/2)(\hat{y} + \hat{z}), \qquad \mathbf{a}_2 = (a/2)(\hat{x} + \hat{z}), \qquad \mathbf{a}_3 = (a/2)(\hat{x} + \hat{y}) \qquad (1.5)$$

Another lattice of interest in semiconductor crystal structures is the hexagonal close-packed lattice. Although not a Bravais lattice, because the lattice sites are not equivalent, it consists of two interpenetrating simple hexagonal lattices which are Bravais lattices. The simple hexagonal lattice consists of lattice sites at each corner of an equilateral triangle of side a, with an additional set of points on a triangle at a distance c above the first.

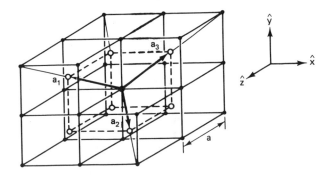

Figure 1.4   Body-centered cubic Bravais lattice with symmetric primitive vectors.

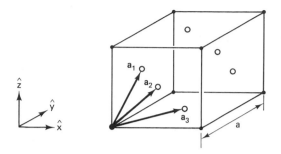

Figure 1.5 Face-centered cubic Bravais lattice with symmetrical primitive vectors.

A set of primitive vectors is

$$\mathbf{a}_1 = a\hat{x}, \qquad \mathbf{a}_2 = (a/2)\hat{x} + \frac{\sqrt{3}}{2}\,a\hat{y}, \qquad \mathbf{a}_3 = c\hat{z} \qquad (1.6)$$

The hexagonal close-packed lattice has two simple hexagonal lattices, with one displaced from the other by the vector $\frac{1}{3}\mathbf{a}_1 + \frac{1}{3}\mathbf{a}_2 + \frac{1}{2}\mathbf{a}_3$, as shown in Fig. 1.6(b). For non-Bravais lattices this is called the *basis vector*. Thus the lattice sites of one lattice are arranged to be halfway between the sites of the other, with each site directly below or above the center of the triangle formed by the sites of the other lattice.

Another important concept in the mathematical description of crystals is that of the primitive unit cell. A *primitive unit cell* is defined as that volume of space which, when translated by all the Bravais lattice vectors of (1.1), exactly fills the space of the Bravais lattice. As with sets of primitive vectors there is no unique primitive unit cell for a given Bravais lattice. However, a primitive unit cell must contain exactly one lattice site, so the volume of the cell is independent of how it is chosen.

The most convenient primitive unit cell to visualize is that for the simple cubic Bravais lattice. This unit cell is a cube of side a determined by the primitive vectors, $\mathbf{a}_1$, $\mathbf{a}_2$, $\mathbf{a}_3$, as shown in Fig. 1.3. It is called the cubic unit cell and is the unit cell used to define the *lattice constant* of a cubic crystal whether or not the structure has a cubic primitive unit cell. That is, the lattice constant for any cubic crystal structure is side a of the cubic unit cell. For noncubic structures, such as the hexagonal close-packed lattice, it is necessary to define more than one lattice constant. Let us examine the number of lattice sites in the cubic unit cell for the simple cubic Bravais lattice. From Fig. 1.3 this unit cell has one lattice site at each of the eight cube corners. Since each corner site is shared by eight cubic unit cells, each unit cell has only one lattice site.

If we used a cubic unit cell for the body-centered cubic lattice, we would have one lattice point from the corners and one point in the center

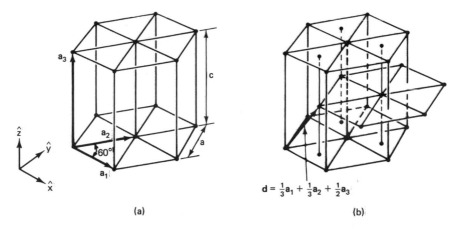

$$d = \tfrac{1}{3}a_1 + \tfrac{1}{3}a_2 + \tfrac{1}{2}a_3$$

(a)   (b)

Figure 1.6 (a) Simple hexagonal Bravais lattice and (b) hexagonal close-backed lattice.

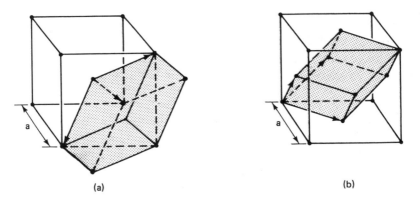

(a)   (b)

Figure 1.7 Cubic and primitive unit cells for (a) the body-centered and (b) the face-centered cubic Bravais lattices.

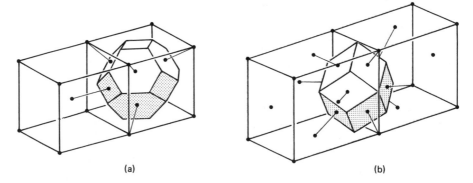

(a)   (b)

Figure 1.8 Wigner–Seitz unit cells for (a) the body-centered and (b) the face-centered cubic Bravais lattices.

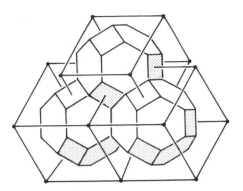

Figure 1.9 Translation of Wigner–Seitz unit cells for a body-centered cubic lattice to exactly fill the complete lattice.

that is not shared for a total of two lattice sites per cubic unit cell. The face-centered cubic lattice would have one site from the corners and one site on each of six faces, each of which is shared by two cells, for a total of four lattice sites in a cubic unit cell. The cubic unit cell, however, is not a primitive unit cell for these lattices.

To form primitive unit cells for the body-centered and face-centered cubic Bravais lattices, we can construct the parallelepipeds determined by the vectors a_i in Figs. 1.4 and 1.5. These are shown in Fig. 1.7. As can be seen, these primitive unit cells are oblique parallelepipeds with one lattice point at each corner. Each of these lattice points is shared by eight primitive unit cells. Thus the primitive cell in each case has one lattice site. These primitive unit cells, however, do not exhibit the complete symmetry of their Bravais lattices. This makes them, conceptually, more difficult to use.

One of the simplest primitive unit cell that exhibits the full symmetry of the lattice is called the Wigner–Seitz unit cell. It is formed by (1) drawing lines from a given lattice point to all nearby lattice points, (2) bisecting these lines with orthogonal planes, (3) and constructing the smallest polyhedron that contains the given point. As shown in Fig. 1.8, this construction produces a truncated octahedral cell for the body-centered lattice and a rhombic dodecahedral cell for the face-centered lattice. Because of the method of construction, the Wigner–Seitz cell translated by all the lattice vectors will exactly fill the Bravais lattice and is thus a primitive unit cell (Fig. 1.9).

The primitive unit cell for the hexagonal close-packed structure is the triangular prism formed by the primitive vectors a_i in Fig. 1.6. This unit cell exhibits the symmetry of the Bravais lattice.

1.2 CRYSTAL STRUCTURES

Thus here we have been discussing the matrix of points that describe the basic geometry of a crystal. A real ideal crystal is described by this underlying set of points, called the Bravais lattice, and the atoms or molecules that are arranged in the primitive unit cell surrounding each lattice point. The atoms or molecules are referred to as the *basis*. The combination of the Bravais lattice and the basis is referred to as the *crystal structure* of the material.

The most common crystal structure for semiconductors is the *sphalerite* or zinc blende structure shown in Fig. 1.10. It consists of two different atoms, each located on the lattice points of a face-centered cubic lattice, and separated by one-fourth of the length of a body diagonal of the cubic unit cell. Each atom is bonded to four atoms of the other type. This configuration is not a Bravais lattice. The sphalerite structure, however, is treated as a face-centered cubic Bravais lattice with a basis of two atoms displaced from each other by $(a/4)(\hat{x} + \hat{y} + \hat{z})$. Thus it has the face-centered cubic primitive unit cell or Wigner–Seitz unit cell with two atoms per cell. Some important properties of this structure result from the fact that the crystal does not appear the same when viewed along a body diagonal from one direction and then the other. Because of this, the sphalerite structure is said to lack inversion symmetry, and polarity effects are observed in the $\hat{x} + \hat{y} + \hat{z}$ direction. When both atoms are the same, the sphalerite structure is called the *diamond* structure, which has inversion symmetry. The diamond structure also has a face-centered cubic primitive unit cell with two atoms per cell, except that in this case, the two atoms are identical. The column IV elemental semiconductors crystallize in the diamond structure, while most of the III–V and II–VI compound semiconductors form in the sphalerite structure. The crystal structures for a number of semiconductors, together with other selected properties, are given in Appendix A.

Several of the more important IV–VI compound semiconductors have the sodium chloride or rock salt crystal structure shown in Fig. 1.11. This structure appears to be a simple cubic lattice with alternating atoms of different type. However, the sodium chloride structure can be most simply described as two interpenetrating face-centered cubic lattices, displaced from each other by one-half the body diagonal, with one type of atom on each lattice. The primitive unit cell is thus face-centered cubic with two atoms per unit cell: one at the origin and one of different type at $(a/2)(\hat{x} + \hat{y} + \hat{z})$. Some of the other IV–VI compounds form in an *orthorhombic* struc-

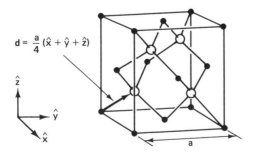

$d = \dfrac{a}{4}(\hat{x} + \hat{y} + \hat{z})$

$\hat{z}$

$\hat{y}$

$\hat{x}$

a

Figure 1.10 Cubic unit cell for the sphalerite structure. The open and solid spheres represent two different atoms. The tetrahedral bonds between atoms within the cubic unit cell are shown.

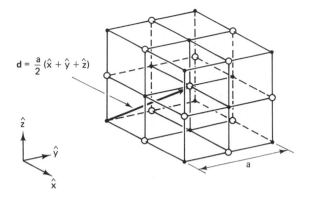

$$d = \frac{a}{2}(\hat{x} + \hat{y} + \hat{z})$$

Figure 1.11 Cubic unit cell for the sodium chloride structure. The open and closed spheres represent two different atoms. The bonds between atoms are indicated.

ture which can be represented by a deformation of the sodium chloride cubic unit cell into a parallelepiped with orthogonal sides.

A few of the III–V and several of the II–VI compounds have the wurtzite structure shown in Fig. 1.12. This structure consists of two interpenetrating hexagonal close-packed lattices, each with different atoms, ideally displaced by $\frac{3}{8} c\hat{z}$ from each other. As with the sphalerite structure, there is no inversion symmetry in this crystal, and polarity effects are observed along the a_3 or $\hat{z}$ axis. It has the primitive unit cell of the simple hexagonal lattice with a basis of four atoms, two of each kind, in each unit cell.

The most complex semiconductor crystal we will consider has the *chalcopyrite* structure of Fig. 1.13. This structure is found in the ternary I–III–VI$_2$ and II–IV–V$_2$ semiconductors. The atoms are arranged on the same lattice sites as in the sphalerite structure, except that in the chalcopyrite structure each nonmetallic atom is surrounded by four metallic atoms, two of each type. The tetragonal chalcopyrite structure is not exactly cubic since the dimension along the $\hat{z}$ axis, c, is generally somewhat less than twice the dimension along the $\hat{x}$ or $\hat{y}$ axis, a: that is, $c \leq 2a$. Because of the two metallic

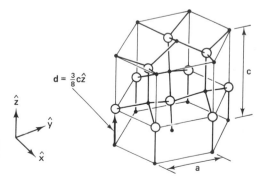

$$d = \frac{3}{8}c\hat{z}$$

Figure 1.12 Wurtzite structure showing the bonds between atoms and the hexagonal symmetry.

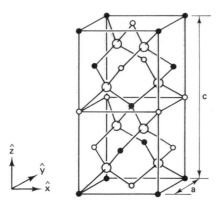

Figure 1.13 The cubic unit cell of the chalcopyrite structure, showing the bonds between atoms. The small closed and open spheres represent two different metallic atoms. The large open spheres represent identical nonmetallic atoms.

atoms, the chalcopyrite cubic unit cell is twice as large as the sphalerite cubic unit cell. The primitive unit cell, however, has eight atoms and is four times as large as the primitive unit cell for the sphalerite structure.

Although there are other semiconductor crystal structures, we have examined the ones of most practical interest. An understanding of these crystal structures should help us to understand some of the basic properties of semiconductors.

1.3 RECIPROCAL LATTICES

The concept of a reciprocal lattice is essential for an understanding of wave propagation and the quantum mechanical properties of electrons in periodic structures. It is also important for a determination of how the law of momentum conservation applies to motion in a periodic potential. As a result, in work on semiconductors, we spend about as much "time" and effort in reciprocal space as in the direct space of the real crystal.

Let us consider the set of Bravais lattice sites generated by the lattice vector $\mathbf{R}$ in (1.1). We wish to define some arbitrary function $f(\mathbf{r})$ in the direct space of the lattice which has the periodicity of the lattice. That is, we require that

$$f(\mathbf{r} + \mathbf{R}) = f(\mathbf{r})$$

for all values of $\mathbf{r}$ and all lattice sites (or vectors), $\mathbf{R}$. In the Cartesian coordinate system

$$\mathbf{r} = x\hat{x} + y\hat{y} + z\hat{z} \tag{1.7}$$

where $\hat{x}$, $\hat{y}$, and $\hat{z}$ are unit vectors. Instead of using (1.1) to define the Bravais lattice vector, $\mathbf{R}$, in terms of the primitive vectors, $\mathbf{a}_i$, we resolve $\mathbf{R}$ into components, a_x, a_y, and a_z, along the Cartesian unit vectors:

$$\mathbf{R} = n_x a_x \hat{x} + n_y a_y \hat{y} + n_z a_z \hat{z} \tag{1.8}$$

In this equation, n_x, n_y, and n_z are the sets of integers required to form the Bravais lattice; and a_x, a_y, and a_z indicate the period of the lattice in the $\hat{x}$, $\hat{y}$, and $\hat{z}$ directions.

We can now expand the function $f(\mathbf{r})$ in a Fourier series,

$$f(\mathbf{r}) = \sum_{K} A_K \exp{(i\mathbf{K}\cdot\mathbf{r})} \tag{1.9}$$

where

$$A_K = \frac{1}{\Omega} \int_{\Omega} f(\mathbf{r}) \exp{(-i\mathbf{K}\cdot\mathbf{r})}\, d\mathbf{r} \tag{1.10}$$

$$\mathbf{K} = 2\pi \left(\frac{l_x}{a_x}\hat{x} + \frac{l_y}{a_y}\hat{y} + \frac{l_z}{a_z}\hat{z} \right) \tag{1.11}$$

The integration in (1.10) is taken over the volume of the unit cell, which for a cubic Bravais lattice is given by

$$\Omega \equiv \mathbf{a}_1\cdot(\mathbf{a}_2 \times \mathbf{a}_3) \tag{1.12}$$

In (1.11) l_x, l_y, and l_z are integers that specify the terms of the Fourier series, while the vectors $\mathbf{K}$ are the wave vectors with dimension of reciprocal length for each component. For these reasons the vectors $\mathbf{K}$ are called reciprocal lattice vectors and are said to define a set of points in *reciprocal space*.

From (1.8) and (1.11) we find for any $\mathbf{K}$ and $\mathbf{R}$ that

$$\mathbf{K}\cdot\mathbf{R} = 2\pi(l_x n_x + l_y n_y + l_z n_z)$$

$$= 2\pi(\text{integer}) \tag{1.13}$$

or

$$\exp{(i\mathbf{K}\cdot\mathbf{R})} = 1 \tag{1.14}$$

Expanding the function $f(\mathbf{r} + \mathbf{R})$ in terms of $\mathbf{K}$ gives us

$$f(\mathbf{r} + \mathbf{R}) = \sum_{K} A_K \exp{[i\mathbf{K}\cdot(\mathbf{r} + \mathbf{R})]}$$

$$= \sum_{K} A_K \exp{(i\mathbf{K}\cdot\mathbf{r})} \exp{(i\mathbf{K}\cdot\mathbf{R})}$$

$$= \sum_{K} A_K \exp{(i\mathbf{K}\cdot\mathbf{r})} = f(\mathbf{r}) \tag{1.15}$$

Since the direct lattice vectors, $\mathbf{R}$, are normally described in terms of nonorthogonal primitive unit cell vectors, $\mathbf{a}_i$, it is useful to define the reciprocal lattice vectors, $\mathbf{K}$, in terms of primitive vectors for the reciprocal lattice, $\mathbf{b}_i$, which satisfy (1.14). Let us construct the vectors

$$\mathbf{b}_1 = \frac{2\pi}{\Omega}\, \mathbf{a}_2 \times \mathbf{a}_3, \qquad \mathbf{b}_2 = \frac{2\pi}{\Omega}\, \mathbf{a}_3 \times \mathbf{a}_1, \qquad \mathbf{b}_3 = \frac{2\pi}{\Omega}\, \mathbf{a}_1 \times \mathbf{a}_2 \tag{1.16}$$

and write

$$\mathbf{K} = h\mathbf{b}_1 + k\mathbf{b}_2 + l\mathbf{b}_3 \tag{1.17}$$

where h, k, and l are integers. From (1.1) and (1.17),

$$\mathbf{K \cdot R} = (h\mathbf{b}_1 + k\mathbf{b}_2 + l\mathbf{b}_3) \cdot (n_1\mathbf{a}_1 + n_2\mathbf{a}_2 + n_3\mathbf{a}_3) \tag{1.18}$$

Since $\mathbf{b}_1 \cdot \mathbf{a}_1 = 2\pi$, $\mathbf{b}_1 \cdot \mathbf{a}_2 = \mathbf{b}_1 \cdot \mathbf{a}_3 = 0$, and so on, we have

$$\mathbf{K \cdot R} = 2\pi(hn_1 + kn_2 + ln_3) = 2\pi(\text{integer}) \tag{1.19}$$

Thus the vectors $\mathbf{b}_i$ constructed in (1.16) satisfy (1.14) and can be taken as primitive vectors that define a primitive unit cell for the reciprocal lattice. From (1.17) we also see that the reciprocal lattice of a Bravais lattice is itself a Bravais lattice.

Let us next determine the volume of this primitive unit cell for any reciprocal lattice. The volume is determined from (1.16) and simple vector analysis as

$$\Omega_K \equiv \mathbf{b}_1 \cdot (\mathbf{b}_2 \times \mathbf{b}_3) = \frac{(2\pi)^3}{\Omega} \tag{1.20}$$

Thus the volume of the unit cell in reciprocal space is inversely proportional to the volume of the unit cell in direct space with units of reciprocal length cubed.

It is also interesting to construct the reciprocal lattice primitive unit cells for the simple direct lattices discussed in Section 1.1. For the simple cubic lattice, the reciprocal unit cell vectors are determined from (1.2) and (1.16) to be

$$\mathbf{b}_1 = \frac{2\pi}{a}\,\hat{x}, \qquad \mathbf{b}_2 = \frac{2\pi}{a}\,\hat{y}, \qquad \mathbf{b}_3 = \frac{2\pi}{a}\,\hat{z} \tag{1.21}$$

where the unit vectors are the same as in real space. Thus the reciprocal lattice unit cell for the simple cubic lattice with a cubic unit cell of side a is a cube with side $2\pi/a$. From (1.4) and (1.16) the reciprocal unit cell for the body-centered cubic lattice is constructed with the vectors

$$\mathbf{b}_1 = \frac{2\pi}{a}\,(\hat{y} + \hat{z}), \qquad \mathbf{b}_2 = \frac{2\pi}{a}\,(\hat{x} + \hat{z}), \qquad \mathbf{b}_3 = \frac{2\pi}{a}\,(\hat{x} + \hat{y}) \tag{1.22}$$

A comparison with (1.5) reveals that these vectors are the same as those of the primitive unit cell for the face-centered cubic direct lattice except that a is replaced by $4\pi/a$. Thus the reciprocal lattice for the body-centered cubic lattice is face-centered cubic with a cubic unit cell of side $4\pi/a$. In a similar manner it can be shown that the reciprocal lattice for the face-centered cubic lattice is body-centered cubic with primitive cell vectors

$$\mathbf{b}_1 = \frac{2\pi}{a}(-\hat{x} + \hat{y} + \hat{z}), \qquad \mathbf{b}_2 = \frac{2\pi}{a}(\hat{x} - \hat{y} + \hat{z}), \qquad \mathbf{b}_3 = \frac{2\pi}{a}(\hat{x} + \hat{y} - \hat{z})$$

$$(1.23)$$

Finally, the reciprocal lattice primitive cell vectors for the simple hexagonal direct lattice are

$$\mathbf{b}_1 = \frac{4\pi}{a\sqrt{3}}\left(\frac{\sqrt{3}}{2}\hat{x} - \frac{1}{2}\hat{y}\right), \qquad \mathbf{b}_2 = \frac{4\pi}{a\sqrt{3}}\hat{y}, \qquad \mathbf{b}_3 = \frac{2\pi}{c}\hat{z} \quad (1.24)$$

We can also construct Wigner–Seitz primitive cells for reciprocal lattices in the same manner that was used for direct lattices. That is, we (1) draw lines from one reciprocal lattice site (taken as the origin) to all nearby sites, (2) bisect each line with an orthogonal plane, and (3) use the smallest polyhedron that encloses the original site. For the reciprocal lattice, however, the primitive cell constructed in this fashion is referred to as the first *Brillouin zone*. The second Brillouin zone is obtained by a similar construction around the first zone, and so on for higher zones. Since most of the important properties of a reciprocal lattice are found in this primitive cell, we will spend most of our time with reciprocal space working in the first Brillouin zone. Let us, therefore, examine the first Brillouin zones for the semiconductor crystal structures discussed in Section 1.2.

The sphalerite, diamond, and sodium chloride structures all have the face-centered cubic direct Bravais lattice. Only the bases are different. As we have seen, the face-centered cubic direct lattice has a body-centered cubic reciprocal lattice, and vice versa. Therefore, the sphalerite, diamond, and sodium chloride structures all have the same reciprocal lattice and first Brillouin zone. Figure 1.14 shows the lattice points in reciprocal space and two Brillouin zones for the face-centered cubic direct lattice. Either cell can be used as the first Brillouin zone.

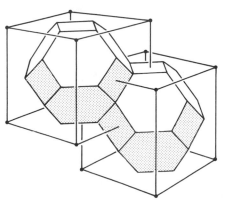

Figure 1.14 Brillouin zones for the face-centered cubic *direct* lattice. These primitive cells are in reciprocal space, which is body-centered cubic, and either could be selected as the first Brillouin zone.

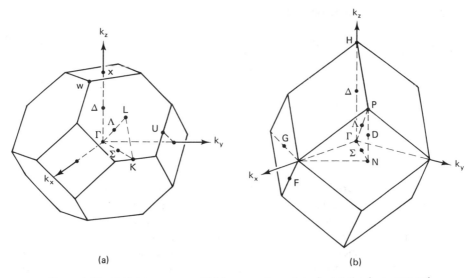

(a)   (b)

Figure 1.15 Brillouin zones and high symmetry points for (a) the face-centered cubic and (b) the body-centered cubic direct lattices.

In discussing various properties of the first Brillouin zone it has become common to use a peculiar notation, from group theory, for high symmetry points. Brillouin zones for the face-centered cubic and body-centered cubic direct lattices with the appropriate notation are shown in Fig. 1.15. The points within the zone are referred to an orthogonal basis k_x, k_y, and k_z with origin at the enclosed reciprocal lattice point. In Fig. 1.15(a), Γ refers to the origin, with X, L, K, W, U being various points on the zone boundary. Δ, Λ, Σ are points along the lines from Γ to X, L, K, respectively. Similar notation is used in Fig. 1.15(b).

The wurtzite structure has the simple hexagonal direct lattice. The appropriate Brillouin zone is a hexagonal prism, as shown in Fig. 1.16. The reciprocal lattice site is at Γ, with the notation for other high-symmetry points as shown.

Since the volume of the chalcopyrite unit cell is four times as large as

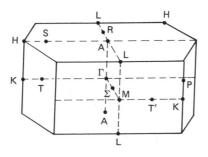

Figure 1.16 Brillouin zone for the simple hexagonal lattice.

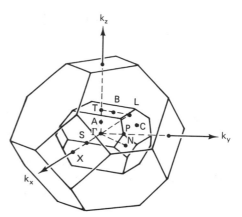

Figure 1.17 Comparison of Brillouin zones for the chalcopyrite and face-centered cubic lattices.

that for the face-centered cubic lattice, its Brillouin zone is four times smaller. The two Brillouin zones are compared in Fig. 1.17 with some of the appropriate notation.

1.4 MILLER INDICES

An important relationship exists between the reciprocal lattice vectors, $\mathbf{K}$, given by (1.17) and the planes of the corresponding direct lattice. A lattice plane is determined by three noncollinear lattice sites. Because of the translational symmetry of the Bravais lattice, however, each plane contains an infinite number of sites (Fig. 1.18). The relationship is that each $\mathbf{K}$ of the reciprocal lattice is normal to some set of planes in the direct lattice and the length of $\mathbf{K}$ is inversely proportional to the spacing between planes of this set.

Consider the reciprocal lattice vector $\mathbf{K}$ in (1.17) with integral components h, k, l *which have no common factor,* and a direct lattice point,

$$\mathbf{R} = n_1\mathbf{a}_1 + n_2\mathbf{a}_2 + n_3\mathbf{a}_3 \tag{1.25}$$

From (1.13) we have

$$\mathbf{K}\cdot\mathbf{R} = 2\pi(hn_1 + kn_2 + ln_3) = 2\pi N \tag{1.26}$$

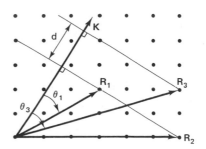

Figure 1.18 Two-dimensional direct lattice showing relationship between lattice planes and reciprocal lattice vectors.

Equation (1.26) tells us that the projection of the vector $\mathbf{R}_1$ along the direction of the vector $\mathbf{K}$ is

$$|\mathbf{R}_1| \cos \theta_1 = \frac{2\pi N}{|\mathbf{K}|} \tag{1.27}$$

Since the Bravais lattice is infinite, we can find an additional point with this same projection. For example, the vector $\mathbf{R}_2$ defined by

$$\mathbf{R}_2 = (n_1 - pl)\mathbf{a}_1 + (n_2 - pl)\mathbf{a}_2 + [n_3 + p(h + k)]\mathbf{a}_3 \tag{1.28}$$

where p is an integer, is such a point. If we then let p range through all integer values, we construct an infinite set of points on the same plane (while keeping N constant). This plane is, of course, orthogonal to $\mathbf{K}$ because of the projection.

In a similar manner, we can define the next adjacent plane by

$$|\mathbf{R}_3| \cos \theta_3 = \frac{2\pi(N + 1)}{|\mathbf{K}|} \tag{1.29}$$

so that the spacing between adjacent lattice planes perpendicular to $\mathbf{K}$ is

$$d = \frac{2\pi}{|\mathbf{K}|} \tag{1.30}$$

Therefore, it is easy to see that sets of planes in a direct lattice can conveniently be characterized by reciprocal lattice vectors or points in the corresponding reciprocal lattice. For this to be a unique prescription, the integers h, k, l in (1.26) must have no common factors.

The use of reciprocal lattice vectors to designate planes in direct space is entirely equivalent to the Miller indices of crystallography. This can be seen by taking a given $\mathbf{K}$ and choosing the n_i of (1.25) so that one plane of the set defined by (1.26) intersects the $\mathbf{a}_i$ axes at $n_i\mathbf{a}_i$, $i = 1, 2, 3$. From (1.26) we then have

$$\mathbf{K} \cdot n_1\mathbf{a}_1 = 2\pi h n_1 = 2\pi N \tag{1.31}$$

and so on, or

$$n_1 = \frac{N}{h}, \qquad n_2 = \frac{N}{k}, \qquad n_3 = \frac{N}{l} \tag{1.32}$$

Thus the intercepts of the plane are inversely proportional to the integral components of the reciprocal lattice vector. This result can also be demonstrated for planes that do not intersect the $\mathbf{a}_i$ axes at discrete lattice sites.

Although the primitive lattice vectors, $\mathbf{a}_i$, are in general not orthogonal, it is customary in all cubic lattices for the Miller indices to refer to the orthogonal simple cubic vectors. The notation used for specific reciprocal lattice vectors (points) and specific sets of direct lattice planes is (hkl). If a plane does not intercept a direct lattice vector (intercept at infinity), the corresponding Miller index is zero. If a plane intercepts in a negative direction, the Miller index has a line drawn over it. Figure 1.19 shows several examples of this notation for lattice planes in cubic lattices. Referring to Figs. 1.15 and 1.17, the corresponding notation for reciprocal lattice vectors is $k_x = (100)$, $k_y = (010)$, $k_z = (001)$. The origin or Γ-point is also referred to as (000). To avoid confusion with planes in the direct lattice and directions in the reciprocal lattice, square brackets are used for directions in the direct lattice and planes in the reciprocal lattice. That is, $a_1 = [100]$, $a_2 = [010]$, and so on, and the point

$$\mathbf{R} = n_1 \mathbf{a}_1 + n_2 \mathbf{a}_2 + n_3 \mathbf{a}_3$$

lies in the direction $[n_1 n_2 n_3]$ from the origin.

There is also a specific notation to indicate families of planes or reciprocal lattice vectors which are equivalent because of the lattice symmetry. For example, {100} is taken to indicate all the planes (100), (010), and (001). The equivalent notation for directions is ⟨100⟩, which is taken to mean all the directions [100], [010], [001] and their reciprocals [$\bar{1}$00], [0$\bar{1}$0], [00$\bar{1}$].

Four Miller indices are used for hexagonal lattices: one for each of three coplanar vectors, which are spaced at 120°, and one in the direction normal to this plane. Thus planes and reciprocal vectors are (hklm), while directions in the direct lattice are [hklm]. Otherwise, the notation is the same as for cubic lattices. Several examples are shown in Fig. 1.20.

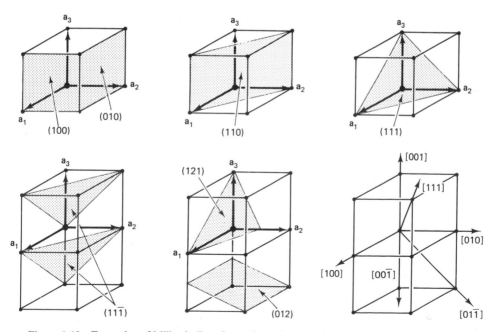

Figure 1.19 Examples of Miller indices for various planes and directions in a cubic lattice.

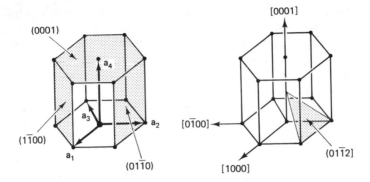

Figure 1.20 Examples of Miller indices for hexagonal lattices.

1.5 DIFFRACTION

An important application for the concepts discussed in this chapter is the analysis of wave diffraction by a crystal. As we will show, diffraction is governed by the Bragg condition,

$$2d \sin \theta = n\lambda \tag{1.33}$$

where d is the spacing between adjacent $\{hkl\}$ planes given by (1.30), θ the angle between the incident wave and the plane, n an integer indicating the diffraction order, and λ the wavelength. Equation (1.33) indicates that the longest wavelength that can be diffracted by a set of planes with spacing d is

$$\lambda_{max} = 2d \tag{1.34}$$

Since d is typically several angstroms, crystal diffraction is limited to wave particles such as neutrons, electrons, and high-energy photons (x-rays).

The de Broglie wavelength associated with neutrons or electrons is related to the energy by

$$\mathscr{E} = \frac{h^2}{2m\lambda^2} \tag{1.35}$$

where h is Planck's constant (6.6262×10^{-34} J·s) and m is the mass of the wave particle. For neutrons, $m = 1.675 \times 10^{-27}$ kg and

$$\lambda(\text{Å}) = \frac{0.28601}{[\mathscr{E}(\text{eV})]^{1/2}} \tag{1.36}$$

where an electron-volt (eV) is equal to 1.60219×10^{-19} J. Thus very low energy neutrons can be diffracted by a crystal. For electrons, $m = 9.10956$

$\times$ 10^{-31} kg and

$$\lambda(\text{Å}) = \frac{12.2643}{[\mathscr{E}(\text{eV})]^{1/2}} \qquad (1.37)$$

For x-rays and other photons the wavelength is related to energy by

$$\mathscr{E} = \frac{hc}{\lambda} \qquad (1.38)$$

where c is the velocity of light (2.9979×10^8 m/s). From (1.38),

$$\lambda(\text{Å}) = \frac{1.23986 \times 10^4}{\mathscr{E}(\text{eV})} \qquad (1.39)$$

and x-rays with energies of order 10^4 eV will be diffracted.

1.5.1 Lattice Structure Factor

To analyze the conditions under which diffraction occurs, consider an x-ray plane wave with electric field,

$$E = E_0 \exp\left[i(\mathbf{k}\cdot\mathbf{r} - \omega t)\right] \qquad (1.40)$$

incident on a multielectron atom. In (1.40) $\mathbf{k}$ is the wavevector, which is related to the wavelength λ by $|\mathbf{k}| = 2\pi/\lambda$ and ω is the angular frequency. The incident electric field accelerates the electrons to higher energy, during which time they emit x-rays in all directions with the electric field,

$$E = \frac{E_0 f_d}{r} \exp\left[i(\mathbf{k}'\cdot\mathbf{r} - \omega t)\right] \qquad (1.41)$$

The electric field in (1.41) is a solution of the radial wave equation where $\mathbf{k}'$ indicates a change in wavevector and phase and f_d is the *atomic scattering factor* of an atom indexed by a basis vector $\mathbf{d}$.

The atomic scattering factors depend on the number of electrons per atom, the scattering angle, and the incident wavelength and are defined as the ratio of the amplitude scattered by the actual electron distribution in an atom to the amplitude scattered by one isolated electron. When all the electrons in an atom scatter in phase (θ) with respect to the observation point, f_d is equal to the atomic number. Other values are tabulated in *International Tables for X-ray Crystallography,* Vol. 3 (Birmingham, England: Kynoch Press, 1968). The diffraction conditions depend on the interference and thus the relative phase of the radiation scattered from the electrons in the many atoms in the crystal.

To examine these interference effects, consider the difference in phase between the radiation scattered by two atoms separated by a lattice vector $\mathbf{R}$ and basis vector $\mathbf{d}$. As indicated in Fig. 1.21, the path difference between

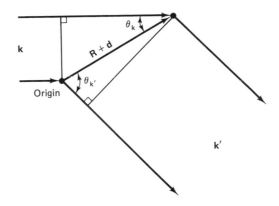

Figure 1.21 Schematic of phase difference between radiation scattered from an atom at the origin and an atom at $\mathbf{R} + \mathbf{d}$.

the incident and scattered waves for the two atoms is

$$|\mathbf{R} + \mathbf{d}| \cos \theta_k - |\mathbf{R} + \mathbf{d}| \cos \theta_{k'} = \frac{(\mathbf{R} + \mathbf{d}) \cdot \mathbf{k}}{|\mathbf{k}|} - \frac{(\mathbf{R} + \mathbf{d}) \cdot \mathbf{k}'}{|\mathbf{k}'|} \quad (1.42)$$

For elastic scattering (no energy loss) the magnitudes of the incident and scattered wavevectors are equal,

$$|\mathbf{k}| = |\mathbf{k}'| \quad (1.43)$$

and the phase difference is

$$(\mathbf{k} - \mathbf{k}') \cdot (\mathbf{R} + \mathbf{d}) = \Delta \mathbf{k} \cdot (\mathbf{R} + \mathbf{d}) \quad (1.44)$$

where $\Delta \mathbf{k}$ is the *scattering vector*.

If we use the phase of the wave scattered from the atom at the origin as our phase reference, the associated phase factor for this atom is

$$\exp (0) = 1 \quad (1.45)$$

The phase factor for the atom at $\mathbf{R} + \mathbf{d}$ is then

$$\exp [i \, \Delta \mathbf{k} \cdot (\mathbf{R} + \mathbf{d})] \quad (1.46)$$

Summing the phase factors for all N unit cells indexed by the lattice vectors $\mathbf{R}$ and all atoms in the unit cell indexed by the basis vectors $\mathbf{d}$, we define a *lattice structure factor*,

$$F_R \equiv \sum_R \sum_d f_d \exp [i \, \Delta \mathbf{k} \cdot (\mathbf{R} + \mathbf{d})] \quad (1.47)$$

which from (1.41) has the phase of and is proportional to the electric field of the resultant scattered radiation. It is also useful to define a *basis structure factor*,

$$F_d \equiv \sum_d f_d \exp{(i \, \Delta \mathbf{k} \cdot \mathbf{d})} \qquad (1.48)$$

in which case (1.47) for the lattice structure factor becomes

$$F_R \equiv F_d \sum_R \exp{(i \, \Delta \mathbf{k} \cdot \mathbf{R})} \qquad (1.49)$$

where $\mathbf{R}$ is given by (1.1).

Since the intensity of the scattered radiation depends on the square of the resultant electric field, we will examine the variation of $F_R^* F_R$ with $\Delta \mathbf{k} \cdot \mathbf{R}$. For simplicity, assume a one-dimensional lattice with N unit cells, so that

$$\mathbf{R} = n\mathbf{a} \qquad (1.50)$$

for

$$n = 0, 1, 2, \ldots, N - 1$$

Equation (1.49) is then

$$\frac{F_R}{F_d} = \sum_{n=0}^{N-1} \exp{(in \, \Delta \mathbf{k} \cdot \mathbf{a})} \qquad (1.51)$$

for

$$x \equiv \Delta \mathbf{k} \cdot \mathbf{a} \qquad (1.52)$$

$$\frac{F_R}{F_d} = \sum_{n=0}^{N-1} \cos nx + i \sum_{n=0}^{N-1} \sin nx \qquad (1.53)$$

$$\frac{F_R}{F_d} = \cos \frac{(N-1)x}{2} \, \sin \frac{Nx}{2} \, \csc \frac{x}{2} + i \sin \frac{(N-1)x}{2} \, \sin \frac{Nx}{2} \, \csc \frac{x}{2} \qquad (1.54)$$

and

$$\frac{F_R}{F_d} = \exp{\left[\frac{i(N-1)x}{2} \right]} \, \sin \frac{Nx}{2} \, \csc \frac{x}{2} \qquad (1.55)$$

Thus

$$\frac{F_R^* F_R}{F_d^* F_d} = \frac{\sin^2 \dfrac{N \, \Delta \mathbf{k} \cdot \mathbf{a}}{2}}{\sin^2 \dfrac{\Delta \mathbf{k} \cdot \mathbf{a}}{2}} \qquad (1.56)$$

1.5.2 Diffraction Conditions

Equation (1.56) for the relative intensity of the scattered radiation as a function of $\Delta\mathbf{k}\cdot\mathbf{a}$ is plotted in Fig. 1.22 for $N = 8$. As can be seen, the scattered radiation has significant value only when

$$\Delta\mathbf{k}\cdot\mathbf{a} = 2\pi n \tag{1.57}$$

where n has integer values (including zero). In fact, for large values of N, (1.56) becomes approximately

$$\frac{F_R^* F_R}{F_d^* F_d} \simeq N^2 \delta(\Delta\mathbf{k}\cdot\mathbf{a} - 2\pi n) \tag{1.58}$$

where $\delta(\Delta\mathbf{k}\cdot\mathbf{a} - 2\pi n)$ is the Dirac delta function. (The secondary maxima in Fig. 1.22 persist, however, even as N approaches infinity.) Thus (1.57) is the condition for constructive interference or diffraction maxima for a one-dimensional crystal. In three dimensions (1.57) is

$$\Delta\mathbf{k}\cdot\mathbf{R} = 2\pi(\text{integer}) \tag{1.59}$$

which from (1.19) defines a reciprocal lattice vector, $\mathbf{K}$. Thus the condition for diffraction is that the scattering vector $\Delta\mathbf{k}$ be a reciprocal lattice vector $\mathbf{K}$, or

$$\mathbf{k} - \mathbf{k}' \equiv \Delta\mathbf{k} = \mathbf{K} \tag{1.60}$$

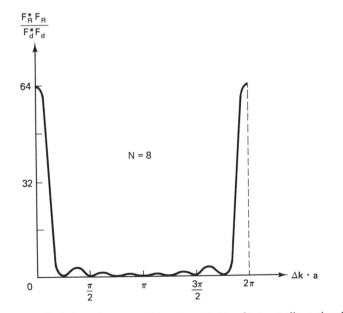

Figure 1.22   Variation of scattered intensity with $\Delta\mathbf{k}\cdot\mathbf{a}$ for a one-dimensional lattice with eight unit cells.

To derive the Bragg condition, notice from Fig. 1.23 that

$$|\mathbf{K}| = |\mathbf{k'}|\sin\theta + |\mathbf{k}|\sin\theta \tag{1.61}$$

For $|\mathbf{k}| = |\mathbf{k'}| = 2\pi/\lambda$, (1.61) is

$$|\mathbf{K}| = \frac{2\pi}{\lambda}(2\sin\theta) \tag{1.62}$$

From (1.30),

$$|\mathbf{K}| = \frac{2\pi}{d} \tag{1.63}$$

where d is the spacing between adjacent $\{hkl\}$ planes. Allowing for higher-order diffraction, (1.62) and (1.63) give

$$2d\sin\theta = n\lambda \tag{1.33}$$

where n is an integer denoting the order of diffraction.

The spacing between adjacent $\{hkl\}$ planes can be calculated from (1.63), (1.17), and the reciprocal lattice unit cell vectors. For a simple cubic unit cell, (1.17) and (1.21) result in

$$\mathbf{K} = \frac{2\pi}{a}(h\hat{x} + k\hat{y} + l\hat{z}) \tag{1.64}$$

where a is the lattice constant or length of the side of the cube. Using (1.63) and (1.64) gives us

$$d = \frac{a}{(h^2 + k^2 + l^2)^{1/2}} \tag{1.65}$$

The diffraction maxima corresponding to this spacing and indicated by the reciprocal lattice points or vectors in (1.64) are referred to as *hkl* maxima.

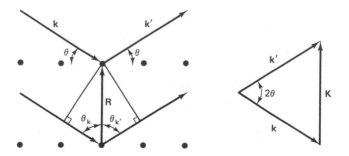

Figure 1.23   Diagram indicating the Bragg diffraction condition.

1.5.3 Basis Structure Factor

In the discussion above we ignored the contents of the unit cell and examined the phase differences in the radiation scattered from N unit cells to establish the conditions under which diffraction occurs. In a real crystal with a basis, however, many of the maxima allowed by

$$\Delta \mathbf{k} \cdot \mathbf{R} = 2\pi(\text{integer}) \tag{1.59}$$

are not observed because of phase cancellation within the unit cell. For this reason we will now examine the basis structure factor,

$$F_d \equiv \sum_d f_d \exp(i\,\Delta \mathbf{k} \cdot \mathbf{d}) \tag{1.48}$$

where the basis vector $\mathbf{d}$ indexes the atoms in the unit cell.

The basis vectors for the various crystal structures are shown in Figs. 1.10, 1.11, and 1.12 and, in general, are given by

$$\mathbf{d} = x_d \mathbf{a}_1 + y_d \mathbf{a}_2 + z_d \mathbf{a}_3 \tag{1.66}$$

where x_d, y_d, and z_d indicate the position of atom d with respect to the primitive vectors of the unit cell. The scattering vector $\Delta \mathbf{k}$ is a reciprocal lattice vector given, in general, by

$$\Delta \mathbf{k} = h\mathbf{b}_1 + k\mathbf{b}_2 + l\mathbf{b}_3 \tag{1.67}$$

For a simple cubic unit cell (1.2) and (1.66) result in

$$\mathbf{d} = a(x_d \hat{x} + y_d \hat{y} + z_d \hat{z}) \tag{1.68}$$

and (1.21) and (1.67) produce

$$\Delta \mathbf{k} = \frac{2\pi}{a}(h\hat{x} + k\hat{y} + l\hat{z}) \tag{1.69}$$

Using (1.68) and (1.69) in (1.48), the basis structure factor is

$$F_d = \sum_d f_d \exp[2\pi i(hx_d + ky_d + lz_d)] \tag{1.70}$$

where h, k, and l are the Miller indices of the diffracting planes. Equation (1.70) determines which diffraction maxima are not allowed because of phase cancellation within the unit cell.

As an example, consider atoms in the face-centered cubic (fcc) unit cell shown in Fig. 1.5. This cell can be represented as a simple cubic unit cell with a basis of four atoms: one at the origin and one at each of the three primitive vectors given by (1.5). That is, the eight atoms in the cube corners are each shared by eight cells for an average of one atom per cell; the six atoms on the cube faces are each shared by two cells for an average of three atoms per cell.

The basis vectors for these four atoms are

$$\mathbf{d} = 0, \quad \frac{a}{2}(\hat{x} + \hat{y}), \quad \frac{a}{2}(\hat{y} + \hat{z}), \quad \text{and} \quad \frac{a}{2}(\hat{z} + \hat{x}) \qquad (1.71)$$

Assuming that all four atoms are the same, (1.70) is

$$\frac{F_d}{f_d} = 1 + e^{i\pi(h+k)} + e^{i\pi(k+l)} + e^{i\pi(l+h)} \qquad (1.72)$$

Thus, for planes such that h, k, and l are either all even or all odd,

$$F_d = 4f_d \qquad (1.73)$$

and diffraction occurs. For planes such that one of the h, k, or l is even or one is odd,

$$F_d = 0 \qquad (1.74)$$

and no diffraction occurs.

1.5.4 X-Ray Methods

A common method of creating x-rays in the laboratory is to bombard a metal target with high-energy ($>10^4$ eV) electrons. Some of these electrons excite electrons from core states in the metal, which then recombine, producing highly monochromatic x-rays. These are referred to as characteristic x-ray lines. Other electrons, which are decelerated by the periodic potential of the metal, produce a broad spectrum of x-ray frequencies. Depending on the experiment, either or both of these x-ray spectra can be used.

The Laue method employs the broad spectrum of x-ray frequencies incident on a single crystal. Each (hkl) plane in the crystal selects a wavelength from the broad spectrum that satisfies the Bragg condition, $2d_{hkl} \sin \theta = n\lambda$. In the back-reflection mode indicated in Fig. 1.24, these wavelengths are reflected back to the film, where they produce a series of exposed spots. Each spot on the film can be identified as a first- or higher-order reflection from the (hkl) planes. If the characteristic lines are not filtered out, they will satisfy the Bragg condition for a few planes, and the corresponding spots will be dominant in the Laue photograph.

Since the resulting pattern exhibits the symmetry of the crystal normal to the primary x-ray beam, this method can be used to determine crystal orientation. Although the Laue method can also be used to determine crystal structure, several wavelengths can reflect in different orders from the same set of planes, with the different order reflections superimposed on the same spot in the film. This makes crystal structure determination by spot intensity difficult.

The rotating crystal method overcomes this problem by using the

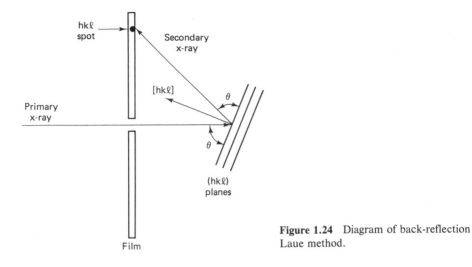

Figure 1.24 Diagram of back-reflection Laue method.

monochromatic part of the x-ray source and rotating the single crystal about its axis until an angle is obtained that satisfies the Bragg condition for a given set of planes. With a suitable reference this method can also be used to determine the lattice constant of the crystal. That is, for a given wavelength λ, if the angle θ at which an *hkl* reflection occurs is known, d_{hkl} can be determined.

Another x-ray technique often used in crystal identification is the powder method. Here a monochromatic x-ray beam is incident on a powdered or polycrystalline sample. Secondary x-rays are observed from those crystallites at an angle θ to the primary beam which happen to satisfy the Bragg condition. This method is obviously useful for samples that are difficult to obtain in single-crystal form.

PROBLEMS

1.1. (a) Show that every edge of the Wigner–Seitz unit cell for the body-centered cubic lattice in Fig. 1.8(a) has the length $\sqrt{2}\,a/4$.

    (b) In the face-centered cubic Wigner–Seitz unit cell of Fig. 1.8(b), show that the two diagonals of each face are in the ratio $\sqrt{2}:1$.

1.2. Show that in the hexagonal close-packed structure of Fig. 1.6(b) the ideal *c/a* ratio is $\sqrt{8/3}$.

1.3. Construct a set of primitive lattice vectors and a primitive unit cell for the chalcopyrite structure of Fig. 1.13.

1.4. Determie the lattice constant for NaCl, which has a molecular weight of 58.44 and a density of 2.167 g/cm³.

1.5. Show that an equation of the form of (1.32) can be obtained for planes that do not intersect the a_i axes at discrete lattice sites.

1.6. GaAs has the sphalerite structure with a lattice constant of 5.65 Å and a molecular weight of 144.64. What is the density in g/cm^3?

1.7. The bcc and fcc lattices can each be described as a simple cubic lattice with a basis of two or four atoms, respectively. *Using this description of these lattices,* determine the reciprocal lattice vectors and the Brillouin zones for the bcc and fcc lattice, and compare and contrast with the results obtained when these lattices are considered Bravais lattices.

1.8. A simple hexagonal lattice with N total atoms has an ideal c/a ratio $\sqrt{8/3}$.
   (a) Determine the volume of the direct lattice primitive unit cell and the volume of the first Brillouin zone in terms of a.
   (b) If the mass of each atom is 8×10^{-26} kg, a is 2 Å, and N is 10^{24} atoms, what is the volume of the crystal in cubic centimeters and the density in g/cm^3?

1.9. For the III–V semiconductor ternary alloys such as $In_xGa_{1-x}P$ and $GaAs_yP_{1-y}$ and others, Vegard's law concerning the lattice constant of the solid solution is followed. That is, the lattice constant varies linearly with composition. Assume that this is true also for quaternary III–V alloys. Given: $a_{InP} = 5.868$ Å, $a_{GaAs} = 5.654$ Å, $a_{GaP} = 5.449$ Å, $a_{InAs} = 6.058$ Å. Suppose that you wish to grow an epitaxial layer of the quaternary $In_{1-x}Ga_xP_{1-z}As_z$ with $z = 0.20$ on a substrate of $GaAs_yP_{1-y}$ with $y = 0.30$. What is the value of x that lattice matches the quaternary and ternary?

1.10. X-ray diffraction conditions are classified by the Miller indices of the planes that "perform the diffraction." That is, the diffraction condition $\Delta \mathbf{k} = \mathbf{K} = m_1\mathbf{b}_1 + m_2\mathbf{b}_2 + m_3\mathbf{b}_3$ is denoted by the (hkl) diffraction condition, where $\mathbf{K}$ is perpendicular to (hkl), the orthogonality condition, and

$$d_{hkl} = \frac{2\pi}{K}$$

the normalization condition. For the following Bravais lattices, state the diffraction condition in terms of constraints on h, k, and l (that is, hkl all even, $h + k + l =$ odd integers, etc.).
   (a) Simple cubic
   (b) Body-centered cubic
   (c) Face-centered cubic

1.11. The secondary maxima of the diffraction pattern for a crystal occur for

$$N \tan \tfrac{1}{2}x = \tan \tfrac{1}{2}Nx$$

for $x = \Delta\mathbf{k}\cdot\mathbf{a}$, where $\mathbf{a}$ is a primitive vector. Approximate solutions to this equation are given by

$$x = \frac{\pi}{N}(2p + 1) - \delta_p, \qquad p = 1, 2, 3, \ldots$$

where

$$\delta_p = \frac{2}{N}\tan^{-1}\left\{\frac{1}{N}\cot\left[\frac{\pi}{2N}(2p + 1)\right]\right\}$$

with an associated intensity

$$\frac{I}{N^2} = \left\{ (N^2 - 1) \sin^2 \left[\frac{\pi}{2N} (2p + 1) \right] + 1 \right\}^{-1}$$

For the case $N = 1000$, determine numerical solutions of the exact relation for the first five secondary maxima and compare these to the approximate solutions. Express the solutions in terms of π/N. [That is, $(N/\pi)x = (2p + 1) - (\pi/N)\delta_p$.] Compare the relative refracted intensity (I/N^2) to the approximate solutions.

1.12. A back-reflection Laue photograph is made using x-rays of wavelength λ directed along a $\langle 100 \rangle$ direction of a simple cubic sample with lattice constant a. For $\lambda = a/5$ and a film width of four times the sample-to-film spacing, determine the pattern on the photograph.

Energy Bands

An electron characterized by its wavefunction, $\psi(\mathbf{r})$, and spin orientation, $\mathbf{s}$, must satisfy the time-independent Schrödinger equation,

$$\mathbf{H}\psi(\mathbf{r}) = \mathscr{E}\psi(\mathbf{r}) \tag{2.1}$$

where $\mathscr{E}$ is the total energy of the electron and $\mathbf{H}$ is the appropriate Hamiltonian operator. The Hamiltonian takes into account all kinetic and potential energy terms, including applied forces and interactions with other particles. If the electron is traveling in a force-free region where it does not interact with other electrons (a free electron), the Hamiltonian contains only a kinetic energy term for the one electron, $\mathbf{p}^2/2m$, where the momentum operator, $\mathbf{p}$, is

$$\mathbf{p} = \frac{\hbar}{i}\frac{\partial}{\partial \mathbf{r}} = \frac{\hbar}{i}\nabla \tag{2.2}$$

and m is the free electron mass. In (2.3) $\hbar = h/2\pi$ where h is Planck's constant. Under these conditions Schrödinger's equation (2.1) reduces to its free one-electron formulation,

$$-\frac{\hbar^2}{2m}\nabla^2\psi(\mathbf{r}) = \mathscr{E}\psi(\mathbf{r}) \tag{2.3}$$

which has solutions of the form

$$\psi_k(\mathbf{r}) = A \exp(i\mathbf{k}\cdot\mathbf{r}) \tag{2.4}$$

where $\mathbf{k}$ is any position-independent vector.

With these solutions we can easily determine the free-electron energy from (2.3) as

$$\mathcal{E}(\mathbf{k}) = \frac{\hbar^2 \mathbf{k}^2}{2m} \tag{2.5}$$

The momentum is determined by operating on (2.4) with (2.2) to give

$$\mathbf{p} = \hbar \mathbf{k} \tag{2.6}$$

Considering the electron as a particle with velocity $\mathbf{v} = \mathbf{p}/m$, we arrive at the simple classical expression for the total energy of a free electron,

$$\mathcal{E} = \frac{\mathbf{p}^2}{2m} = \frac{1}{2} m\mathbf{v}^2 \tag{2.7}$$

From (2.4) we can also consider the electron as a plane wave with wave-vector $\mathbf{k}$ and de Broglie wavelength,

$$\lambda = \frac{2\pi}{|\mathbf{k}|} \tag{2.8}$$

The problem we consider in this chapter is how this free-electron description is modified for electrons in a periodic crystal structure. We expect a substantial modification for the following reasons. The atoms in the crystal, consisting of valence electrons, core electrons, and nuclei, produce a potential energy $U(\mathbf{r})$ with the periodicity of the direct Bravais lattice,

$$U(\mathbf{r}) = U(\mathbf{r} + \mathbf{R}) \tag{2.9}$$

for all direct lattice vectors, $\mathbf{R}$. Equation (2.8) tells us that a free-electron wavelength is of the same order of magnitude as the lattice periodicity. Thus we expect electrons to be strongly diffracted by the lattice.

To solve this problem, one would, in principle, have to include in the Hamiltonian of (2.1) terms that take into account interactions among the nuclei, core electrons, and valence electrons. Such a problem would be difficult to formulate, let alone solve. Since, in semiconductors the valence electrons are shared among atoms, a useful approximation is to treat the valence electrons as noninteracting entities that move through the crystal under the influence of an effective potential which includes the combined effects of the nuclei, the core electrons, and other valence electrons. Although not obvious a priori, this *one-electron approximation* provides a good description of semiconductor properties. In this manner we formulate a one-electron Schrödinger equation from (2.1) and (2.9), where $U(\mathbf{r})$ is taken as an effective one-electron potential: that is, the potential that the nuclei, core electrons, and all the other valence electrons produce for one valence electron. The problem is then a manner of solving (2.1) for the allowed one-electron energy levels.

There are several reasons why this one-electron approximation works as well as it does.

1. Electrons tend to be spatially removed from one another by Coulomb repulsion and by Fermi exclusion when they have the same spin. This reduces the interaction between the one electron and the rest of the valence electrons taken as a whole.

2. The valence electrons tend to cluster around the ion cores (nuclei and core electrons) due to Coulomb attraction. This effectively screens the Coulomb attraction of the ionic cores for the one electron and reduces this interaction.

3. Electrons passing near the ionic cores are accelerated by the Coulomb attraction. Because of this, electrons spend less time in the neighborhood of a core, effectively reducing the Coulomb attraction. (It is this effective *repulsion* that produces the *pseudopotential* discussed in Section 2.5.)

For these reasons we will examine in detail relatively simple one-electron models that illustrate some of the more important properties of electrons in periodic structures. We then discuss the results of more detailed computations on specific crystal structures.

2.1 BLOCH ELECTRONS

Let us first examine those properties of electrons in periodic structures which are independent of the specific nature of the potential $U(\mathbf{r})$. The term *Bloch electron* is used to refer to an electron that obeys the one-electron Schrödinger equation in a periodic potential. Bloch found that such electrons have wavefunctions in the form of a plane wave multiplied by a function that has the periodicity of the direct lattice. That is,

$$\psi_k(\mathbf{r}) = \exp(i\mathbf{k}\cdot\mathbf{r})u_k(\mathbf{r}) \tag{2.10}$$

where $\mathbf{k}$ is a wavevector and

$$u_k(\mathbf{r}) = u_k(\mathbf{r} + \mathbf{R}) \tag{2.11}$$

for all direct lattice vectors $\mathbf{R}$. This result is known as *Bloch's theorem* [F. Bloch, *Z. Phys. 52*, 555 (1928)]. From (2.10) we also have

$$\psi_k(\mathbf{r} + \mathbf{R}) = \exp[i\mathbf{k}\cdot(\mathbf{r} + \mathbf{R})]u_k(\mathbf{r} + \mathbf{R}) \tag{2.12}$$

or using (2.11),

$$\psi_k(\mathbf{r} + \mathbf{R}) = \exp(i\mathbf{k}\cdot\mathbf{R})\psi_k(\mathbf{r}) \tag{2.13}$$

for any value of $\mathbf{k}$ and every $\mathbf{R}$ in the direct lattice.

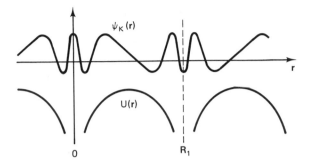

Figure 2.1 Possible Bloch electron wavefunction and periodic potential showing phase difference from one unit cell to the next. $\mathbf{r}$ and $\mathbf{R}_1$ are taken to be parallel.

Equation (2.13) is an alternative form of Bloch's theorem. It tells us that the electron wavefunction in any primitive unit cell of the direct lattice differs from that in any other cell only by the factor exp ($i\mathbf{k}\cdot\mathbf{R}$). For real $\mathbf{k}$ this represents a difference in phase as shown in Fig. 2.1. We can see that this factor is similar to the expression obtained in (1.14), which was

$$\exp(i\mathbf{K}\cdot\mathbf{R}) = 1 \tag{2.14}$$

for all reciprocal lattice vectors $\mathbf{K}$. The wavevector, $\mathbf{k}$, thus has dimensions of reciprocal length and belongs in reciprocal space with the vectors $\mathbf{K}$. Let us assume, for instance, that some electron wavefunction has a wavevector that is equal to a reciprocal lattice vector. From (2.13),

$$\psi_K(\mathbf{r} + \mathbf{R}) = \exp(i\mathbf{K}\cdot\mathbf{R})\psi_K(\mathbf{r})$$

$$= \psi_K(\mathbf{r}) \tag{2.15}$$

for all $\mathbf{R}$. That is, the electron wavefunctions ψ_K are periodic in $\mathbf{R}$.

Let us assume that an electron has a wavevector $\mathbf{k}$ given by

$$\mathbf{k} = \mathbf{K} + \mathbf{k}' \tag{2.16}$$

where $\mathbf{k}'$ is some other vector in reciprocal space. From (2.13) and (2.14) we find that

$$\psi_k(\mathbf{r} + \mathbf{R}) = \exp\{i[(\mathbf{K} + \mathbf{k}')\cdot\mathbf{R}]\}\psi_k(\mathbf{r})$$

$$= \exp(i\mathbf{k}'\cdot\mathbf{R})\psi_k(\mathbf{r}) \tag{2.17}$$

or the wavefunctions ψ_k obey Bloch's theorem as if they had wavevector $\mathbf{k}'$. Thus the wavefunction does not have a unique, wavevector $\mathbf{k}$, but a set of wavevectors that differ from each other by the set of reciprocal lattice vectors.

As indicated in Fig. 2.2, we can define a wavevector uniquely by reducing it with the appropriate reciprocal lattice vector to the first Brillouin zone. The prescription for this reduction is as follows. We choose the value

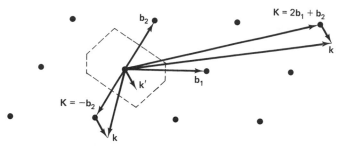

Figure 2.2 A two-dimensional reciprocal lattice indicating that any wavevector **k** in a higher Brillouin zone can be reduced to a value **k'** in the first Brillouin zone by choosing the appropriate reciprocal lattice vector **K**.

of **K** that will make the point **k'** lie as close to the origin as is possible. Since the value of **K** can be selected in increments of $\mathbf{b}_i$, the primitive vectors, the point **k'** can be made to lie closer to the origin than to any other lattice point in reciprocal space. This, of course, is the first Brillouin zone. Therefore, we have shown that any wavevector in higher Brillouin zones in reciprocal space is equivalent to one in the first Brillouin zone. It is for this reason that the first Brillouin zone is to be preferred over other primitive unit cells in the reciprocal lattice.

Let us now consider the number of allowed values of **k** in the first Brillouin zone. This can be determined by introducing boundary conditions at the outer surfaces of the crystal. If there are a total of N primitive unit cells of volume Ω in a crystal of volume V so that $N\Omega = V$, the N unit cells can be divided into N_i unit cells in the directions of the $\mathbf{a}_i$ primitive vectors, $i = 1, 2, 3$, so that

$$N = N_1 N_2 N_3 \qquad (2.18)$$

Thus the boundary of the crystal in the $\mathbf{a}_i$ direction is at $N_i\mathbf{a}_i$. To avoid standing electron waves we impose a cyclic or Born–von Kármán [M. Born and T. von Kármán, *Z. Phys. 13*, 297 (1912)] condition at these boundaries,

$$\psi(\mathbf{r}) = \psi(\mathbf{r} + N_i\mathbf{a}_i), \qquad i = 1, 2, 3 \qquad (2.19)$$

According to Bloch's theorem (2.13), we have

$$\psi_k(\mathbf{r} + N_i\mathbf{a}_i) = \exp(iN_i\mathbf{k}\cdot\mathbf{a}_i)\psi_k(\mathbf{r}) \qquad (2.20)$$

or

$$\exp(iN_i\mathbf{k}\cdot\mathbf{a}_i) = 1 \qquad (2.21)$$

Since the $\mathbf{a}_i$ are real, the **k** must also be real to satisfy (2.21). If we define the wavevectors k in terms of the primitive vectors for the reciprocal lattice b_i, we can write

$$\mathbf{k} = k_1\mathbf{b}_1 + k_2\mathbf{b}_2 + k_3\mathbf{b}_3 \qquad (2.22)$$

where the components k_i are to be determined. By inserting (1.16) for the $\mathbf{b}_i$ into (2.22) and (2.22) into (2.21), we obtain

$$\exp(i2\pi N_i k_i) = 1 \tag{2.23}$$

or

$$N_i k_i = m_i, \qquad i = 1, 2, 3 \tag{2.24}$$

where the m_i takes on all integer values. Thus the allowed values of $\mathbf{k}$ in reciprocal space are

$$\mathbf{k} = \frac{m_1}{N_1}\mathbf{b}_1 + \frac{m_2}{N_2}\mathbf{b}_2 + \frac{m_3}{N_3}\mathbf{b}_3 \tag{2.25}$$

Let us recall from (1.17) that the reciprocal lattice vector which defines the reciprocal lattice primitive unit cell is

$$\mathbf{K} = \mathbf{b}_1 + \mathbf{b}_2 + \mathbf{b}_3 \tag{2.26}$$

and from (1.20) that the volume of this cell is

$$\Omega_K = \mathbf{b}_1 \cdot (\mathbf{b}_2 \times \mathbf{b}_3) = \frac{(2\pi)^3}{\Omega} \tag{2.27}$$

In a similar manner, the volume of reciprocal space occupied by an allowed value of k is defined by

$$\mathbf{k} = \frac{\mathbf{b}_1}{N_1} + \frac{\mathbf{b}_2}{N_2} + \frac{\mathbf{b}_3}{N_3} \tag{2.28}$$

and given as

$$\Omega_k = \frac{\mathbf{b}_1}{N_1} \cdot \left(\frac{\mathbf{b}_2}{N_2} \times \frac{\mathbf{b}_3}{N_3} \right) = \frac{(2\pi)^3}{N\Omega} = \frac{(2\pi)^3}{V} \tag{2.29}$$

Therefore,

$$\Omega_K = N\Omega_k \tag{2.30}$$

and since the volume of a primitive cell is independent of how it is chosen, there are N allowed values of $\mathbf{k}$ in the first Brillouin zone. These results are summarized in Fig. 2.3 for a two-dimensional reciprocal lattice.

Since the number of unit cells N in a crystal of volume V is equal to or has the same order of magnitude as the number of atoms (10^{22} to 10^{23} per cubic centimeter), the number of allowed values of $\mathbf{k}$ in the first Brillouin zone is quite large and Ω_k is very small. Because of this it is sometimes convenient to treat reciprocal space and the first Brillouin zone as a continuum for $\mathbf{k}$ values. However, when $\mathbf{k}$ is used to index the energy levels in each energy band, it is treated as discrete.

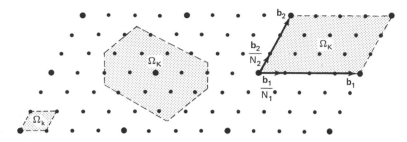

Figure 2.3   Allowed values of **k** (small points) in a two-dimensional reciprocal lattice. The figure is drawn with $N_1 = 4$, $N_2 = 3$. Both Ω_K equal 12 Ω_k.

It is interesting to compare the wavevector **k** for Bloch electrons to the wavevector for free electrons. From (2.6) we know that for free electrons **k** is proportional to the electron momentum,

$$\mathbf{p} = \hbar\mathbf{k} \tag{2.31}$$

For Bloch electrons, however, this is not the case. To determine the relationship between **p** and **k** for Bloch electrons, we operate on (2.10) with (2.2) to give

$$\mathbf{p}\psi_k(\mathbf{r}) = \hbar\mathbf{k}\psi_k(\mathbf{r}) + \exp(i\mathbf{k}\cdot\mathbf{r})\frac{\hbar}{i}\frac{\partial}{\partial\mathbf{r}} u_k(\mathbf{r})$$

which for a periodic potential, is not a constant times the wavefunction. Thus $\hbar\mathbf{k}$ is not the momentum of a Bloch electron. It is, nevertheless, useful and convenient to define a *crystal momentum* for Bloch electrons as

$$\mathbf{P} = \hbar\mathbf{k} \tag{2.32}$$

We will find in Section 2.8 that this crystal momentum, **P** behaves as a momentum only for externally applied forces. The "real" momentum, **p**, must take into account the response of the Bloch electrons to externally applied forces *and* the internal periodic potential of the crystal.

2.2 EMPTY LATTICE MODEL

We have seen that a substantial amount of information can be obtained about the wavevector of a Bloch electron without reference to the specific nature of the effective one-electron potential energy, $U(\mathbf{r})$. We can also obtain the general form and degeneracy (states at the same energy) of Bloch electron energy bands by solving the one-electron Schrödinger equation (2.3) for $U(\mathbf{r})$ = 0. The solutions, of course, are identical to those obtained for free electrons (2.4) and consist of plane waves with wavevectors, **k**, which are continuous throughout the reciprocal lattice. For this reason, the resulting energy values (2.5), given by

$$\mathscr{E}(k) = \frac{\hbar^2}{2m} \mathbf{k}^2 \tag{2.33}$$

are referred to as *free-electron* or *empty lattice* energy bands. The relationship between $\mathscr{E}$ and $\mathbf{k}$ in (2.33) is referred to as a *parabolic* energy band.

To determine the general form of the energy bands for Bloch electrons, we wish to reduce the free-electron wavevector, $\mathbf{k}$, to the first Brillouin zone. As we saw in Fig. 2.2 this can be done for any wavevector (including a free-electron wavevector) by a suitable choice of reciprocal lattice vector,

$$\mathbf{k} = \mathbf{k}' + \mathbf{K} \tag{2.34}$$

The wavefunction (2.4) is then

$$\psi_k(\mathbf{r}) = A \exp [i(\mathbf{k}' + \mathbf{K}) \cdot \mathbf{r}] \tag{2.35}$$

which is a Bloch wavefunction (2.10) with

$$u_k(\mathbf{r}) = A \exp (i\mathbf{K} \cdot \mathbf{r}) \tag{2.36}$$

From (2.34) and (2.33) the empty lattice energy bands are

$$\mathscr{E}(\mathbf{k}) = \frac{\hbar^2}{2m} (\mathbf{k}' + \mathbf{K})^2 \tag{2.37}$$

where $\mathbf{K}$ is given by (1.17) as

$$\mathbf{K} = h\mathbf{b}_1 + k\mathbf{b}_2 + l\mathbf{b}_3 \tag{2.38}$$

and h, k, l are integers. Thus it can be seen that $\mathbf{K}$ serves as an index for the different energy bands. Since there are an infinite number of h, k, l, there are an infinite number of energy bands.

We can demonstrate these free-electron energy bands most simply by looking at values of $\mathbf{k}$ in the direction of $\mathbf{K}$ without reference to a specific crystal structure. This gives the one-dimensional energy bands shown in Fig. 2.4. The lowest-lying band in the first Brillouin zone is obtained by setting $\mathbf{K} = 0$ in (2.37) and letting $\mathbf{k}$ range from $-\frac{1}{2}\mathbf{K}$ to $+\frac{1}{2}\mathbf{K}$. Continuing this procedure, an infinite number of energy bands are generated in the first Brillouin zone, each indexed with a separate reciprocal lattice vector. These free-electron energy bands, however, are of most interest in three dimensions, where they illustrate the form and degeneracy of electron states for a specific lattice.

As an example, we consider the general form and degeneracies of the empty lattice energy bands for the face-centered cubic lattice. The reciprocal space for this lattice is shown in Fig. 2.5. Values of wavevector, $\mathbf{k}$, for some

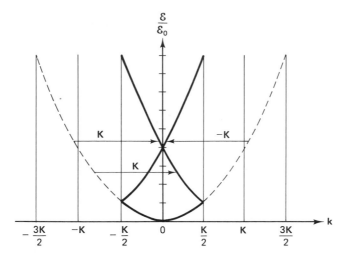

Figure 2.4 Empty lattice energy bands for **k** in the direction of **K**. The dashed line shows the parabolic free-electron description. $\mathscr{E}_0 = (\hbar^2/2m)(\tfrac{1}{2}\,\mathbf{K})^2$ is the energy at $\pm\tfrac{1}{2}\,\mathbf{K}$, the boundaries of the first Brillouin zone.

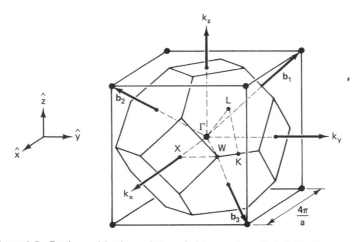

Figure 2.5 Reciprocal lattice points, primitive vectors, first Brillouin zone, and several high symmetry points for the face-centered cubic direct lattice. This is reciprocal space for crystals with the diamond, sphalerite, and sodium chloride structures.

of the high-symmetry points in the Brillouin zone are:

Γ-point: $\mathbf{k} = 0$

X-points: $\mathbf{k} = \pm\dfrac{2\pi}{a}\,\hat{x}, \quad \pm\dfrac{2\pi}{a}\,\hat{y}, \quad \pm\dfrac{2\pi}{a}\,\hat{z}$

L-points: $\mathbf{k} = \pm\dfrac{\pi}{a}\,(\hat{x} + \hat{y} + \hat{z}), \quad \pm\dfrac{\pi}{a}\,(-\hat{x} + \hat{y} + \hat{z}),$

$$\pm\dfrac{\pi}{a}\,(\hat{x} - \hat{y} + \hat{z}), \quad \pm\dfrac{\pi}{a}\,(\hat{x} + \hat{y} - \hat{z}) \qquad (2.39)$$

K-points: $\mathbf{k} = \pm\dfrac{3}{2}\dfrac{\pi}{a}\,(\hat{x} + \hat{y}),$ etc.

W-points: $\mathbf{k} = \pm\dfrac{\pi}{a}\,(2\hat{x} + \hat{y}),$ etc.

In the discussion to follow it is convenient to express the reciprocal lattice vectors (points) in terms of their components along orthogonal unit vectors. Using (1.23) in (2.38), we obtain

$$\mathbf{K} = \dfrac{2\pi}{a}\,[(-h + k + l)\hat{x} + (h - k + l)\hat{y} + (h + k - l)\hat{z}] \qquad (2.40)$$

Let us first consider the lowest-lying energy band given by (2.37). This is given by $h = k = l = 0$ or $\mathbf{K} = 0$. The minimum of the band is $\mathscr{E} = 0$ at $\mathbf{k} = 0$ or at the Γ-point. The energy of this band increases parabolically with increasing $\mathbf{k}$ in the Brillouin zone until it reaches the values, $\mathscr{E} = (\hbar^2/2m)(2\pi/a)^2$ at the X-points, $\mathscr{E} = (\hbar^2/2m)(\sqrt{3}\,\pi/a)^2$ at the L-points, and so on in different directions. From (2.35) there is only one wavefunction that corresponds to this range of energies up to the zone boundaries, so this lowest-lying energy band is nondegenerate (except for the twofold spin degeneracy).

We see from (2.37) that there are an infinite number of energy values for $\mathbf{k} = 0$, corresponding to the infinite number of reciprocal lattice points. The next energy value above zero at the Γ-point occurs for $h = \pm 1$, or $k = \pm 1$, or $l = \pm 1$ in (2.38). From (2.40) these reciprocal lattice vectors are given by

$$\mathbf{K} = \dfrac{2\pi}{a}\,(-\hat{x} + \hat{y} + \hat{z}) \qquad (2.41)$$

for $h = \pm 1$, and so on. Since the magnitude of all these $\mathbf{K}$ vectors is the same, $(2\pi/a)\sqrt{3}$, the energy level is

$$\mathscr{E}(0) = \dfrac{\hbar^2}{2m}\left(\dfrac{2\sqrt{3}\,\pi}{a}\right)^2 \qquad (2.42)$$

From (2.35) there are eight wavefunctions that have this energy at $\mathbf{k} = 0$, so this point is eightfold degenerate. The next highest level above (2.42) occurs for $h = k = l, l = 0$ and so on, or for

$$\mathbf{K} = \frac{2\pi}{a} (2\hat{z}) \qquad (2.43)$$

and so on. The energy of this point is sixfold degenerate and has the value

$$\mathscr{E}(0) = \frac{\hbar^2}{2m} \left(\frac{4\pi}{a} \right)^2 \qquad (2.44)$$

By generating a sufficient number of energy levels at the Γ-point and following them out to the desired high-symmetry points on the zone boundary with (2.37), we obtain the empty lattice energy bands for the face-centered cubic lattice shown in Fig. 2.6. As can be seen, the free-electron energy bands are fairly complicated, with multiple degeneracy at high-symmetry points, and the electrons can range through all energy values. When we apply a periodic potential to the one-electron Schrödinger equation, however, some of the degeneracy at the zone boundaries will be removed and the electrons will be constrained to certain energy values. These effects are demonstrated in the nearly free electron model.

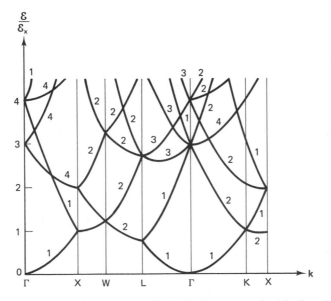

Figure 2.6   Empty lattice energy bands for the face-centered cubic direct lattice. The numbers indicate the degeneracy of each band. $\mathscr{E}_x = (\hbar^2/2m)(2\pi/a)^2$ is the energy at the X-points. [After F. Herman in *An Atomistic Approach to the Nature and Properties of Materials,* ed. J. A. Pask (New York: Wiley, 1967).]

2.3 NEARLY FREE ELECTRON MODEL

When the kinetic energy of the electrons is large compared to the periodic energy of the lattice, the behavior of the electrons can be approximated by nearly free electron wavefunctions. Although often considered a pedagogical exercise, with the advent of pseudopotential theory [J. C. Phillips and L. Kleinman, *Phys. Rev. 116*, 287 (1959)] this method has become important in the calculation of semiconductor energy bands. In the nearly free electron model, gaps occur in allowed electron energy values for the following reasons. Traveling electron waves reflected from adjacent atoms interfere constructively to produce standing waves. Some of these standing waves pile up charge at the atomic sites, where their energy is lowered, while other standing waves pile up charge between atomic sites, where their energy is increased over the free-electron values. This shift in energy between the standing-wave states produces an *energy gap*.

Let us consider a Bloch electron with a wavefunction given by (2.10) and (2.11). From (1.9) the periodic part of the wavefunction can be expanded in a Fourier series with reciprocal lattice vector, $\mathbf{K}$, as index:

$$u_k(\mathbf{r}) = \sum_K A_K \exp (i\mathbf{K}\cdot\mathbf{r}) \qquad (2.45)$$

where

$$A_K = \frac{1}{\Omega} \int_\Omega u_k(\mathbf{r}) \exp (-i\mathbf{K}\cdot\mathbf{r}) \, d\mathbf{r} \qquad (2.46)$$

The wavefunction then has the form

$$\psi_k(\mathbf{r}) = \sum_K A_K \exp [i(\mathbf{k} + \mathbf{K})\cdot\mathbf{r}] \qquad (2.47)$$

The one-electron Schrödinger equation (2.1) is then

$$\left[-\frac{\hbar^2}{2m} \nabla^2 + U(\mathbf{r}) - \mathscr{E} \right] \sum_K A_K \exp [i(\mathbf{k} + \mathbf{K})\cdot\mathbf{r}] = 0 \qquad (2.48)$$

or

$$\sum_K \left[\frac{\hbar^2}{2m} (\mathbf{k} + \mathbf{K})^2 + U(\mathbf{r}) - \mathscr{E} \right] A_K \exp [i(\mathbf{k} + \mathbf{K})\cdot\mathbf{r}] = 0 \qquad (2.49)$$

Multiplying on the left by $(1/\Omega) \exp [-i(\mathbf{k} + \mathbf{K}')\cdot\mathbf{r}]$, where $\mathbf{K}'$ is another reciprocal lattice vector and integrating over a primitive unit cell, (2.49) becomes

$$\frac{1}{\Omega} \sum_K A_K \int_\Omega \left[\frac{\hbar^2}{2m} (\mathbf{k} + \mathbf{K})^2 + U(\mathbf{r}) - \mathscr{E} \right] \exp [i(\mathbf{K} - \mathbf{K}')\cdot\mathbf{r}] \, d\mathbf{r} = 0 \qquad (2.50)$$

But

$$\frac{1}{\Omega} \int_{\Omega} \exp\left[i(\mathbf{K} - \mathbf{K}')\cdot\mathbf{r}\right] d\mathbf{r} = \delta_{K,K'} \tag{2.51}$$

where the Kronecker delta is

$$\delta_{K,K'} = \begin{cases} 1 & \text{for } \mathbf{K} = \mathbf{K}' \\ 0 & \text{for } \mathbf{K} \neq \mathbf{K}' \end{cases} \tag{2.52}$$

We now have

$$\sum_{K} A_K \left[\frac{\hbar^2}{2m} (\mathbf{k} + \mathbf{K})^2 - \mathscr{E} \right] \delta_{K,K'}$$

$$= -\frac{1}{\Omega} \sum_{K} A_K \int_{\Omega} U(\mathbf{r}) \exp\left[i(\mathbf{K} - \mathbf{K}')\cdot\mathbf{r}\right] d\mathbf{r} \tag{2.53}$$

or

$$A_{K'} \left[\mathscr{E} - \frac{\hbar^2}{2m} (\mathbf{k} + \mathbf{K}')^2 \right]$$

$$= \frac{1}{\Omega} \sum_{K} A_K \int_{\Omega} U(\mathbf{r}) \exp\left[i(\mathbf{K} - \mathbf{K}')\cdot\mathbf{r}\right] d\mathbf{r} \tag{2.54}$$

Since from (2.9) $U(\mathbf{r})$ is periodic in $\mathbf{R}$, we can expand it in a Fourier series with reciprocal lattice vector, $\mathbf{K}''$, as index:

$$U(\mathbf{r}) = \sum_{K''} B_{K''} \exp\left(i\mathbf{K}''\cdot\mathbf{r}\right) \tag{2.55}$$

where

$$B_{K''} = \frac{1}{\Omega} \int_{\Omega} U(\mathbf{r}) \exp\left(-i\mathbf{K}''\cdot\mathbf{r}\right) d\mathbf{r} \tag{2.56}$$

Using (2.55) in (2.54), we obtain

$$A_{K'} \left[\mathscr{E} - \frac{\hbar^2}{2m} (\mathbf{k} + \mathbf{K}')^2 \right]$$

$$= \frac{1}{\Omega} \sum_{K} \sum_{K''} A_K B_{K''} \int_{\Omega} \exp\left[i(\mathbf{K} - \mathbf{K}' + \mathbf{K}'')\cdot\mathbf{r}\right] d\mathbf{r} \tag{2.57}$$

But just as in (2.51),

$$\frac{1}{\Omega} \int_{\Omega} \exp\left[i(\mathbf{K} - \mathbf{K}' + \mathbf{K}'')\cdot\mathbf{r}\right] d\mathbf{r} = \delta_{K,K'-K''} \tag{2.58}$$

where

$$\delta_{K,K'-K''} = \begin{cases} 1 & \text{for } \mathbf{K} = \mathbf{K}' - \mathbf{K}'' \\ 0 & \text{for } \mathbf{K} \neq \mathbf{K}' - \mathbf{K}'' \end{cases} \tag{2.59}$$

Thus (2.57) becomes

$$A_{K'} \left[\mathscr{E} - \frac{\hbar^2}{2m} (\mathbf{k} + \mathbf{K}')^2 \right] = \sum_K \sum_{K''} A_K B_{K''} \delta_{K,K'-K''}$$

$$= \sum_{K''} A_{K'-K''} B_{K''} \tag{2.60}$$

which, with a change of notation, gives us

$$A_K \left[\mathscr{E} - \frac{\hbar^2}{2m} (\mathbf{k} + \mathbf{K})^2 \right] = \sum_{K'} A_{K-K'} B_{K'} \tag{2.61}$$

Equation (2.61) is an exact expression relating the Fourier coefficients A_K for the expansion of the Bloch function $u(\mathbf{r})$ to the Fourier coefficients $B_{K'}$ for the expansion of the effective one-electron lattice potential energy $U(\mathbf{r})$. Formally, we could now solve this problem by taking self-consistent expressions for $U(\mathbf{r})$ and $u(\mathbf{r})$ and determining the coefficients A_K and $B_{K'}$ from (2.46), (2.56), and (2.61) in an iterative fashion. We can obtain approximate analytical solutions, however, by making certain assumptions regarding the relative sizes of the various coefficients. The simplest assumption to make is that only the terms for $\mathbf{K}$ and $\mathbf{K}'$ equal to zero are important. This assumption just leads to free-electron solutions.

The next simplest assumption is that only the terms for $\mathbf{K}$ and $\mathbf{K}'$ equal to zero and $\mathbf{K}$ equal to $\mathbf{K}'$ are important. Under this assumption we obtain two equations for the coefficients from (2.61):

$$A_0 \left[\mathscr{E} - \frac{\hbar^2}{2m} \mathbf{k}^2 \right] = A_0 B_0 + A_{-K} B_K \qquad \text{for } \mathbf{K} = 0$$

$$A_K \left[\mathscr{E} - \frac{\hbar^2}{2m} (\mathbf{k} + \mathbf{K})^2 \right] = A_0 B_K + A_K B_0 \qquad \text{for } \mathbf{K} = \mathbf{K}' \tag{2.62}$$

For simplicity we have ignored the degenerate $-\mathbf{K}'$ term in each of these equations. If you wish to include them, simply replace every B_K in the rest of the analysis by $2B_K$. Since from (2.46) and (2.56) $A_{-K} B_K = A_K B_{-K}$, we can put this coupled set of equations in the matrix form

$$\begin{bmatrix} \mathscr{E} - \mathscr{E}_0 & -B_{-K} \\ -B_K & \mathscr{E} - \mathscr{E}_K \end{bmatrix} \begin{bmatrix} A_0 \\ A_K \end{bmatrix} = 0 \tag{2.63}$$

where

$$\mathscr{E}_0 \equiv \frac{\hbar^2 \mathbf{k}^2}{2m} + B_0 \quad \text{and} \quad \mathscr{E}_K \equiv \frac{\hbar^2}{2m} (\mathbf{k} + \mathbf{K})^2 + B_0 \tag{2.64}$$

Note that $\mathscr{E}_0$ and $\mathscr{E}_K$ have the form of free-electron energies in a constant potential energy B_0 and correspond to the first two terms in the wavefunction (2.47),

$$\psi_k(\mathbf{r}) = A_0 \exp(i\mathbf{k}\cdot\mathbf{r}) + A_K \exp[i(\mathbf{k} + \mathbf{K})\cdot\mathbf{r}] \tag{2.65}$$

To keep the coefficients of the wavefunction expansion, A_0 and A_K, from vanishing identically, the determinant of the matrix in (2.63) must be zero. Since $B_{-K}B_K = B_K^*B_K = B_K^2$ for crystals with inversion symmetry, we obtain a quadratic equation for the energy of the electrons,

$$\mathscr{E}^2 - (\mathscr{E}_0 + \mathscr{E}_K)\mathscr{E} + (\mathscr{E}_0\mathscr{E}_K - B_K^2) = 0 \tag{2.66}$$

which has the solutions

$$\mathscr{E}^\pm = \tfrac{1}{2}(\mathscr{E}_0 + \mathscr{E}_K) \pm \tfrac{1}{2}[(\mathscr{E}_0 - \mathscr{E}_K)^2 + 4B_K^2]^{1/2} \tag{2.67}$$

By comparing (2.67) with (2.64) and (2.65), we see that the states with energy $\mathscr{E}_0$ and $\mathscr{E}_K$ are combined into two states ψ^+ and ψ^- with energies $\mathscr{E}^+$ and $\mathscr{E}^-$ by the energy perturbation B_K.

We can examine these nearly free electron bands most simply for $\mathbf{k}$ in the direction of $\mathbf{K}$. The lowest-lying band is obtained for $\mathbf{K} = 0$ in (2.67) and looking at $\mathscr{E}^-$ for $\mathbf{k}$ near zero. This gives us

$$\mathscr{E}^- = \frac{h^2\mathbf{k}^2}{2m} \tag{2.68}$$

so the $\mathscr{E}^-$ versus $\mathbf{k}$ is parabolic near $\mathbf{k} = 0$. At the zone boundary, $\mathbf{k} = \tfrac{1}{2}\mathbf{K}$, and for $\mathbf{K} = -\mathbf{K}$, we have, from (2.67),

$$\mathscr{E}^\pm = \mathscr{E}_0 \pm |B_K| \tag{2.69}$$

Thus we see that at the zone boundary the energy of the ψ^- state is lower than the free-electron energy by $|B_K|$, and the energy of the ψ^+ state is higher by $|B_K|$. It is evident, then, that the periodic lattice potential has created an *energy gap* of magnitude $2|B_K|$ between these two states. In Fig. 2.7 the higher-energy bands are shown inside the first Brillouin zone and also outside. The bands outside are referred to as an extended zone scheme. They tend to show more clearly how the nearly free electron bands are perturbed from the free-electron parabola with discontinuities at the zone boundaries. Note that because of this perturbation, K is no longer a band index.

It is interesting to examine the wavefunctions that correspond to these energy bands. From (2.65) the wavefunctions at the zone boundary are

$$\psi(r) = A_0 \exp(i\tfrac{1}{2}\mathbf{K}\cdot\mathbf{r}) + A_K \exp(-i\tfrac{1}{2}\mathbf{K}\cdot\mathbf{r}) \tag{2.70}$$

The ratio of the coefficients can be obtained from (2.62) and (2.64) as

$$\frac{A_K}{A_0} = \frac{B_K}{\mathscr{E} - \mathscr{E}_K} \tag{2.71}$$

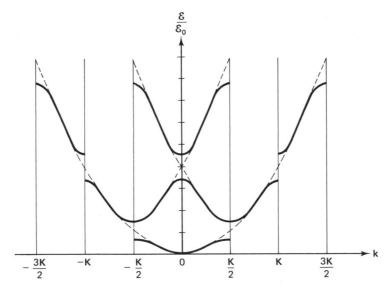

Figure 2.7 Nearly free electron energy bands for **k** in the direction of **K**. The dashed line shows the parabolic free-electron $\mathscr{E}$ versus **k** dependence. The higher-lying bands outside the first Brillouin zone is referred to as an extended zone scheme; $\mathscr{E}_0 = (\hbar^2/2m)(\tfrac{1}{2}\mathbf{K})^2 + B_0$.

or using (2.67),

$$\frac{A_K}{A_0} = \frac{2B_K}{(\mathscr{E}_0 - \mathscr{E}_K) \pm [(\mathscr{E}_0 - \mathscr{E}_K)^2 + 4B_K^2]^{1/2}} \tag{2.72}$$

At the zone boundary $\mathscr{E}_0 = \mathscr{E}_K$, and

$$\frac{A_K}{A_0} = \mp \frac{B_K}{|B_K|} \tag{2.73}$$

Let us assume that B_K is negative. This corresponds to a periodic potential that is negative in the neighborhood of each atom and thus tends to attract electrons. Then

$$\psi^-(r) = A_0[\exp(i\tfrac{1}{2}\mathbf{K}\cdot\mathbf{r}) + \exp(-i\tfrac{1}{2}\mathbf{K}\cdot\mathbf{r})]$$

$$= 2A_0 \cos(\tfrac{1}{2}\mathbf{K}\cdot\mathbf{r}) \tag{2.74}$$

and

$$\psi^+(r) = 2A_0 i \sin(\tfrac{1}{2}\mathbf{K}\cdot\mathbf{r}) \tag{2.75}$$

Therefore, we find that the periodic potential has converted the traveling electron plane waves of (2.65) into the standing waves of (2.74) and (2.75) at the zone boundaries. This corresponds to Bragg reflection of the electrons.

    If we determine the electron charge density for these two standing waves, as indicated in Fig. 2.8, we find that the charge for ψ^- is concentrated near the sites or atoms in the direct lattice, while the charge for ψ^+ is concentrated between atoms. We conclude, then, that the energy gap is caused by the potential well (B_K negative) at the atoms, which attracts the charge of ψ^- (lowers its energy) and repulses the charge of ψ^+ (raises its energy). Because of this position dependence, the ψ^- state is said to be "s-like," in analogy to atomic s-levels, which do not vanish at the ion, and the ψ^+ state is referred to as "p-like," since its charge vanishes at the atoms.

With the nearly free electron model we have developed the concept of allowed and "forbidden" energy bands. In a relatively simple manner, two other very important properties of electrons in a periodic potential can also be obtained from this model. Since $\mathcal{E}_0$ is about equal to $\mathcal{E}_K$ near the zone boundary, we can approximate the term in square brackets in (2.67) by

$$(1 + x)^{1/2} \simeq 1 + \tfrac{1}{2}x$$

or

$$\mathcal{E}^{\pm} \simeq \frac{1}{2}(\mathcal{E}_0 + \mathcal{E}_K) \pm |B_K| \left[1 + \frac{1}{2}\left(\frac{\mathcal{E}_0 - \mathcal{E}_K}{2|B_K|} \right)^2 \right] \qquad (2.76)$$

Using (2.64) and $\mathbf{k} \simeq \pm\tfrac{1}{2}\mathbf{K}$ near a boundary, (2.76) can be put in the form

$$\mathcal{E}^{\pm} \simeq B_0 \pm |B_K| + \frac{\hbar^2 \mathbf{k}^2}{2m^*} \qquad (2.77)$$

where

$$m^* = \frac{m}{1 \pm \hbar^2 \mathbf{K}^2 / m |B_K|} \qquad (2.78)$$

Equations (2.77) and (2.78) show us that the electron energies near the zone boundaries have an approximately parabolic dependence on wavevector in a manner similar to free electrons. The Bloch electrons, however, behave as though they have mass different from the free-electron mass. This behavior is characterized by an *effective mass, m^**, as given in (2.78). Equa-

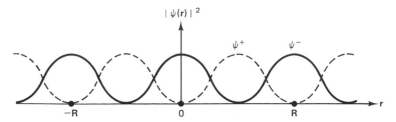

Figure 2.8 Distribution of electron charge density in direct space for **r** in the direction of **R**, the direct lattice vectors, for values of **k** at the zone boundaries in reciprocal space.

tion (2.78) also shows us that the effective mass for ψ^- states near the top of the lowest-lying band in Fig. 2.7 is *negative,* while the effective mass for ψ^+ states near the bottom of the next-highest band is *positive.*

The concept of an electron in a periodic potential energy having an effective mass different from its free mass should not be surprising. We found in Section 2.1 that the wavevector, $\mathbf{k}$, is not proportional to the real electron momentum, which takes into account the response of the electron to external forces and the internal forces due to the periodic potential. It reflects only the response of the electron to external forces. The response of the electron to the periodic potential energy of the crystal is accounted for with the effective mass, m^*, which can be greater or smaller than the free mass, m. A *negative* effective mass for an electron simply means that in the process of increasing its $\mathbf{k}$ vector under the influence of an external force, the momentum transfer from the crystal to the electron is larger than and in the opposite direction to the applied external force. We consider the effects of external forces in more detail in Chapter 5.

In the discussion above we took into account only the periodic potential due to the direct Bravais lattice or, more accurately, a crystal structure with one atom located at each site of the Bravais lattice. An interesting effect is obtained in a crystal with a basis of more than one atom per unit cell. Let us assume that we can approximate the periodic potential in (2.55) by taking the sum of potentials coming from each atom in the unit cell and each unit cell in the crystal structure. That is, let

$$U(\mathbf{r}) = \sum_{R,d} U_d(\mathbf{r} - \mathbf{R} - \mathbf{d}) \tag{2.79}$$

where U_d represents the atomic potentials with the index $\mathbf{d}$ allowing for different atoms in the unit cell. If we substitute (2.79) into (2.56), we obtain for the Fourier coefficients of $U(r)$,

$$B_K = \frac{1}{\Omega} \int_\Omega \sum_{R,d} U_d(\mathbf{r} - \mathbf{R} - \mathbf{d}) \exp(-i\mathbf{K}\cdot\mathbf{r}) \, d\mathbf{r}$$

$$= \frac{1}{\Omega} \sum_d \int_V U_d(\mathbf{r} - \mathbf{d}) \exp(-i\mathbf{K}\cdot\mathbf{r}) \, d\mathbf{r} \tag{2.80}$$

where V is the volume of the whole crystal. Making the substitution $\mathbf{r} - \mathbf{d} = \mathbf{r}'$, we have

$$B_K = \sum_d \exp(-i\mathbf{K}\cdot\mathbf{d}) \frac{N}{\Omega} \int_\Omega U_d(\mathbf{r}') \exp(-i\mathbf{K}\cdot\mathbf{r}') \, d\mathbf{r}' \tag{2.81}$$

$$= \sum_d B_{Kd} \exp(-i\mathbf{K}\cdot\mathbf{d}) \tag{2.82}$$

where B_{Kd} are the Fourier coefficients of U_d.

The exponential in (2.82) has a simple physical interpretation: It shows the relative phase of the traveling electron waves reflected from the various atoms in the unit cell. Depending on the basis and the direction of the electron wave, these reflections can add constructively or destructively, changing the values of B_K or possibly reducing it to zero. From (2.69), B_K, of course, determines the gap of forbidden energies between allowed energy bands. Thus the addition of a basis to the nearly free electron model can significantly alter the energy gap.

2.4 TIGHTLY BOUND ELECTRON MODEL

In the nearly free electron model it was assumed that the kinetic energy of the electrons was large compared to the periodic potential energy due to the lattice. Under these conditions the wavefunctions were found to be plane waves with a perturbation from the lattice, and the allowed energy bands were large compared to the forbidden energy regions. In the tightly bound electron model this situation is reversed. The periodic potential energy due to the lattice is assumed to be large compared to the kinetic energy of the electrons, so that the electrons are largely bound to the atomic cores. In this situation we expect the electron wavefunctions to be more like atomic orbitals than plane waves. That is, we expect the wavefunction overlap between adjacent atoms in the lattice to be sufficiently small that the band structure will be closely related to the wavefunctions and discrete energies of electrons in isolated atoms.

In this manner we consider N isolated atoms with electron wavefunctions ψ_a and discrete energy levels $\mathscr{E}_a$ located at the lattice sites $\mathbf{R}$. The one-electron Schrödinger equation for these atoms is

$$\mathbf{H}_a\psi_a(\mathbf{r} - \mathbf{R}) = \mathscr{E}_a\psi_a(\mathbf{r} - \mathbf{R}) \tag{2.83}$$

where

$$\mathbf{H}_a = -\frac{\hbar^2}{2m}\nabla^2 + U_a(\mathbf{r} - \mathbf{R}) \tag{2.84}$$

The Bloch electrons satisfy (2.1):

$$\mathbf{H}\psi_k(\mathbf{r}) = \mathscr{E}\psi_k(\mathbf{r}) \tag{2.85}$$

where

$$\mathbf{H} = -\frac{\hbar^2}{2m}\nabla^2 + U(\mathbf{r}) \tag{2.86}$$

and $U(\mathbf{r})$ satisfies (2.9). We construct a wavefunction for the Bloch electrons from a *l*inear *c*ombination of *a*tomic *o*rbitals (L.C.A.O.).

$$\psi_k(\mathbf{r}) = \sum_R \exp(i\mathbf{k}\cdot\mathbf{R})\psi_a(\mathbf{r} - \mathbf{R}) \tag{2.87}$$

which must satisfy Bloch's theorem (2.13). From (2.87) we have

$$\psi_k(\mathbf{r} + \mathbf{R}') = \sum_R \exp(i\mathbf{k}\cdot\mathbf{R})\psi_a(\mathbf{r} + \mathbf{R}' - \mathbf{R})$$

$$= \exp(i\mathbf{k}\cdot\mathbf{R}') \sum_R \exp[i\mathbf{k}\cdot(\mathbf{R} - \mathbf{R}')]\psi_a[\mathbf{r} - (\mathbf{R} - \mathbf{R}')]$$

$$= \exp(i\mathbf{k}\cdot\mathbf{R}')\psi_k(\mathbf{r}) \tag{2.88}$$

which is Bloch's theorem.

The Bloch electron Hamiltonian (2.86) can be put in the form

$$\mathbf{H} = -\frac{\hbar^2}{2m}\nabla^2 + U_a(\mathbf{r} - \mathbf{R}) + \Delta U(\mathbf{r} - \mathbf{R})$$

$$= \mathbf{H}_a + \Delta U(\mathbf{r} - \mathbf{R}) \tag{2.89}$$

where

$$\Delta U(\mathbf{r} - \mathbf{R}) \equiv U(\mathbf{r}) - U_a(\mathbf{r} - \mathbf{R}) \tag{2.90}$$

indicates the extent to which the periodic potential energy of the crystal deviates from the isolated atomic potential (Fig. 2.9). If the electrons are tightly bound, this deviation is small. Operating on (2.87) with (2.84), we find that

$$\mathbf{H}_a\psi_k(\mathbf{r}) = \left[-\frac{\hbar^2}{2m}\nabla^2 + U_a(\mathbf{r} - \mathbf{R})\right]\sum_R \exp(i\mathbf{k}\cdot\mathbf{R})\psi_a(\mathbf{r} - \mathbf{R})$$

$$= \sum_R \exp(i\mathbf{k}\cdot\mathbf{R})\mathbf{H}_a\psi_a(\mathbf{r} - \mathbf{R})$$

$$= \mathscr{E}_a\psi_k(\mathbf{r}) \tag{2.91}$$

Using (2.89) in (2.85), we have

$$\mathscr{E}\psi_k(\mathbf{r}) = [\mathbf{H}_a + \Delta U(\mathbf{r} - \mathbf{R})]\psi_k(\mathbf{r}) \tag{2.92}$$

$$(\mathscr{E} - \mathscr{E}_a)\psi_k(\mathbf{r}) = \Delta U(\mathbf{r} - \mathbf{R})\psi_k(\mathbf{r}) \tag{2.93}$$

Multiplying on the left by $\psi_k^*(\mathbf{r})$ and integrating over the crystal volume yields

$$(\mathscr{E} - \mathscr{E}_a)\int_V \psi_k^*(\mathbf{r})\psi_k(\mathbf{r})\,d\mathbf{r} = \int_V \psi_k^*(\mathbf{r})\,\Delta U(\mathbf{r} - \mathbf{R})\psi_k(\mathbf{r})\,d\mathbf{r} \tag{2.94}$$

Since the wavefunctions are normalized over the volume of the unit cell Ω, the integral on the left-hand side of (2.94) is just equal to the number of atoms in the crystal,

$$\mathscr{E} - \mathscr{E}_a = \frac{1}{N}\int_V \psi_k^*(\mathbf{r})\,\Delta U(\mathbf{r} - \mathbf{R})\psi_k(\mathbf{r})\,d\mathbf{r} \tag{2.95}$$

We now substitute the atomic wavefunctions of (2.87) for the Bloch

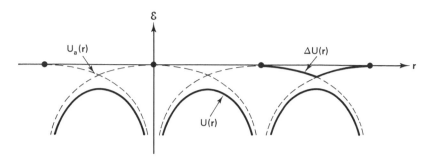

Figure 2.9   The periodic lattice potential energy, atomic potential energies, and their difference for **r** in the direction of **R**.

wavefunctions in (2.95), with the result

$$\mathcal{E} - \mathcal{E}_a = \frac{1}{N} \sum_R \sum_{R'} \exp\left[i\mathbf{k}\cdot(\mathbf{R} - \mathbf{R}')\right] \int_V \psi_a^*(\mathbf{r} - \mathbf{R}') \,\Delta U(\mathbf{r} - \mathbf{R})\psi_a(\mathbf{r} - \mathbf{R}) \, d\mathbf{r}$$

$$(2.96)$$

Because of the periodicity of the lattice, we can set the origin at **R** for each term in the summation over **R** (set **R** = 0). It is then easy to see that all N terms in the summation over **R** are identical and (2.96) becomes

$$\mathcal{E} - \mathcal{E}_a = \sum_{R'} \exp\left(-i\mathbf{k}\cdot\mathbf{R}'\right) \int_V \psi_a^*(\mathbf{r} - \mathbf{R}') \,\Delta U(\mathbf{r})\psi_a(\mathbf{r}) \, d\mathbf{r} \quad (2.97)$$

If we take out the term for **R**' = 0, we have

$$\mathcal{E} = \mathcal{E}_a - \alpha - \sum_{R'} \beta_{R'} \exp\left(i\mathbf{k}\cdot\mathbf{R}'\right) \qquad \text{for } \mathbf{R}' \neq 0 \qquad (2.98)$$

where

$$\alpha \equiv - \int_V \psi_a^*(\mathbf{r}) \,\Delta U(\mathbf{r})\psi_a(\mathbf{r}) \, d\mathbf{r} \qquad\qquad (2.99)$$

$$\beta_{R'} \equiv - \int_V \psi_a^*(\mathbf{r} - \mathbf{R}') \,\Delta U(\mathbf{r})\psi_a(\mathbf{r}) \, d\mathbf{r} \qquad (2.100)$$

and **R**' is now a vector joining an atom at the origin to all other atoms in the crystal. Since the integrals in (2.99) and (2.100) are difficult to calculate, we will express our results in terms of α and $\beta_{R'}$.

In (2.98) α is referred to as the Coulomb energy. It determines the shift in the atomic core levels, $\mathcal{E}_a$, caused by the interactions among atoms. $\beta_{R'}$ is referred to as the exchange energy and it determines the extent of broadening of the atomic levels into energy bands (Fig. 2.10). Since $\Delta U(\mathbf{r})$ is negative and becomes larger with decreasing atomic spacing, both α and $\beta_{R'}$ are positive and increase in magnitude with decreasing atomic spacing. Thus

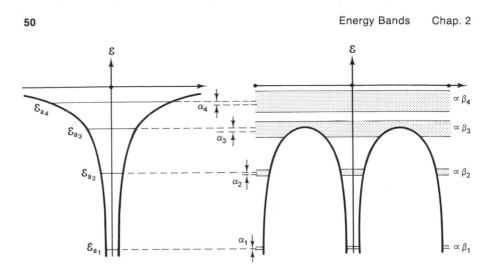

Figure 2.10 Diagram showing how energy bands are formed from the terms in (2.98). Each band is constructed from one atomic level.

these terms contain the overlap of atomic orbitals and account for the difference in potential energy between isolated atoms and a crystal of interacting atoms. The series in (2.98) allows for a summation of the overlap between an atom at the origin and all other atoms in the crystal. From Fig. 2.9, however, it can be seen that most of $\Delta U(\mathbf{r})$ can be accounted for by summing over nearest-neighbor atoms only.

As an example, let us consider the tight-binding energy bands for a face-centered cubic direct lattice with a basis of one atom. For spherically symmetric atomic wavefunctions, $\beta_{R'} = \beta$ is the same for all nearest neighbors. Figure 2.11 shows that there are 12 nearest neighbors in this structure located at

$$\mathbf{R} = \frac{a}{2}(\pm\hat{x} \pm \hat{y}), \qquad \frac{a}{2}(\pm\hat{y} \pm \hat{z}), \qquad \frac{a}{2}(\pm\hat{x} \pm \hat{z}) \qquad (2.101)$$

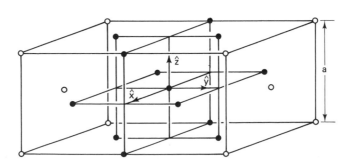

Figure 2.11 The 12 nearest-neighbor atoms (solid spheres) in face-centered cubic structure.

For $\mathbf{k}$ in the form

$$\mathbf{k} = k_x\hat{x} + k_y\hat{y} + k_z\hat{z} \tag{2.102}$$

(2.98) becomes

$$\mathcal{E} = \mathcal{E}_a - \alpha - 4\beta[\cos \tfrac{1}{2}k_x a \cos \tfrac{1}{2}k_y a + \cos \tfrac{1}{2}k_y a \cos \tfrac{1}{2}k_z a$$

$$+ \cos \tfrac{1}{2}k_z a \cos \tfrac{1}{2}k_x a] \tag{2.103}$$

Equation (2.103) shows that the energy of a Bloch electron, derived from one atomic state, consists of some constant value and an expression that varies with wavefunction between well-defined limits. Thus we find that for every electron state in the free atom, there exists a band of energies in the crystal.

Figure 2.12 shows this tight-binding energy band for several high-symmetry directions in the first Brillouin zone. This figure can be compared directly to the lowest-lying empty lattice energy band in Fig. 2.6. It can be noted that the width of the band depends on β and varies for different directions of wavevector in a manner similar to the lowest-lying empty lattice band. This similarity shows the strong dependence of energy band structure on the crystal lattice. The number of nondegenerate electronic states in the band is equal to N, the number of atoms in the crystal, and each state can be occupied by two electrons of opposite spin. To obtain the complete energy band diagram in the tight-binding approximation, it is necessary to add additional atomic states to this picture, as indicated in Figs. 2.10 and 2.13.

Figure 2.13 illustrates conceptually how the complete energy band diagram for a crystal can be obtained from discrete atomic levels. We take N

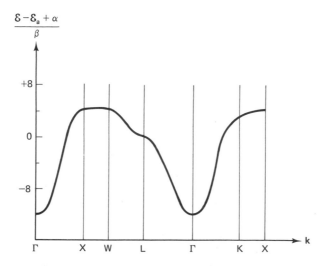

Figure 2.12   Tight-binding energy band, corresponding to one atomic state, for a face-centered cubic lattice with a basis of one atom.

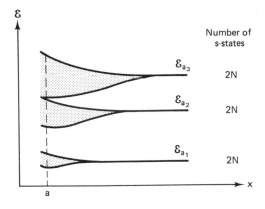

Figure 2.13 Diagram showing how energy bands in tightly bound electron model are formed from discrete atomic energy levels as the distance between atoms, x, is decreased to the lattice constant, a.

atoms and locate them on the sites of the appropriate Bravais lattice, except that the distance between atoms is so large that they do not interact. Each atomic level is then N-fold degenerate (excluding spin and angular momentum degeneracy). As we bring the atoms closer together, the higher-energy atomic states begin to interact, because of their large orbitals, and the N degenerate states split into a band of N discrete states. As the distance between atoms decreases further, lower-lying atomic levels begin to split into bands, and so on. In this manner, band states can be classified according to their atomic origin. That is, bands can be referred to as s-bands, p-bands, and so on. For higher-lying bands that overlap, however, this classification is not as obvious.

To take into account a basis in the tightly bound electron model, (2.87) for the Bloch wavefunctions can be changed to

$$\psi_k(\mathbf{r}) = \sum_R \exp{(i\mathbf{k}\cdot\mathbf{R})} \sum_d \psi_{ad}(\mathbf{r} - \mathbf{R} - \mathbf{d}) \qquad (2.104)$$

If the molecular wavefunctions are known, however, the Bloch wavefunctions could be formed from a linear combination of *molecular* orbitals.

2.5 OTHER BAND MODELS

We have used the general models discussed up to now to demonstrate the existence of energy bands and to gain some qualitative understanding of their properties. These simple analytical models, however, are not sufficiently accurate to obtain results that can be compared with experiment. The nearly free electron model, for example, tends to overemphasize the plane wave aspects of the wavefunction, while the tightly bound electron model overemphasizes the atomic core aspects. Also, we were not able to obtain an analytical expression for the periodic potential.

To overcome some of these problems, other methods for calculating energy bands have been devised. Even with these more sophisticated models it is necessary to base some of the parameters, such as energy gap or effective mass, on experimental results. In addition, the calculation of semiconductor energy bands is sufficiently complex that numerical methods must be used to obtain reliable results. We will discuss some of these methods and look at the resulting energy bands for specific semiconductors.

2.5.1 Orthogonalized Plane Wave Method

The orthogonalized plane wave or OPW method [C. Herring, *Phys. Rev.* *57*, 1169 (1940)] resolves some of the problems associated with the nearly free electron (plane wave) approximation. Near the atomic core, we expect the Bloch wavefunctions to differ substantially from plane waves, and many plane waves with large **k** values are required in the series expansion of (2.47) to approximate its behavior. Thus (2.54) or (2.61) must be solved for a large number of coefficients to obtain accurate results. The OPW method helps to resolve this problem by constructing a wavefunction that behaves like a plane wave between atoms and approximates more closely an atomic wavefunction near the atomic core. The wavefunction consists of an atomic orbital part which is made orthogonal to a plane wave part. This wavefunction is used instead of (2.47), and fewer coefficients have to be considered in (2.54) or (2.61) to obtain reliable results.

2.5.2 Pseudopotential Method

Another technique used to improve the nearly free electron model is the pseudopotential method [J. C. Phillips and L. Kleinman, *Phys. Rev.* *116*, 287 (1959)]. The concept of a pseudopotential arises in the use of OPW wavefunctions instead of the plane waves in (2.47). When these wavefunctions are operated on with the one-electron Hamiltonian, the orthogonalized atomic orbital part produces a term on the right-hand side of (2.54) in addition to that involving the real potential $U(\mathbf{r})$. The combination of the two terms defines an effective or pseudopotential for the valence electrons. An examination of these two terms individually shows that the term due to the real potential is negative, while the term due to the orthogonalization of the wavefunction is positive. This cancellation produces a net pseudopotential for the electrons which is weaker than the real potential. Physically, the reason for this is that the electron wavefunctions in the atomic core oscillate rapidly with a high kinetic energy. This kinetic energy acts as a repulsive potential in the attractive potential well of the core and tends to cancel part of the negative potential energy. In this manner, the behavior of the valence electrons is less sensitive to the form chosen for $U(\mathbf{r})$, such as the number of terms in (2.55). If the pseudopotential is regarded as an adjustable parameter to be determined from experimental data, this method can produce reliable results for specific semiconductor energy bands.

2.5.3 Cellular Methods

Some of the other techniques used to calculate energy bands can be classified as cellular methods [E. P. Wigner and F. Seitz, *Phys. Rev. 43,* 804 (1933)]. The basic idea behind these methods is that due to the Bloch equation (2.13), it is only necessary to solve Schrödinger's equation for the wavefunction within a primitive unit cell (Wigner–Seitz unit cell) of the direct lattice. The wavefunction for any other cell can then be obtained from (2.13). The problem is that not every solution to Schrödinger's equation within a primitive cell is valid for the entire crystal. That is, the wavefunctions and their gradients must be continuous and approximate plane waves at the boundaries of the cell. These conditions can be met by the use of a periodic lattice potential which has the topology of a pan used for baking rolls. This so-called "muffin-tin potential" consists of an attractive atomic potential within a sphere of arbitrary radius (usually less than half the nearest-neighbor distance) around each lattice site and zero everywhere else.

Two methods often used for calculating energy bands with a muffin-tin potential are the augmented plane wave or APW method [J. C. Slater, *Phys. Rev. 51,* 846 (1937)] and the KKR method [J. Korringa, *Physica 13,* 392 (1947); W. Kohn and N. Rostoker, *Phys. Rev. 94,* 1411 (1954)]. In the APW method, a solution to Schrödinger's equation is obtained by first matching plane wave solutions outside the core radius to atomic solutions within the core. A wavefunction is then constructed from these augmented plane waves. Unfortunately, these augmented plane waves have a discontinuous gradient at the boundary between the core and flat potential regions, so that substantial calculation is required to obtain reliable results. The KKR method is essentially a Green's function approach to the solution of Schrödinger's equation with a muffin-tin potential. It has certain similarities to the APW method and the results from the two are in substantial agreement when the same potential is used in each. Both, however, are first-principles calculations which should be viewed with caution.

2.5.4 Spin-Orbit Coupling

In the discussion of energy band calculations we have ignored any effects due to the electron spin operator, **s**. We have simply assumed that two electrons of opposite spin can occupy any state defined by a Bloch wavefunction. In free atoms we know, however, that the electrons moving through the electric field of the ion core experience a potential that depends on the scalar product of the spin magnetic moment with the vector product of its velocity and the electric field. This *spin-orbit coupling* tends to remove the degeneracy of states with the same wavefunction and opposite spin. We should then expect similar effects for electrons moving in the periodic potential of a lattice. The effects of spin-orbit coupling on energy bands can be determined by using a Hamiltonian that includes this potential and solving the one-electron Schrödinger equation with a perturbation technique known as the **k·p** method [E. O. Kane, *J. Phys. Chem. Solids 1,* 83 (1956)]. The results indicate that even strong spin-orbit coupling will not remove the spin degeneracy for crystals with inversion symmetry, provided that a band is not degenerate with one of different wavefunction. Even in crystals that lack inversion symmetry, the spin-orbit splitting is small. For any crystal, however, if several bands with different wavefunction are degenerate, the spin-orbit splitting can be substantial.

2.6 VALENCE BANDS AND BONDS

Before we consider the details of semiconductor energy bands, it is worthwhile to pursue the relationship between energy bands and valence bonds. We learned in Section 2.1 that the first Brillouin zone for a direct Bravais lattice has N allowed values of **k**, where N is the number of primitive unit cells in the direct lattice. Each allowed k value corresponds to a wavefunction or electron state which can have two electrons of opposite spin. Including spin, we then have $2N$ electron states per energy band or two electron states per unit cell per energy band for a Bravais lattice. If we now add a basis (atoms) to the lattice to obtain a crystal structure, we can count the number of electrons per unit cell and determine the occupancy of the energy bands.

As an example, let us consider the occupancy of energy bands in silicon. Silicon has the diamond crystal structure, which is a face-centered cubic Bravais lattice with a basis of two atoms per primitive unit cell. The atomic structure is $(1s)^2(2s)^2(2p)^6(3s)^2(3p)^2$. Since the first two shells are completely filled and tightly bound to the nucleus, we will consider only the third shell of valence electrons. Here the two $3s$ states (including spin states) are completely filled with two electrons, and the six $3p$ states have only two electrons. If we use a tight-binding approach to form the crystal from the atoms, we will need $2N$ silicon atoms. As shown in Fig. 2.14, the resulting $12N$ $3p$-states and the $4N$ $3s$-states interact and cross over when the atoms are brought together to the appropriate lattice constant. The result for 0 K is $8N$ lower-lying states occupied by $8N$ electrons separated by an energy gap from $8N$ higher-lying states with zero electrons. Since there are $2N$ states per energy band, the crystal has three $3p$ bands and one $3s$ band (all spin degenerate) in both the low- and high-energy regions. The four bands appear to be only one in an $\mathscr{E}$ versus **r** diagram, as in Fig. 2.14.

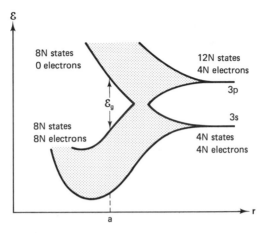

Figure 2.14 The formation of energy bands in silicon from the two-highest occupied atomic levels. The $8N$ available electrons completely fill the $8N$ states in the lowest bands (valence bands), leaving the $8N$ states in the highest bands (conduction bands) completely empty.

In molecules such as CH_4 this combination of s and p states is known to give a tetrahedral bonding arrangement, called *covalent bonding,* where two electrons of opposite spin are shared between atoms. For the silicon crystal, however, there are $8N$ electrons in the four lower-lying bands, or eight electrons per unit cell. This means that every silicon atom in the crystal is surrounded by eight outer electrons. From the crystal structure of silicon (Fig. 1.10), we see that every silicon atom has four nearest neighbors, so that each silicon atom must be covalently bonded to its nearest neighbors, with the $8N$ electrons in the lowest-lying energy bands. For this reason these occupied bands are referred to as *valence bands.* The decrease in the energy of these bands over the atomic states, as shown in Fig. 2.14, reflects the binding or cohesive energy. In a similar manner the four higher-lying bands correspond to an antibonding molecular state and are referred to as *conduction bands.* It should be pointed out, however, that this analogy with covalently bonded molecules is not precise. The wavefunction overlap, which produces the energy bands in the crystal, demonstrates that the *valence electrons are not localized in the bonds.* There is, nevertheless, a high concentration of electron charge in the regions between atoms, which is equivalent to a covalent electron pair.

This covalent bonding arrangement can conveniently be illustrated by the two-dimensional schematic shown in Fig. 2.15. Each solid line between atoms represents one of the paired electrons in the covalent bond. Adding up the number of electrons in the Wigner–Seitz unit cell, we find a total of eight. These are the eight electrons that completely fill the four valence bands

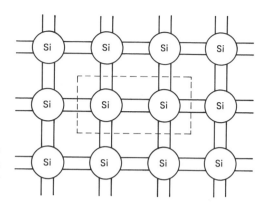

Figure 2.15 Two-dimensional schematic of covalent bonding arrangement in silicon. The dashed lines show the Wigner–Seitz unit cell.

in Fig. 2.14. The other column IV semiconductors can be represented in a similar fashion.

To illustrate the origin of the various bands, we can look at the $\mathscr{E}$ versus **k** curve near the Γ-point as in Fig. 2.16(a). The lowest-lying valence band, labeled Γ_1, is twofold (spin) degenerate and is derived from the bonding s orbitals for the two atoms in the unit cell. The other three valence bands ($\Gamma_{25'}$) are derived from p orbitals and are sixfold degenerate at Γ. When spin orbit effects are included, these bands are split into two fourfold degenerate bands, labeled $\Gamma_{25'}^{3/2}$, and one lower-lying twofold degenerate band, labeled $\Gamma_{25'}^{1/2}$, at Γ. These three valence bands are also referred to as the heavy mass,

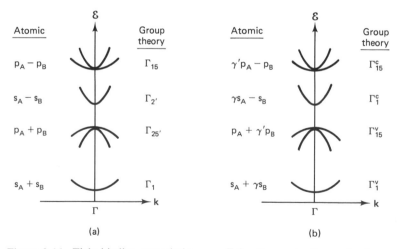

Figure 2.16 Tight-binding prescription near Γ for the conduction and valence bands of crystals with (a) the diamond structure and (b) the sphalerite structure. The eight bands are labeled with both atomic and group theory notation. Spin-orbit coupling effects are neglected. A and B refer to the two atoms in the unit cell.

light mass, and split-off bands, in order of decreasing energy. As indicated in Fig. 2.16, the four conduction bands are derived from antibonding s and p orbitals, with properties that reflect these atomic states.

It is interesting to see what happens to the bonding arrangement when the column IV elements in the diamond structure are replaced by two different elements in the sphalerite structure. As an example, let us replace the two Ge atoms in a unit cell by one Ga (column III) and one As (column V) atom. These atoms have the same core as Ge, with an outer atomic configuration of $(4s)^2(4p)$, $(4s)^2(4p)^2$, and $(4s)^2(4p)^3$ for Ga, Ge, and As, respectively. In the sphalerite structure each As atom is surrounded by four Ga atoms, and vice versa. In the unit cell the Ga atom can contribute three electrons to the bonds, and the As atom can contribute five electrons, for a total of eight. This is the same number of electrons a unit cell of Ge has, so we can expect the same kind of bonding arrangement. If these electrons are distributed uniformly in the unit cell, however, the $+5$ charge of the As core is not quite neutralized by its surrounding average electron charge of -4, and the $+3$ charge of the Ga core is in a similar situation. We thus have a residual positive charge in the region of the As core, which must be neutralized. As indicated in Fig. 2.17, this is easily realized by having a greater electronic charge around the As atom than around a Ga atom. The unit cell, however, still has eight electrons.

If we proceed to the II–VI compounds with the sphalerite structure, this accumulation of electronic charge around the nonmetallic ions becomes even more pronounced and the bonding arrangement looks less and less covalent. The result of this transfer of electrons from the metallic to the nonmetallic ion is that the bonding becomes more ionic in nature. Thus compound semiconductors have mixed covalent and ionic bonding.

The degree of ionic bonding can be estimated by the difference in electronegativity between the atoms forming the compound: that is, the greater the electronegativity difference, the larger the ionic component of bonding. Electronegativity describes the ability of an atom to attract electrons. Elec-

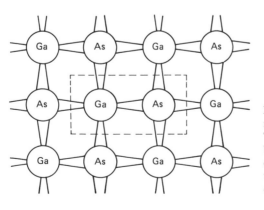

Figure 2.17 Two-dimensional schematic of bonding arrangement in GaAs. The tilt in the lines, which indicate the bonding electrons, is meant to show more charge around the As atoms than around the Ga atoms. The unit cell still has eight electrons.

TABLE 2.1   Electronegativity Values for the Elements in Tetrahedral Environments[a]

Li	Be	B	C	N	O	F
1.0	1.5	2.0	2.5	3.0	3.5	4.0
Na	Mg	Al	Si	P	S	Cl
0.72	0.95	1.18	1.41	1.64	1.87	2.1
(0.9)	(1.2)	(1.5)	(1.8)	(2.1)	(2.5)	(3.0)
Cu	Zn	Ga	Ge	As	Se	Br
0.79	0.91	1.13	1.35	1.57	1.79	2.01
(1.9)	(1.6)	(1.6)	(1.8)	(2.0)	(2.4)	(2.8)
Ag	Cd	In	Sn	Sb	Te	I
0.57	0.83	0.99	1.15	1.31	1.47	1.63
(1.9)	(1.7)	(1.7)	(1.8)	(1.9)	(2.1)	(2.5)
Au	Hg	Tl	Pb	Bi		
0.64	0.79	0.94	1.09	1.24		
(2.4)	(1.9)	(1.8)	(1.8)	(1.9)		

[a] The values in parentheses are from L. Pauling, *The Nature of the Chemical Bond* (Ithaca, N.Y.: Cornell University, 1960).

TABLE 2.2   Ionic Component of Bonding in Percent for Selected Semiconductors

Si	0	GaAs	32	CdO	79	ZnO	62
Ge	0	GaSb	26	CdS	69	ZnS	62
SiC	18	InP	44	CdSe	70	ZnSe	63
		InAs	35	CdTe	67	ZnTe	61
		InSb	32				

Source: J. C. Phillips, *Phys. Rev. Lett. 22,* 705 (1969).

tronegativity values for the elements in tetrahedrally coordinated environments [J. C. Phillips, *Bands and Bonds in Semiconductors* (New York: Academic Press, 1973)] are shown in Table 2.1. In general, the farther the elements are removed from one another in the periodic table, the greater their electronegativity difference is and their ionic component of bonding in compound formation. Table 2.2 shows the extent of ionic bonding for a number of semiconductors, calculated on this basis.

2.7 ENERGY BAND STRUCTURES

Figure 2.18 shows the results of an empirical pseudopotential calculation for the energy bands of the elemental semiconductors Si and Ge. The results of this calculation are in good agreement with experiments, except that spin-orbit effects are not included. These materials have the diamond crystal

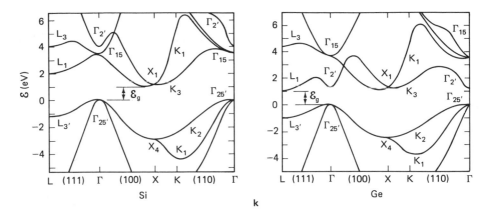

Figure 2.18 Empirical pseudopotential results for the energy bands of the elemental semiconductors Si and Ge. [After M. L. Cohen and T. K. Bergstresser, *Phys. Rev. 141*, 789 (1966).]

structure. As can be seen, the $\Gamma_{25'}$ valence bands for these two materials are quite similar, with the maximum at the Γ-point. The Γ_1-band is off-scale on the lower end of these figures. Although the valence bands are similar, there are significant differences in the conduction band structure. For many applications the most important energy band parameter is the energy gap, $\mathscr{E}_g$, defined as the minimum separation between the conduction and valence bands. For Si the energy gap (1.17 eV at 0 K) is between the valence band maximum at Γ and the conduction band minimum at about 0.8X in the (100) direction. This energy gap is referred to as an *indirect bandgap,* since a change in **k** value is required for an electron to make a transition from this conduction band minimum to the valence band maximum. Also, we note that there are six {100} directions in reciprocal space, so that the minimum at 0.8X in Si is actually *six equivalent minima.* For Ge the minimum in the conduction band is at L, on the outer surface of the first Brillouin zone. Thus it also has an indirect band gap of value 0.74 eV at 0 K. We can see from Fig. 2.5 that there are eight L-points (or eight {111} directions) in the first Brillouin zone. Since these points are shared with the next Brillouin zone, we find that Ge has *four equivalent minima* at L in the first Brillouin zone.

Most of the III–V compound semiconductors have the sphalerite crystal structure, so we expect the form of the energy bands to be similar to those for the diamond structure. The energy bands for GaAs and GaP, shown in Fig. 2.19, are fairly typical of this class of materials. As can be seen, the valence band structures are quite similar to Si and Ge. The minimum conduction band energy for GaP is at the X-point, so that GaP has an indirect bandgap with three equivalent conduction band minima. GaAs, however, has a minimum in the conduction band at Γ. Since no change in **k** value is

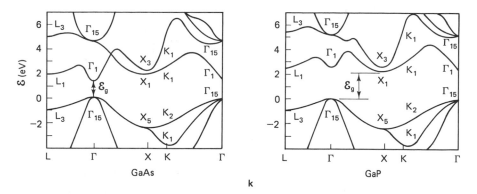

Figure 2.19 Energy bands for the III–V compound semiconductors GaAs and GaP. [After M. L. Cohen and T. K. Bergstresser, *Phys. Rev. 141*, 789 (1966).]

required for an electron to go from this conduction band to the maximum in the valence band, GaAs has a *direct bandgap*. Notice, however, that the minima at L_1 and X_1 are not far above Γ_1. This has important implications which will be discussed later.

The II–VI compound semiconductors crystallize into sphalerite or wurtzite structures. Figure 2.20 shows typical band structures for these materials. Sphalerite CdTe has a band structure similar to the III–V compounds with a direct bandgap. Wurtzite ZnS is also shown. The band structure in this case is more complex. The bandgap, however, is direct, which is typical of II–VI compounds in either the sphalerite or wurtzite crystal structure.

The II–IV–V$_2$ compounds crystallize in the chalcopyrite structure (Fig. 1.13), which is very similar to the sphalerite structure. Since the primitive unit cell is four times as large as for sphalerite, the Brillouin zone shown in Fig. 1.17 is four times smaller. These ternary compounds have simple an-

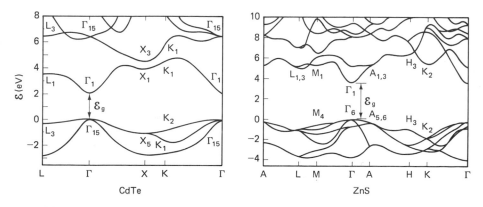

Figure 2.20 Energy bands for sphalerite CdTe and wurtzite ZnS. [After T. K. Bergstresser and M. L. Cohen, *Phys. Rev. 164*, 1069 (1967).]

alogs in the III–V binary compounds, and their energy band structure can be obtained by mapping the analog III–V bands into the chalcopyrite Brillouin zone. Because of the 4-to-1 size ratio, Γ, X, and W for sphalerite map into Γ for chalcopyrite; L and Σ map into N; and X and L map into T. Results for ZnGeAs$_2$, which has GaAs as its analog, and ZnGeP$_2$, with GaP as its analog, are shown in Fig. 2.21. ZnGeAs$_2$ has a direct energy gap at Γ, with conduction band minima Γ_1 derived from Γ_1 in GaAs (Fig. 2.19); Γ_3 and $T_1 + T_2$ from X_1; Γ_2 and T_5 from X_3; and N_1 from L_1. The three degenerate Γ_{15} valence bands in GaAs form the one nondegenerate Γ_4 and the two degenerate Γ_5 valence bands in ZnGeAs$_2$. In ZnGeP$_2$ the X_1 minima, which cause GaP to have an indirect bandgap, map into Γ_3 and $T_1 + T_2$ minima. If Γ_3 occurs at the lowest energy, ZnGeP$_2$ would have a direct bandgap while its III–V analog, GaP, is indirect. Experiments indicate that this could be the situation for several II–IV–V$_2$ compounds with indirect gap III–V analogs.

The IV–VI compounds have the NaCl structure or an orthorhombic structure very close to that of NaCl. Figure 2.22 shows the energy bands for PbTe, which is fairly typical of these compounds. The maximum in the valence bands is at L_6^+, with the minimum in the conduction bands at L_6^-. Thus PbTe has a direct energy gap at L which is small, 0.18 eV, and both bands have four equivalent extrema. Small, direct energy gaps are typical of these materials. However, the ordering of the bands is not the same in all IV–VI compounds. For example, the minimum in the conduction band of SnTe is L_6^+ and the maximum valence band is L_6^-, the reverse of

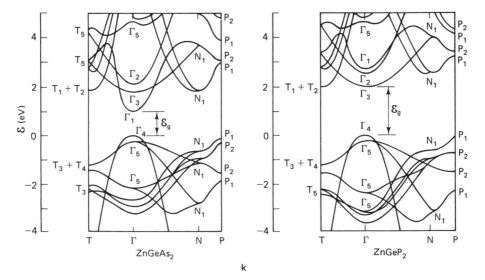

Figure 2.21 Energy bands for chalcopyrite ZnGeAs$_2$ and ZnGeP$_2$. [After A. Shileika, *Surf. Sci. 37*, 730 (1973).]

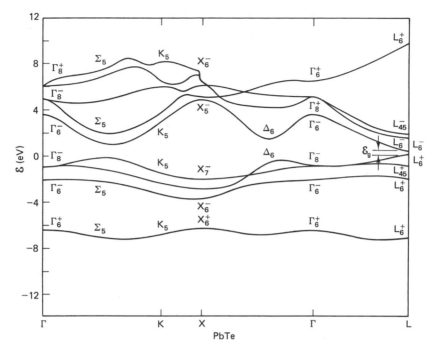

Figure 2.22   Energy bands for PbTe, which has the sodium chloride structure. [After Y. W. Tung and M. L. Cohen, *Phys. Rev. 180,* 823 (1969).]

PbTe [J. O. Dimmock et al., *Phys. Rev. Lett. 16,* 1193 (1966)]. For this reason, the energy gap of the alloy PbSnTe can approach zero at some intermediate composition [I. Melngailis, *J. Phys. Paris 29,* C4-84 (1968)].

In the energy band calculations above, spin-orbit effects are not included. When these effects are taken into account, some of the degeneracy of the highest-lying valence bands is removed, resulting in a lower split-off band. The energy separation between this split-off band and the degenerate light- and heavy-hole bands is referred to as the spin-orbit energy, Δ, shown

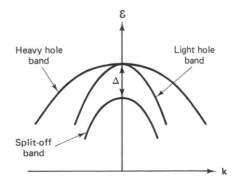

Figure 2.23   Heavy-hole, light-hole, and split-off valence bands.

TABLE 2.3   Spin-Orbit Splitting Energies

	Δ (eV)		Δ (eV)
C	0.006	InP	0.11
Si	0.044	InAs	0.38
Ge	0.29	InSb	0.82
α-Sn	0.80	ZnO	-0.005
AlN	0.012	ZnS	0.07
AlP	0.060	ZnSe	0.43
AlAs	0.29	ZnTe	0.93
AlSb	0.75	CdS	0.066
GaN	0.011	CdSe	0.42
GaP	0.127	CdTe	0.92
GaAs	0.34	HgS	0.13
GaSb	0.80	HgSe	0.48
InN	0.08	HgTe	0.99

in Fig. 2.23. Values of this parameter for a number of semiconductors are listed in Table 2.3.

2.8 ELECTRON DYNAMICS

There is an intimate relationship between band structure and the transport properties of electrons. By *transport properties* we mean the response of the electrons to applied forces. The proper approach to the problem of electronic motion in a periodic lattice would be to solve the time-dependent Schrödinger equation. A simpler solution, however, can be obtained, by first constructing a wave packet from plane wave solutions to the time-dependent Schrödinger equation. Then since it is well known that this wave packet behaves as a classical particle (correspondence principle), we can treat the problem semiclassically.

Let us then construct a wave packet (particle) from solutions to the time-dependent equation of the form

$$\psi(\mathbf{r}, t) = \psi(\mathbf{r}) \exp(-i\omega t) \tag{2.105}$$

where $\psi(\mathbf{r})$ is given in (2.10). The group velocity of the wave packet (or the average velocity of the particle) is

$$\mathbf{v} = \boldsymbol{\nabla}_k \omega = \frac{\partial \omega}{\partial \mathbf{k}} \tag{2.106}$$

If we use the quantum scalar operator

$$\mathscr{E} \equiv -\frac{\hbar}{i} \frac{\partial}{\partial t} \tag{2.107}$$

on the wave packet formed from (2.105), we obtain

$$\mathscr{E} = \hbar\omega \qquad (2.108)$$

which is just Planck's relationship. With this, (2.106) becomes

$$v = \frac{1}{\hbar}\,\nabla_k\mathscr{E} = \frac{1}{\hbar}\frac{\partial\mathscr{E}}{\partial\mathbf{k}} \qquad (2.109)$$

It should be pointed out that (2.109) is a rather surprising result. It tells us that an electron in a state specified by wavevector $\mathbf{k}$ in a given energy band has a velocity determined by the slope of the $\mathscr{E}$ versus $\mathbf{k}$ curves. An examination of the $\mathscr{E}$ versus $\mathbf{k}$ diagrams in Section 2.7 shows that for most allowed values of $\mathbf{k}$, the electrons will have a finite velocity. Thus an electron moving in a perfect periodic potential with no applied forces has a constant velocity and is *not scattered* by the atoms of the crystal. In other words, a perfect crystal offers no resistance to the motion of an electron. Only crystal imperfections serve to scatter electrons. We examine these scattering processes in Chapter 6.

2.8.1 Effective Mass Concept

In the meantime, let us apply an external force $\mathbf{F}$ to the wave packet. This force changes the energy of the electron according to

$$\mathbf{v}\cdot\mathbf{F} = \frac{d\mathscr{E}}{dt} = \frac{1}{\hbar}\,\nabla_k\mathscr{E}\cdot\hbar\,\frac{d\mathbf{k}}{dt} \qquad (2.110)$$

Then from (2.109),

$$\hbar\,\frac{d\mathbf{k}}{dt} = \mathbf{F} \qquad (2.111)$$

This is an important relationship. It tells us that $\hbar\mathbf{k}$ behaves as a momentum for external forces applied to a Bloch electron. It is for this reason that we defined a crystal momentum in (2.32). Let us now take the time derivative of the velocity in (2.109), to get

$$\frac{d\mathbf{v}}{dt} = \frac{1}{\hbar}\frac{d}{dt}\,\nabla_k\mathscr{E} = \frac{1}{\hbar}\,\nabla_k\left[\nabla_k\mathscr{E}\cdot\frac{d\mathbf{k}}{dt}\right] \qquad (2.112)$$

Using (2.111), equation (2.112) becomes

$$\mathbf{a} = \frac{1}{\hbar^2}\,\nabla_k[\nabla_k\mathscr{E}\cdot\mathbf{F}] \qquad (2.113)$$

where $\mathbf{a}$ is the electron acceleration. This equation tells us that $\mathbf{F}$ can produce a change in $\mathbf{v}$ in directions other than the direction of $\mathbf{F}$. Comparing this to the Newtonian force equation, we see that the closest thing to a mass for the electron is an inverse tensor which depends on the curvature of the $\mathscr{E}$ versus $\mathbf{k}$ diagrams. This can be determined by resolving (2.113) into components along three arbitrary axes. Then

$$a_i = \frac{1}{\hbar^2} \sum_j \frac{\partial^2 \mathscr{E}}{\partial k_i \partial k_j} F_j \qquad \text{for } i, j = 1, 2, 3 \qquad (2.114)$$

and we obtain an inverse effective mass tensor with components

$$\frac{1}{m_{ij}^*} = \frac{1}{\hbar^2} \frac{\partial^2 \mathscr{E}}{\partial k_i \partial k_j} \qquad (2.115)$$

If we examine the curvature of some of the energy bands in Section 2.7, we see that at band minima the effective mass is positive and at band maxima the effective mass is negative. This is the same as the result that we found for the nearly free electron energy bands in Section 2.3. Another conclusion that results from (2.115) is that the effective mass becomes infinite at some point within a band. This merely indicates the point in the band where the externally applied force stops accelerating and begins decelerating the electron. From (2.115) we can also see why the $\Gamma_{25'}$ and Γ_{15} valence bands in Figs. 2.18 and 2.19, respectively, are referred to as light and heavy bands.

Generally, the effective mass tensor can be diagonalized by a suitable choice of axes. If $\mathbf{k}$ is given by (2.22), a general expression for a band extrema at Γ or $\mathbf{k} = 0$ would be

$$\mathscr{E} = \mathscr{E}_0 \pm \sum_i A_i \mathbf{k}_i^2 \qquad \text{for } i = 1, 2, 3 \qquad (2.116)$$

where the $+$ and $-$ are for conduction and valence bands, respectively. Using (2.115) this becomes

$$\mathscr{E} = \mathscr{E}_0 \pm \frac{\hbar^2}{2} \mathbf{k} \cdot \mathbf{M} \cdot \mathbf{k} \qquad (2.117)$$

where $\mathbf{M}$ is the inverse effective mass tensor. For parabolic bands of spherical symmetry (2.117) reduces to

$$\mathscr{E} = \mathscr{E}_0 \pm \frac{\hbar^2 \mathbf{k}^2}{2m^*} \qquad (2.118)$$

the equation for a sphere in three dimensions.

For cylindrical symmetry, equal-energy surfaces are ellipsoidal with the form

$$\mathscr{E} = \mathscr{E}_0 \pm \frac{\hbar^2}{2} \left(\frac{\mathbf{k}_1^2}{m_1^*} + \frac{\mathbf{k}_2^2 + \mathbf{k}_3^2}{m_2^*} \right) \qquad (2.119)$$

For a band extrema at $\mathbf{k} = \mathbf{k}_0$, the equal-energy surfaces would have the form

$$\mathscr{E} = \mathscr{E}_0 \pm \frac{\hbar^2}{2} \sum_i \frac{(\mathbf{k}_i - \mathbf{k}_0)^2}{m_i^*} \qquad (2.120)$$

and so on. Cubic crystals typically have extrema at $\mathbf{k} = 0$ with spherical symmetry and extrema at $\mathbf{k} = \mathbf{k}_0$ with cylindrical symmetry. Hexagonal crystals have extrema at $\mathbf{k} = 0$ with cylindrical symmetry.

2.8.2 Effective Mass Limitations

Effective mass is one of the more important concepts in the analysis of semiconductors and semiconductor devices, so it is desirable to examine the conditions under which it can be used. Basically, it is obtained by not using or neglecting terms in the one-electron Schrödinger equation which can produce interband transitions. In Chapter 3, for example, where impurity energy levels are derived in the effective mass approximation, we tacitly assume that the impurity Coulomb potential in the Hamiltonian of (3.10) is not large enough to produce band-to-band transitions by Zener (field) tunneling. In Chapter 7, however, where optical absorption is developed, we include a photon vector potential in the Hamiltonian of (7.8) to provide for band-to-band transitions. The resulting equations for absorption depend on the free-electron mass.

Specifically, the effective mass approximation fails when the potentials in the one-electron Hamiltonian change so rapidly in time or in space that interband transitions occur in the semiconductor. For variations of potential with time, this sets an upper bound on the radial frequency of

$$\omega \simeq \frac{\mathscr{E}_g}{\hbar} \qquad (2.121)$$

where $\mathscr{E}_g$ is the minimum energy gap of the semiconductor. For variations of potential with space, this sets an upper bound on the electric field of

$$E \simeq \frac{\mathscr{E}_g}{qa} \qquad (2.122)$$

where a is the atomic spacing. When frequencies or fields of the magnitudes indicated by (2.121) and (2.122) are involved, the effective mass approximation is clearly not valid. There are, however, other limitations on the effective mass approximation that may be more restrictive than these.

For many applications we would like the charge carriers to be small enough (in some sense) compared to the smallest device dimension that they are localized within the device structure, and fast enough compared to the oscillation period that they can respond to the applied field. That is, the application may require that the charge carriers behave more like particles than like waves. (The effective mass approximation is valid only for a wave packet.) The limitation is simply that particles are constructed from a superposition of states indexed by k, within a range Δk, that are thermally distributed at some temperature, T. The range of energy $\Delta\mathscr{E}$ within the wave packet corresponding to Δk is then

$$\Delta\mathscr{E} = \frac{\hbar^2(\Delta k)^2}{2m^*} \simeq kT \qquad (2.123)$$

With (2.123) and the Heisenberg uncertainty relations,

$$\Delta x \Delta p \geq h, \qquad \Delta t \Delta\mathscr{E} \geq h \qquad (2.124)$$

we can estimate how localized in time and space a particle can be with an energy range of kT. The results are

$$\Delta t \geq \frac{h}{kT} \qquad (2.125)$$

and

$$\Delta x \geq \frac{h}{(2m^*kT)^{1/2}} \qquad (2.126)$$

For a semiconductor at 300 K with an energy gap of 1 eV, (2.125) gives a bandwidth of 6.3×10^{12} Hz, whereas (2.121) gives 2.4×10^{14} Hz, more than an order of magnitude larger. However, even 6.3×10^{12} Hz is much higher than the anticipated operating frequency of any current semiconductor device, so there should be no problem with the effective mass approximation in this aspect. At 300 K (2.126) gives

$$\Delta x \geq 76 \left(\frac{m}{m^*}\right)^{1/2} \text{Å} \qquad (2.127)$$

as the minimum size of a wave packet with an energy spread of kT. This value can be interpreted as the smallest dimension a device can have and still be analyzed using the effective mass approximation. For GaAs with an effective mass of $0.067m$, (2.127) gives a minimum device dimension of 294 Å. Since this is much larger than the dimensions of many quantum well devices, the effective mass approximation would not be expected to be a very reliable tool in evaluating these devices.

PROBLEMS

2.1. Plot the two lowest-lying empty lattice energy bands for the simple hexagonal lattice with ideal c/a ratio from A to L to M to Γ to A to H to K to Γ. Indicate the degeneracy of each band.

2.2. Use the nearly free electron approximation to obtain an expression in terms of a, b, and U_0 for the forbidden energy gaps of the periodic square-well potential shown in Fig. P2.2.

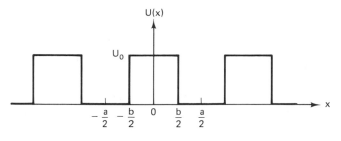

Figure P2.2

2.3. **(a)** Use the tight-binding approximation to obtain an expression for the energy bands of a simple cubic direct lattice with a lattice constant of a. Use nearest neighbors only.

  (b) Determine the electron velocity, $\mathbf{v}$, and the effective mass tensor, m_{ij}, for these bands.

2.4. Obtain an expression similar to (2.103) for a tight-binding energy band in the simple hexagonal lattice and plot as in Fig. 2.12 from L to M to Γ.

2.5. **(a)** Assuming no collisions, obtain an expression for the time it takes an electron to completely traverse the first Brillouin zone in a face-centered cubic crystal in the direction of L under an applied electric field $\mathbf{E}$.

  (b) If the average time between collisions is 10^{-12} s, what fraction of the zone can an electron travel in an electric field of 10 V/cm? Use a lattice constant of 5 Å.

2.6. Consider a hypothetical solid made up of N hydrogen atoms at positions R_j on a simple cubic lattice of spacing a.

  (a) Write the expression for a one-electron wave function ψ_k which is a linear combination of $1s$ atomic orbitals ϕ_{1s}.

  (b) Show that if the ψ_k is properly constructed, it satisfies the Bloch theorem.

  (c) Using the tight-binding method, obtain $\mathcal{E}(k) = \mathcal{E}_0 - \mathcal{E}_0' - 2\mathcal{E}_\pm' \cos ka$.

2.7. Calculate the four lowest empty lattice free-electron energy bands for a simple square lattice in the $\Delta(\Gamma - X)$ and $\Sigma(\Gamma - M)$ directions and the two lowest bands in the $Z(X - M)$ directions, and plot in terms of energy in units of $\hbar^2/ma^2$ and wave vector in units of $2\pi/a$.

2.8. The sp^3 bond hybridization is very important in the bonding of diamond, Si, and Ge. These bonds can be described by linear combinations of the s, p_x, p_y, and p_z wavefunctions of the valence electrons. The appropriate linear combinations for diamond are

$$\psi_1 = \frac{1}{2}(\phi_s + \phi_{px} + \phi_{py} + \phi_{pz}), \quad \text{where } \phi_s = \frac{1}{(4\pi)^{1/2}}$$

$$\psi_2 = \frac{1}{2}(\phi_s - \phi_{px} - \phi_{py} + \phi_{pz}), \quad \text{where } \phi_{px} = \left(\frac{3\pi}{4}\right)^{1/2} \sin\theta\cos\phi$$

$$\psi_3 = \frac{1}{2}(\phi_s + \phi_{px} - \phi_{py} - \phi_{pz}), \qquad \text{where } \phi_{py} = \left(\frac{3\pi}{4}\right)^{1/2} \sin\theta \sin\phi$$

$$\psi_4 = \frac{1}{2}(\phi_s - \phi_{px} + \phi_{py} - \phi_{pz}), \qquad \text{where } \phi_{pz} = \left(\frac{3\pi}{4}\right)^{1/2} \cos\theta$$

(a) Considering only the angular parts of the wave functions, find the directions of the maximum charge densities $\{q \mid \psi_1 \mid^2, q \mid \psi_2 \mid^2, q \mid \psi_3 \mid^2,$ and $q \mid \psi_4 \mid^2\}$ of these bonds.

(b) What are the angles between the directions of maximum charge concentration for ψ_1 and ψ_2, ψ_3 and ψ_4, and ψ_1 and ψ_3?

2.9. Calculate and plot the constant-energy contours in the $k_x k_y$-plane of the Brillouin zone (reduced zone scheme) of the simple cubic lattice in

(a) The tight-binding approximation

(b) The nearly free electron approximation

(c) The free-electron approximation

2.10. Use degenerate perturbation theory to calculate the splitting and form for small k of the second and third energy bands at the center of the zone of a one-dimensional crystal in the nearly free electron model. What is the effective mass of an electron at the bottom of the third energy band? What is the effective mass of an electron at the top of the second energy band?

2.11. Calculate (estimate) and plot the electron velocity for $\mathbf{k}$ from Γ–X along Δ for GaAs. The effective mass at the bottom of the Γ minimum is $0.0665m_0$, and the effective mass at the bottom of the X minima can be assumed to be spherical and equal to $0.4m_0$.

Crystal
Imperfections

Up to now we have considered only the properties of perfect crystals. We assumed that the atoms of the crystal were arranged at the sites of a mathematically precise Bravais lattice. This resulted in a perfectly periodic lattice potential energy for the electrons and we were able to establish relationships between the energy and wavevector of an electron on this basis. We also found, incidentally, that electrons can have a constant average velocity in a perfect periodic potential without the application of external forces. This, of course, is contrary to experimental observation. Because of this and other important experimental results, we realize that the perfect crystal is an overly simplified picture of a more complicated situation. Thus, to take into account many of the more important properties of semiconductors, we must examine possible crystal imperfections. We will base our analysis of these imperfections on the assumption that they can be, in general, regarded as perturbations in a perfect crystal.

3.1 ELECTRONS AND HOLES

Let us first consider the situation where one of the bonds between atoms in a perfect crystal is broken. From a chemical point of view, sufficient energy has been given to an electron to move it from a bonding state to an antibonding state. In the energy band picture, an electron is excited from somewhere near the top of a completely filled valence band to somewhere near the bottom of a completely empty conduction band. After the excitation,

of course, the valence band is missing one electron and the conduction band
has acquired one. We will assume for the time being that the conduction
band electron is *spatially* removed from the empty state it left in the valence
band a sufficient distance that interaction between them can be neglected.
This broken-bond situation represents an excited state of the one-electron
energy bands discussed in Chapter 2.

To understand what will happen in this situation, let us first look at
the electrons in a completely filled valence band under the influence of an
external force applied at $t = 0$. As indicated in Fig. 3.1, we assume only
eight allowed values of **k** for each energy band in the first Brillouin zone.
According to (2.111),

$$\hbar \frac{d\mathbf{k}}{dt} = \mathbf{F} \tag{3.1}$$

which tells us that the **k** vectors for all the electrons in the valence band
will increase continually in the direction of **F** for t greater than zero. Thus
the electron in state 0 will move to state 1, the electron in state 1 will move
to state 2, and so on. What will happen to the electron in state 8? It can
move to the first state in the second Brillouin zone. However, since we want
to have a unique wavevector for every electron state, we flip it over
(Umklapp process) to state 0 in the first Brillouin zone by adding to it a
reciprocal lattice vector $-\mathbf{K}$. As long as the force is applied, this process
is repeated for each electron and they cycle continuously through the first
Brillouin zone.

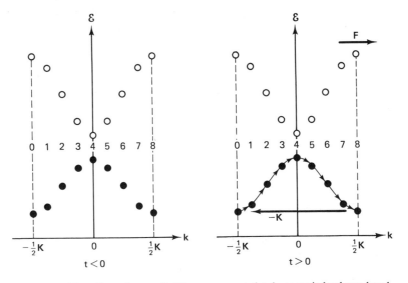

Figure 3.1   The effect of an applied force on a completely occupied valence band.
For simplicity we take $N = 8$ or 8 allowed values of **k** in the first Brillouin zone.
The open circles represent empty states and the solid circles represent filled states.

Let us look at the velocity and acceleration of an electron as it goes through one of these cycles. From (2.109) the velocity is given by

$$\mathbf{v} = \frac{1}{\hbar} \frac{\partial \mathscr{E}}{\partial \mathbf{k}} \tag{3.2}$$

If we consider a parabolic band, the effective mass tensor of (2.115) reduces to a scalar, and from (2.113) the acceleration is

$$\mathbf{a} = \frac{1}{m^*} \mathbf{F} \tag{3.3}$$

From (3.2) and (3.3) an electron starting at state 0 has zero velocity, and since m^* is positive at the bottom of a band, maximum acceleration in the direction of the force. It thus proceeds to state 1 and then state 2, increasing its velocity. At state 2 it has maximum velocity, and since m^* is infinite, zero acceleration. As the electron goes to state 3 its velocity starts to decrease, since m^* becomes negative, and the acceleration is opposite to the direction of the force. At state 4 the velocity is zero, m^* is negative, and the deceleration is maximum. From state 0 to state 4 the electron has gained energy from the applied force. As the electron goes from state 4 to 5, its velocity is in the direction opposite to the force. Thus, it starts to give energy back to the applied force, and so on. These variations of electron energy, velocity, acceleration, and effective mass during one cycle through the first Brillouin zone are summarized in Fig. 3.2 for a valence band and also for a conduction band.

It should be pointed out that the discussion above has been for pedagogical purposes only. In a real solid, an electron will normally undergo many collisions before it can traverse the first Brillouin zone. The effects of these collisions on carrier transport are discussed in Chapters 5 and 6. Without the collisions, however, it would be possible for a direct-current (dc) field to induce an alternating current in a semiconductor with a partially filled band [C. Zener, *Proc. R. Soc. London A145,* 523 (1934)].

From Figs. 3.1 and 3.2 we can see that for a filled valence band, there is no net energy transfer from the applied force to the electrons through either part or all of a cycle. If we sum all the values of **k** in Fig. 3.1, we find $k_4 = 0$, $k_3 = -k_5$, $k_2 = -k_6$, $k_1 = -k_7$, and $\frac{1}{2}k_0 = -\frac{1}{2}k_8$, so there is also no net gain in **k** from the applied force. (To avoid counting states twice, the sum is over only one-half of states 0 and 8.) It should be pointed out that these results are independent of the specific form of the energy bands, because of the equivalence of zone edge points in a given direction. Therefore, the only effect of an applied force on a filled valence band is to cause the electrons to accelerate and decelerate cyclically from one state to another. Without the force the electrons would stay in their states defined by constant wavevector **k** at a constant average velocity **v**.

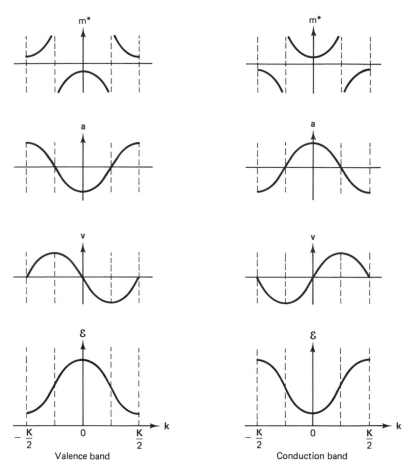

Figure 3.2   Variation of electron energy, $\mathscr{E}$, velocity, $\mathbf{v}$, acceleration, $\mathbf{a}$, and effective mass, m^*, with wavevector, $\mathbf{k}$, in the first Brillouin zone for both valence and conduction bands.

Let us now examine the effect of an applied force at $t = 0$ on a conduction band with one electron and a valence band with one missing electron at $\mathbf{k} = 0$. This situation is shown in Fig. 3.3. For $t < 0$ we assume that the electron in valence band state 4 has been excited into conduction band state 4. When the force is turned on, according to (3.1), the $\mathbf{k}$ vectors for *all* electrons increase simultaneously. Thus, at t_1 the electron in conduction band state 4 has moved to state 5, the electron in valence band state 8 has been Umklapped to state 0, the electron in state 0 has moved to state 1, and so on. The net effect of the valence band electron motion is that the empty state has moved from 4 to 5. At t_2 the conduction band electron has moved from state 5 to state 6, and the empty valence band state has moved from 5 to 6. Therefore, we see that the conduction band electron and the empty

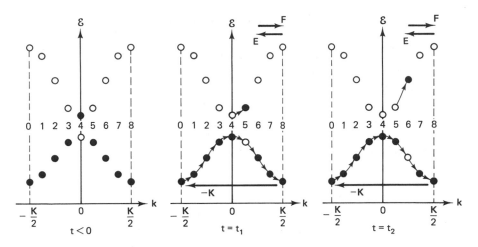

Figure 3.3 The effect of an applied force on a conduction band with one electron and a valence band with one missing electron.

valence band state both increase their **k** value in the direction of the applied force **F**.

If we sum the **k** vectors for the conduction band at t_1, the result is obviously k_5. For the valence band, the sum of the **k** vectors is $-k_3$. At time t_2 the electron in the conduction band has wavevector k_6, while the sum of the electron wavevectors in the valence band is $-k_2$. The interpretation of this behavior for the electron in the conduction band is straightforward: Its **k** vector increases with time under an applied force, **F**, according to (3.1). An interpretation of the behavior for the electron assembly in the valence band, however, is not as obvious: the sum of **k** vectors for the assembly decreases (becomes more negative) with time under an applied force, **F**. If we consider the force on an electron wave packet in an applied electric field, **E**,

$$\mathbf{F} = -q\mathbf{E} \tag{3.4}$$

$$\hbar \frac{d\mathbf{k}}{dt} = -q\mathbf{E}$$

we see that the direction of the force depends on the charge of the particle, $-q$. Since **E** is in the $-\mathbf{k}$ direction in Fig. 3.3 (**F** in the $+\mathbf{k}$ direction), the electron in the conduction band follows (3.4) nicely. The assembly of electrons in the valence band, however, has a **k** vector that decreases with time, so that (3.4) for them is

$$-\hbar \frac{d\mathbf{k}}{dt} = -q\mathbf{E}$$

$$h \frac{d\mathbf{k}}{dt} = +q\mathbf{E} \tag{3.5}$$

This is the equation of motion for a positively charged particle provided that the sum of velocities for the assembly of valence band electrons is in the same direction as the velocity of the conduction band electron. From Figs. 3.2 and 3.3, the conduction band electron has velocity $+v_5$ at t_1 and $+v_6$ at t_2. When we sum over the valence band electrons with the aid of Figs. 3.2 and 3.3, we find that the valence band assembly has a net velocity $+v_3$ at t_1, which is equal to $+v_5$ for the conduction band, and $+v_2$ at t_2, which is equal to $+v_6$. Therefore, the conduction band electron and the assembly of valence band electrons have the same velocity. If we look at the acceleration, the net effective mass for the assembly of valence band electrons behaves in a manner similar to the effective mass of the conduction band electron: that is, they are both positive, negative, or infinite at the same time. We can also see from Fig. 3.3 that the energy of the valence band electrons with a vacant state increases with time under an applied force.

If we have followed this discussion carefully, we can see that there is nothing particularly unusual (other than the effective mass) about the behavior of the electron in the conduction band under an applied external force. These are usually referred to as conduction electrons, electrons, or sometimes, inappropriately, as "free" electrons. The assembly of valence band electrons with an empty state, however, is another matter. In an applied force we have seen that this entity has an increasing energy when the empty state goes to lower energy, a decreasing wavevector when the empty state goes to higher wavevector, a positive velocity when the empty state is in a negative velocity region, a positive effective mass when the empty state is in the top of the valence band, and a negative effective mass when the empty state is at the bottom of the band. In fact, the assembly of valence band electrons acts as a positively charged particle in an otherwise empty band. This particle would have positive m^* near the top of the band and negative m^* near the bottom. It would also travel from state 4 to 3 to 2 in Fig. 3.3 as the empty site traveled from 4 to 5 to 6. To avoid a tedious summation over N values of wavevector every time we wish to examine the properties of a valence band with empty states, it is convenient to invent an equivalent particle [W. Heisenberg, *Ann. Phys. Leipzig 10,* 388 (1931)]. Such a particle with the properties of the assembly of valence band electrons is called a *hole*.

At time t_2 in the electron energy band diagram of Fig. 3.3, the empty valence band state was at 6, moving in the direction of increasing **k**, while the fictitious hole appeared to be at 2, moving in the direction of decreasing **k**. This is the situation shown in Fig. 3.4(a). Equation (3.5) tells us, however, that the wavevector for the hole should increase in the direction of **E** because of its positive charge. Also, we saw that the valence band energy increased as the vacant state went toward decreasing electron energy. Thus the direction of increasing energy for the hole is opposite to the electron energy. For these reasons it is convenient to use two separate $\mathscr{E}$ versus **k** diagrams

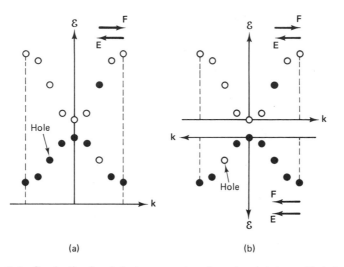

(a)   (b)

Figure 3.4 Conduction band electrons, empty valence band state, and hole in (a) an electron representation and (b) a mixed electron and hole representation.

simultaneously, as indicated in Fig. 3.4(b): one to represent the behavior of electrons in the conduction band and another to represent the behavior of holes in the valence band.

Although we have demonstrated that the assembly of valence band electrons with an empty state behaves as a positively charged particle, it is somewhat difficult to see the physical origin of this charge in the energy band scheme. In the bond diagram for a covalent semiconductor in Fig. 3.5, however, the origin of the positive charge is obvious. An examination of the unit cell that contains the broken bond or hole has a net charge of $+1$ from the atomic cores.

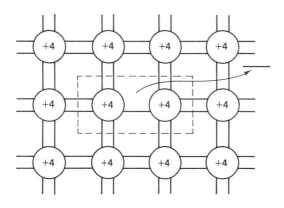

Figure 3.5 Bond diagram of a covalent semiconductor positive charge in the unit cell that contains the hole or broken bond.

At the beginning of this discussion on electrons and holes, we indicated that the broken-bond situation was an excited state of the one-electron energy bands. It is not, however, the first excited state. We assumed that the electron and hole were sufficiently removed from one another that interaction between them could be neglected. When they are sufficiently close we expect a Coulombic attraction between them because of their opposing effective charges. This interaction can result in a situation where the electron and hole circle each other around their joint center of mass. This bound electron–hole pair is referred to as an *exciton* since it represents the first excited state of the one-electron energy bands. The Coulomb attraction lowers the energy of an electron, which would, otherwise, be in the conduction band, and produces a series of allowed energy states just below the conduction band in the forbidden bandgap. When an exciton travels throughout the crystal, it carries no net charge and is thus not important in conduction processes. Excitons, however, play an important role in optical processes, and we consider them in more detail in Chapter 7.

3.2 IMPURITIES

One of the most important technological properties of semiconductors is the ability to make substantial changes in conductivity and other physical processes by the addition of impurity atoms. The controlled addition of impurities to alter the properties of a semiconductor is called *doping*. To determine how an impurity affects certain properties, it is necessary to consider several factors:

1. Does the impurity atom replace an atom of the host crystal, or does it take up an interstitial position in the host lattice? The former are referred to as *substitutional* impurities and the latter as *interstitial* impurities.
2. If the impurity is substitutional, what is its valence relative to the host atom it replaces? The valence is also important for an interstitial impurity.
3. If a substitutional impurity has the same valence as the host atom, what is its electronegativity relative to the host atom it replaces? Although these impurities are commonly referred to as *isoelectronic,* the proper terminology is *isovalent*. The effect of such impurities in an otherwise perfect crystal is most easily seen in the bond picture.

3.2.1 Bonding Concepts

When a column V impurity substitutes for a column IV Ge atom in the diamond structure as in Fig. 3.6(a), four of the five valence electrons from

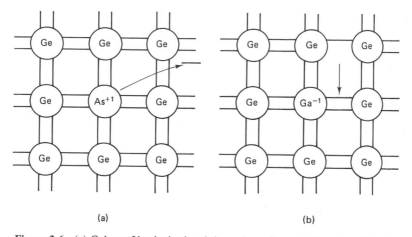

Figure 3.6   (a) Column V substitutional donor impurity and (b) column III substitutional acceptor impurity in Ge.

the impurity replace the four electrons of the Ge atom to complete the bonding arrangement. Since all valence bonds or valence band states are filled, the fifth electron must go into conduction band states. However, the fifth electron is attracted to the net positive charge in the impurity core, and this produces a series of allowed energy levels in the forbidden bandgap just below the conduction band. The energy separation between these levels and the conduction band is sufficiently small that at room temperature, the attraction can be overcome and the electron will be free to move in the conduction band. Since this column V impurity can donate one electron to the conduction band, it is referred to as a monovalent *donor*. When the conductivity of a semiconductor is controlled by donor impurities, it is called *n-type*. Such a donor impurity can exist in two charge states: When the electron is bound to the impurity, it is neutral; when the electron is ionized to the conduction band, it is positive.

With the column III impurity shown in Fig. 3.6(b), the situation is significantly different. There are only three valence electrons to replace the four valence electrons of the Ge atom, so that one valence bond is empty. If an electron is removed from a valence bond elsewhere to complete the bonding arrangement around the impurity atom, the missing electron elsewhere is a positively charged hole and the impurity atom has a net negative charge from the extra electron it *accepted* to complete its bonds. The attraction between the hole and the negatively charged *acceptor* impurity produce bound states for valence band electrons in the forbidden bandgap just above the valence band. When acceptors control the conductivity the material is *p-type*. If the hole acquires sufficient energy to overcome this attraction, it is free to conduct net charge in the valence band. When this monovalent acceptor has a hole bound to it (empty bond), it is neutral. When the hole is free (filled bonds), it is negative.

Similar behavior occurs in compound semiconductors. For example, in a III–V compound such as GaAs a column VI impurity replacing an As atom would be a donor, while a column II impurity replacing a Ga atom would be an acceptor. Some impurities, however, can be donors or acceptors, depending on which host atom in the compound semiconductor they replace. In Fig. 3.7(a) we show a column IV atom replacing a Ga atom in GaAs. Three of the four Ge valence electrons replace the three Ga valence electrons, leaving the fourth to conduct charge in the conduction band. In Fig. 3.7(b), a column IV impurity replaces an As atom. Since there are only four valence electrons to replace five from the As atom, a hole can be formed to conduct charge in the valence band. The ability of an impurity to produce either *n*- or *p*-type conductivity is referred to as *amphoteric* behavior. For Ge in GaAs it is possible to make a *p-n* junction utilizing only this single dopant!

When an impurity atom differs in valence from the host atom it replaces by more than one, it can donate more than one electron to the conduction band or accept more than one electron from the valence band. Figure 3.8 shows the effects of a column VI donor in Ge. In this case there are two electrons left over after the bond is completed. If one electron is removed to the conduction band as in Fig. 3.8(a), the impurity has a net charge of $+1$. When the second electron is removed as in Fig. 3.8(b), the impurity has a net charge of $+2$. This impurity would be a *divalent donor*.

An impurity that has the same valence as the host atom it replaces must be looked at differently. In this situation, the behavior of the isovalent impurity is determined by the electronegativity difference between it and

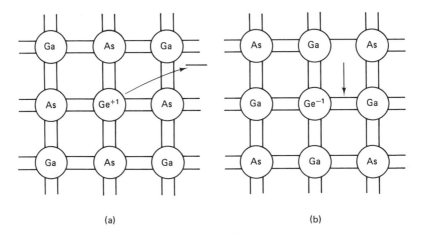

(a)   (b)

Figure 3.7   Column IV impurity in GaAs can be (a) a donor on a Ga lattice site and (b) an acceptor on a As lattice site.

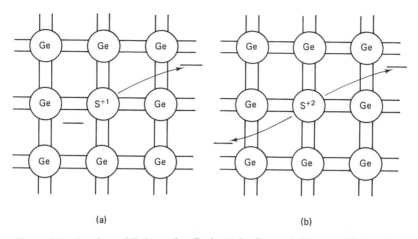

Figure 3.8   A column VI donor for Ge in (a) its first and (b) second ionization states.

the host atom. Values of electronegativity for the elements are given in Table 2.1. When the electronegativity of the impurity is sufficiently less than that of the host atom, it can lose an electron and become positively charged. The positive charge can then trap an electron in an outer orbit. These impurities are called *isovalent donors*. Some examples are Bi substituting for P in GaP and Te substituting for S in CdS. On the other hand, if the electronegativity of the impurity is sufficiently greater than that of the host atom, it can accept an electron from a host atom bond, become negatively charged, and trap the hole from the host bond in an outer orbit. These are referred to as *isovalent acceptors*. Examples are N for P in GaP and O for Te in ZnTe.

Interstitial impurities are apparently not as important as substitutional impurities for most semiconductors. The atoms have to be very small to occupy the interstitial sites in semiconductor crystal structures. Generally, their behavior depends on the relative ease with which they can completely fill or completely empty an electronic shell. Thus column I impurities would tend to be donors, while column VII impurities would be acceptors.

3.2.2 Effective Mass Approximation

If the binding energy of an electron or hole to a donor or acceptor impurity is relatively small, the energy levels can be determined by considering the impurity potential energy as a perturbation on the periodic potential energy of an otherwise perfect lattice. This calculation is called an *effective mass approximation* [W. Kohn, *Solid State Phys. 5*, 257 (1957)] and is illustrated schematically in Fig. 3.9. Because of Coulombic attraction, an electron (or hole) will move in orbital motion around the fixed donor (or acceptor). When the binding energy is small, the orbital radius will be very

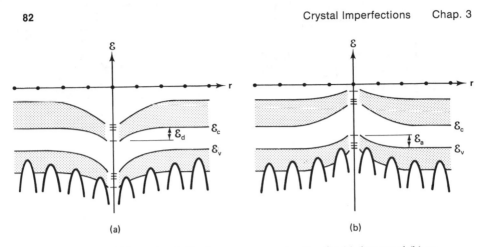

Figure 3.9 Schematic of effective mass approximation for (a) donor and (b) acceptor energy levels in the forbidden bandgap.

large. An orbiting particle will then see the Coulombic potential energy of the charged core of the impurity screened through the polarizability of a large number of host valence and core electrons. When the number of these polarized electrons is sufficiently large, we can use the macroscopic dielectric constant, ϵ_r, to reduce the Coulombic potential energy of the charged impurity core.

Let us first consider the problem for the unperturbed lattice. The one-electron Schrödinger equation is given by (2.1) as

$$\mathbf{H}\psi_k(\mathbf{r}) = \mathscr{E}\psi_k(\mathbf{r}) \tag{3.6}$$

where

$$\mathbf{H} = \frac{-\hbar^2}{2m}\nabla^2 + U(\mathbf{r}) \tag{3.7}$$

From (2.10) the unperturbed wavefunction has the form

$$\psi_k(\mathbf{r}) = A \exp(i\mathbf{k}\cdot\mathbf{r})u_k(r) \tag{3.8}$$

The problem for the lattice perturbed by an impurity atom is set up in a similar manner. The Schrödinger equation for the electron from the impurity is

$$\mathbf{H}_i\psi_i(\mathbf{r}) = \mathscr{E}_i\psi_i(\mathbf{r}) \tag{3.9}$$

where

$$\mathbf{H}_i = \frac{-\hbar^2}{2m}\nabla^2 + U(\mathbf{r}) - \frac{Zq^2}{4\pi\epsilon r}$$

$$= \mathbf{H} - \frac{Zq^2}{4\pi\epsilon r} \tag{3.10}$$

In (3.10), q is the electronic charge, ϵ the permittivity of the crystal ($\epsilon = \epsilon_r\epsilon_0$), $\mathbf{r}$ the distance of the electron from the impurity core, and Z accounts for a multiply charged impurity: $Z = 1$ for monovalent donor, $Z = 2$ for divalent donor, and so on.

We first construct a wavefunction for the impurity from the Bloch electron wavefunctions in (3.8) as

$$\psi_i(\mathbf{r}) = \sum_k A_k\psi_k(\mathbf{r})$$

$$= \sum_k A_k \exp\,(i\mathbf{k\cdot r})u_k(\mathbf{r}) \tag{3.11}$$

Note that in (3.11) the sum is over all $\mathbf{k}$ in the first Brillouin zone. Using (3.11) in (3.9) we have

$$\left[\mathbf{H} - \frac{Zq^2}{4\pi\epsilon r}\right]\sum_k A_k\psi_k(\mathbf{r}) = \mathcal{E}_i \sum_k A_k\psi_k(\mathbf{r}) \tag{3.12}$$

We now multiply (3.12) on the left by $\psi_{k'}^*(\mathbf{r})$ and integrate over the entire volume of the crystal, V, to obtain

$$\frac{1}{N}\sum_k A_k \int_V \psi_{k'}^*(\mathbf{r})\left[\mathbf{H} - \frac{Zq^2}{4\pi\epsilon r}\right]\psi_k(\mathbf{r})\,d\mathbf{r} = \frac{\mathcal{E}_i}{N}\sum_k A_k \int_V \psi_{k'}^*(\mathbf{r})\psi_k(\mathbf{r})\,d\mathbf{r}$$

$$\tag{3.13}$$

The integration must be over the entire crystal because the Coulombic potential and the electron wavefunction extend over a large number of unit cells. We know, however, that for normalization

$$\frac{1}{N}\int_V \psi_{k'}^*(\mathbf{r})\psi_k(\mathbf{r})\,d\mathbf{r} = \delta_{k',k} \tag{3.14}$$

and

$$\frac{1}{N}\int_V \psi_{k'}^*(\mathbf{r})\mathbf{H}\psi_k(\mathbf{r})\,d\mathbf{r} = \mathcal{E}\delta_{k',k} \tag{3.15}$$

where $\delta_{k',k}$ is the Kronecker delta. Equation (3.13) then simplifies to

$$(\mathcal{E} - \mathcal{E}_i)A_k - \sum_{k'} U_{kk'}A_{k'} = 0 \tag{3.16}$$

where

$$U_{kk'} \equiv \frac{1}{N}\int_V \psi_{k'}^*(\mathbf{r})\frac{Zq^2}{4\pi\epsilon r}\psi_k(\mathbf{r})\,d\mathbf{r}$$

$$= \frac{1}{N}\int_V u_{k'}^*(\mathbf{r})u_k(\mathbf{r})\exp\,[i(\mathbf{k} - \mathbf{k'})\cdot\mathbf{r}]\frac{Zq^2}{4\pi\epsilon r}\,d\mathbf{r} \tag{3.17}$$

Since the Bloch functions are periodic in **R**, we can expand their products in Fourier series with index **K** and

$$u_{k'}^*(\mathbf{r})u_k(\mathbf{r}) = \sum_K B_K \exp(i\mathbf{K}\cdot\mathbf{r}) \qquad (3.18)$$

where

$$B_K = \frac{1}{V}\int_V u_{k'}^*(\mathbf{r})u_k(\mathbf{r}) \exp(-i\mathbf{K}\cdot\mathbf{r}) \, d\mathbf{r} \qquad (3.19)$$

Using (3.18) in (3.17), we have

$$U_{kk'} = \frac{Zq^2}{4\pi\epsilon N}\sum_K B_K \int_V \exp[i(\mathbf{k} - \mathbf{k}' + \mathbf{K})\cdot\mathbf{r}] \frac{d\mathbf{r}}{\mathbf{r}} \qquad (3.20)$$

The integral in (3.20) can be evaluated by putting it in spherical coordinates and integrating over all three-dimensional direct space. Since

$$\int_0^\infty r \exp(-ar) \, dr = \frac{1}{|a|^2} \qquad \text{for Re}(a) > 0 \qquad (3.21)$$

(a being complex), (3.20) becomes

$$U_{kk'} = \frac{Zq^2}{\epsilon N}\sum_K \frac{B_K}{|\mathbf{k} - \mathbf{k}' + \mathbf{K}|^2} \qquad (3.22)$$

Substituting (3.22) into (3.16), we have the result

$$(\mathscr{E} - \mathscr{E}_i)A_k - \frac{Zq^2}{\epsilon N}\sum_{k'} A_{k'} \sum_K \frac{B_K}{|\mathbf{k} - \mathbf{k}' + \mathbf{K}|^2} = 0 \qquad (3.23)$$

Equation (3.23) is an exact expression relating the expansion coefficients of the impurity electron wavefunction, A_k, to the coefficients of the Bloch function, B_K. With this equation the impurity energy levels associated with any conduction or valence band extrema can be obtained.

Let us first look at the simplest case, which is a parabolic conduction band minimum at Γ. For small values of **k** the most important term in the summation over **K** in (3.23) is **K** = 0 and the Bloch electron energy is

$$\mathscr{E} = \mathscr{E}_0 + \frac{\hbar^2 k^2}{2m^*} \qquad (3.24)$$

For this case (3.23) is

$$\left(\mathscr{E}_c - \mathscr{E}_i + \frac{\hbar^2 k^2}{2m^*}\right)A_k - \frac{Zq^2}{\epsilon N}\sum_{k'} \frac{A_{k'}}{|\mathbf{k} - \mathbf{k}'|^2} = 0 \qquad (3.25)$$

since from (3.19) and (3.14), $B_0 = 1$. Equation (3.25) is similar to a well-known equation which appears unfamiliar because it is in reciprocal space. To transform (3.25) into direct space we define a Fourier transform, $A(\mathbf{r})$, for the coefficients of the impurity wavefunction, A_k, as

$$A(\mathbf{r}) = \sum_k A_k \exp(i\mathbf{k}\cdot\mathbf{r}) \tag{3.26}$$

Let us now multiply (3.25) by $\exp(i\mathbf{k}\cdot\mathbf{r})$ and sum over $\mathbf{k}$:

$$\sum_k \left(\mathscr{E}_c - \mathscr{E}_i + \frac{\hbar^2 k^2}{2m^*}\right) A_k \exp(i\mathbf{k}\cdot\mathbf{r}) - \frac{Zq^2}{\epsilon N} \sum_k \sum_{k'} \frac{A_{k'} \exp(i\mathbf{k}\cdot\mathbf{r})}{|\mathbf{k} - \mathbf{k}'|^2} = 0$$

$$\tag{3.27}$$

The first term on the left in (3.27) can be transformed fairly easily. We just put $\mathbf{k}$ in operator form and take it outside the summation. This is just the inverse of operating on a wavefunction with a momentum operator. The first term is then

$$\sum_k \left(\mathscr{E}_c - \mathscr{E}_i + \frac{\hbar^2 k^2}{2m^*}\right) A_k \exp(i\mathbf{k}\cdot\mathbf{r}) = \left(\mathscr{E}_c - \mathscr{E}_i - \frac{\hbar^2}{2m^*}\nabla^2\right) A(\mathbf{r})$$

$$\tag{3.28}$$

In the last term of (3.27), let $\mathbf{k} = \mathbf{k}' + \mathbf{k}''$:

$$-\frac{Zq^2}{\epsilon N} \sum_k \sum_{k'} \frac{A_{k'} \exp(i\mathbf{k}\cdot\mathbf{r})}{|\mathbf{k} - \mathbf{k}'|^2} = -\frac{Zq^2}{\epsilon N} \sum_{k'} A_{k'} \exp(i\mathbf{k}'\cdot\mathbf{r}) \sum_{k''} \frac{\exp(i\mathbf{k}''\cdot\mathbf{r})}{|\mathbf{k}''|^2}$$

$$= -\frac{Zq^2}{4\pi\epsilon r} A(\mathbf{r})$$

$$\tag{3.29}$$

since the Fourier transform of $1/\mathbf{r}$ is

$$\frac{1}{\mathbf{r}} = \frac{4\pi}{N} \sum_k \frac{\exp(i\mathbf{k}\cdot\mathbf{r})}{k^2} \tag{3.30}$$

With (3.28) and (3.29) (3.27) is transformed into

$$\left(-\frac{\hbar^2}{2m^*}\nabla^2 - \frac{Zq^2}{4\pi\epsilon r}\right) A(\mathbf{r}) = (\mathscr{E}_i - \mathscr{E}_c) A(\mathbf{r}) \tag{3.31}$$

where the $A(\mathbf{r})$ are called Wannier functions.

Equation (3.31) has the form of the Schrödinger equation for a hydrogen atom. Thus we can use the solutions for the hydrogen atom with appropriate substitutions of effective mass for free mass and semiconductor permittivity. The result is a series of hydrogenic impurity energy levels given by

$$(\mathscr{E}_c - \mathscr{E}_{in}) = \frac{q^4 Z^2 m^*}{2n^2(4\pi\epsilon\hbar)^2} \tag{3.32}$$

where n has all positive integer values. For a donor associated with the conduction band minimum at Γ, (3.32) gives a series of donor ionization energies,

$$\Delta \mathscr{E}_{dn} \equiv \mathscr{E}_c - \mathscr{E}_{dn}$$

$$= 13.6 \left(\frac{Z}{n\epsilon_r}\right)^2 \left(\frac{m^*}{m}\right) \text{ eV} \tag{3.33}$$

where ϵ_r is the dielectric constant of the semiconductor. The orbital radii of an electron in these levels are

$$r_{dn} = \frac{4\pi\epsilon\hbar^2 n^2}{m^* q^2 Z}$$

$$= 0.53 \frac{n^2 \epsilon_r}{Z} \left(\frac{m}{m^*}\right) \text{ Å} \tag{3.34}$$

In analogy with a hydrogen atom, the donor has a 1s ground state given by $n = 1$ in (3.33), a 2s and three 2p excited states given by $n = 2$ in (3.33), and so on. For materials with nearly spherical conduction bands at Γ such as GaAs, this hydrogenic model gives results that are in very good agreement with experiment [G. E. Stillman et al., *Solid State Commun. 9*, 2245 (1971)].

Equation (3.32) can also be used to determine the ionization energy for acceptors associated with spherical valence band maxima at Γ. As indicated in Section 2.7, most of these maxima are degenerate. The light and heavy hole maxima at Γ can be accounted for by using an equivalent effective mass in (3.32) given as

$$m^* = (m_l^{3/2} + m_h^{3/2})^{2/3} \tag{3.35}$$

In materials with ellipsoidal indirect conduction band minima, such as Si and Ge, or with an ellipsoidal direct conduction band minimum, such as wurtzite II–VI compounds, (3.25) must be modified to take into account the anisotropic effective masses. Equation (3.31) then becomes

$$\left[\frac{-\hbar^2}{2m_t}\left(\frac{\partial^2}{\partial x^2} + \frac{\partial^2}{\partial y^2}\right) - \frac{\hbar^2}{2m_1}\frac{\partial^2}{\partial z^2} - \frac{Zq^2}{4\pi\epsilon r}\right] A(\mathbf{r}) = (\mathscr{E}_i - \mathscr{E}_c)A(\mathbf{r}) \tag{3.36}$$

where m_t and m_1 are the effective masses transverse and longitudinal to the principal axis of revolution of the minima. The z axis is chosen to lie along the direction of the **k** vector for indirect minima. Unlike the hydrogenic equation, however, (3.36) is not separable and must be solved numerically. Figure 3.10 shows how the cylindrical anisotropy of the bands for Si and Ge removes some of the degeneracy of the p states and produces a reordering of levels as compared to the hydrogenic levels on the left.

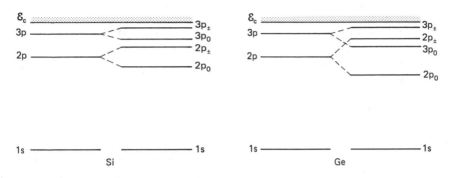

Figure 3.10   Effect of ellipsoidal conduction band minima on "hydrogenic" donor states in Si and Ge.

3.2.3 Ionization Energies

The effective mass approximation predicts that all donors or acceptors in a given material should have the same ionization energy. This is essentially true for impurity excited states. An examination of (3.34) shows that electrons in these states have large orbital radii and spend little time in the unit cell that contains the impurity. However, there are substantial differences in the $1s$ ground state from one impurity to another in a given material. Since an electron in the $1s$ state has spherical symmetry and the smallest orbital radius, it will spend more time in the unit cell containing the impurity and be more affected by impurity differences. Deviations in the ground-state energy from the hydrogenic value are referred to as *central cell corrections*. Qualitatively, these central cell corrections can be explained in many instances by differences in electronegativity between the substitutional impurity and the atom it replaces. If the impurity is more electronegative than the host, the ionization energy of the ground state should be larger than the hydrogenic value. When the impurity is less electronegative, the ground-state energy should be smaller than hydrogenic.

We can see from (3.33) and (3.34) that the higher excited states of an impurity have small ionization energies and large orbital radii because of the screening of the impurity potential by the valence electrons. If the concentration of an impurity is increased to the point where the average distance between impurity atoms is equal to twice the orbital radius of an excited state, we expect the wavefunctions of the impurities to interact. This interaction produces a banding of the excited state similar to the banding of atomic states in the tight-binding approximation. Because of the small ionization energy, when the banding of the impurity state is sufficiently large, it can merge with its associated valence or conduction band. As the impurity concentration is increased further, successively lower and lower excited states will interact, band, and eventually merge with the next-higher impurity band.

At some concentration, the ground state will broaden and merge with the banded first excited state, which is merged with the second, and so on. When this occurs, the impurity states are indistinguishable from the states of their associated conduction or valence bands. (In n-type GaAs this occurs at donor concentrations of about 10^{16} cm^{-3}.)

To distinguish between the ionization states of an impurity, it is convenient to introduce the charge notation 0, $+$, and $-$ to indicate the effective charge of the impurity atom. That is, D^0 represents a nonionized neutral donor, D^+ represents a donor that has lost its electron (no electron in the ground or excited states), A^0 a nonionized neutral acceptor, and A^- an acceptor that has captured an electron from a valence bond. A divalent donor that has lost both electrons would be D^{2+}. The ionization process can be represented as a quasi-chemical reaction of the form

$$D^0 = D^+ + e^- : \Delta\mathscr{E}_d$$
$$A^0 = A^- + h^+ : \Delta\mathscr{E}_a$$

(3.37)

where the ionization energies, $\Delta\mathscr{E}_d$ and $\Delta\mathscr{E}_a$, are the energy values required to make the reactions proceed from left to right. The energy required to ionize an impurity can be obtained from the ambient thermal energy of the crystal, by optical means, or from externally applied forces. The constraints on forming quasi-chemical equations, such as (3.37), are that mass and net charge be equal. An analogous equation for the formation of an electron–hole pair would be

$$0 = e^- + h^+ : \mathscr{E}_g$$

(3.38)

Ionization energies for common impurities in several semiconductors are listed in Table 3.1.

3.3 STOICHIOMETRIC DEFECTS

Crystalline imperfections where the atoms of the basis do not have a perfectly periodic arrangement on the sites of a Bravais lattice are called stoichiometric defects. These defects can be electrically or optically active and, in some materials, are more important than impurities in determining the properties of the material. The prevalence and importance of stoichiometric defects depend to a large extent on the ionicity of the bonding. That is, compounds with mostly ionic bonding can have large concentrations of stoichiometric defects, while the elemental semiconductors with covalent bonding have very few defects of this sort. This observation reflects the stability of the covalent bond.

There are several types of stoichiometric defects: vacant lattice sites, interstitial atoms, and misplaced atoms. If we consider a binary compound MX, where M represents the metal atom and X the nonmetal atom, we can

TABLE 3.1 Impurity Ionization Energies in meV for Several Semiconductors

	Si	Ge		GaAs	GaP		CdS	ZnSe
Li+	32.81	9.89	Si+	5.854	82.1	Li+	28.6	
P+	45.31	12.76	Ge+	5.908	201.5	Na+	31.5	
As+	53.51	14.04	Sn+	5.817	65.5	Al+		26.3
Sb+	42.51	10.19	S+	5.890	104.2	Ga+	33.1	27.9
Bi+	70.47	12.68	Se+	5.808	102.6	In+	33.8	28.9
B−	45	10.47	Te+	5.892	89.5	F+	35.1	29.3
Al−	57	10.80	Be−	30	48.7	Cl+	32.7	26.9
Ga−	65	10.97	Mg−	30	53.5	Br+	32.5	
In−	160	11.61	Zn−	31.4	64.0	I+	32.1	
Tl−		13.10	Cd−	35.4	96.5	Li−	165	114
			C−	26.7	48.0	Na−	169	100
			Si−	35.2	203	P−	~1000	700
			Ge−	41.2	257	As−	~1000	
			Sn−	171				

develop a shorthand notation for these defects. In a perfect unit cell an M atom will reside on an M site and an X atom will reside on an X site. The notation is $M_M X_X$: that is, the upper part of the symbol represents the atom and the subscript represents the site. An atom on an interstitial site would have a subscript i. An atom missing from a site would have an upper part V. Thus the native defects can be represented as:

1. Vacant lattice sites: V_M or V_X
2. Interstitial atoms: M_i or X_i
3. Misplaced atoms: M_X or X_M

These species can produce several important defect situations called:

1. *Schottky disorder:* V_M and/or V_X
2. *Frenkel disorder:* $V_M + M_i$ and/or $V_X + X_i$
3. *Antistructure disorder:* M_X and/or X_M

These defect situations involve reactions among defects and phases. For example, a Frenkel pair is formed by an atom going from a normal lattice site to an interstitial site. A vacancy can be formed by an atom leaving the solid and entering the vapor phase. To describe these reactions, we can use quasi-chemical equations similar to those used to describe the ionization of impurities. The rules for formulating these equations are as follows:

1. *Mass conservation.* The number of atoms of each constituent must be the same before and after the reaction. The mass of a vacancy is zero.
2. *Electrical neutrality.* Both sides of the equations must have the same net charge. Only neutral atoms (or molecules) are exchanged between external phases (gases or liquids) and the solid.
3. *Site stoichiometry.* For a binary compound MX, the number of M sites must equal the number of X sites. Thus, if an equation creates an M site, it must also create an X site. Adding a vacancy creates a site. Adding an interstitial does not create a site.
4. *Surface sites.* When an atom goes from the bulk to the surface, the number of sites the atom normally occupies increases.
5. *Interstitial atoms.* When dealing with interstitial atoms, empty interstitial sites (interstitial vacancies) must be included in the equations.

As examples of typical reactions among stoichiometric defects, let us first consider some changes that occur in the solid without exchange with an external phase. The formation of Schottky disorder would be

$$0 = V_M + V_X \qquad (3.39)$$

The atoms that occupied the M and X sites are removed to the surface. The formation of Frenkel disorder would be

$$M_M + V_i = M_i + V_M \tag{3.40}$$

and so on. When an atom from a surrounding vapor phase is incorporated into a binary compound, several reactions can occur:

$$M(g) = M_M + V_X \tag{3.41}$$

$$M(g) + V_M = M_M \tag{3.42}$$

$$M(g) + X_X = M_X + X(g) \tag{3.43}$$

$$M(g) + V_i = M_i \tag{3.44}$$

To incorporate substitutional or interstitial impurities from a vapor phase, the same rules apply. Let L represent the impurity atom. Several possibilities are

$$L(g) + M_M = L_M + M(g) \tag{3.45}$$

$$L(g) + V_i = L_i \tag{3.46}$$

$$L(g) + V_M = L_M \tag{3.47}$$

When a stoichiometric defect is electrically active, ionization reactions similar to those for impurities [reaction (3.37)] can also be formulated. In this case it is first necessary to know whether a given defect acts as a donor or acceptor. This is not too difficult to determine in ionic crystals where deviations from stoichiometry are large. In II–VI and IV–VI compounds, for example, the conductivity usually becomes more n-type when the samples are annealed in excess metal vapor. This indicates that either metal interstitials or nonmetal vacancies act as donors, since the concentrations of both of these species are expected to increase with this annealing process. An interstitial host atom would be expected to behave electrically like an interstitial impurity atom, that is, to donate or accept electrons in an effort to completely empty or fill the outer shell. On this basis, a metal interstitial in a II–VI compound should be a divalent donor. In the II–VI compounds, however, other measurements indicate that the increase in n-type conductivity is due to a nonmetal vacancy. The behavior of a vacancy is a little more difficult to visualize.

Figure 3.11(a) shows a Te vacancy in CdTe. In ionic materials most of the electronic charge in the bonds is concentrated around the more electronegative atom, in this case Te. Around a Te vacancy there are two electrons contributed by one or two Cd atoms which do not contribute to the bonding (dangling bonds). Since the Cd atoms around the vacancy share very little of the charge in the bonds, they can empty their outer electronic shell with relative ease by donating one or two electrons. In this manner we

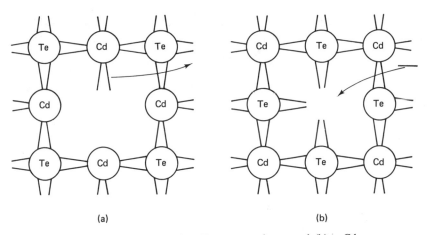

Figure 3.11 Bond diagram for (a) a Te-vacancy donor and (b) a Cd-vacancy acceptor in CdTe.

expect a Te vacancy to be a bivalent donor. A similar argument applies to the Cd vacancy in Fig. 3.11(b). Two or three Te atoms around the vacancy have completed outer electronic shells. The other one or two Te atoms can complete their shells by accepting one or two electrons.

In materials that are more covalent than ionic, such as the III–V compounds and the column IV elements (which are completely covalent), the concentrations of stoichiometric defects are much smaller than in ionic materials. The extra electrons around the defects tend to pair up to produce centers that can be neutral, positive, or negative, depending on other factors in the material. Because of their small concentrations, native defects typically play secondary roles in determining the properties of covalently bonded semiconductors. For example, they may interact with impurities to produce electrically compensating or nonradiative recombination centers.

3.4 COMPLEXES

When stoichiometric defects interact with impurities, they can form associated centers called complexes. Impurities can also interact with other impurities or defects with other defects to form complexes. Some of the most commonly observed complexes are those that involve association between oppositely charged donors and acceptors. When the donor and the acceptor are both singly charged, the quasi-chemical equation that describes the formation of the pair would be

$$D^+ + A^- = (DA)^0 \qquad (3.48)$$

This neutral complex can then be ionized by donating an electron to the conduction band or accepting a hole from the valence band,

$$(DA)^0 = (DA)^+ + e^- : \mathcal{E}_p$$
$$(DA)^0 = (DA)^- + h^+ : \mathcal{E}_p \tag{3.49}$$

In Equation (3.49) $\mathcal{E}_p$ is the ionization energy of the donor–acceptor pair which can be determined if we know the ionization energies of the isolated donor and acceptor.

Let us consider what happens when the positively charged donor and negatively charged acceptor are brought together. We expect a Coulombic attraction between the two ions which depends on their separation, r_{da}. As they are brought together, the energy levels of the donor are raised due to the negative charge on the acceptor, and the energy levels of the acceptor are lowered due to the positive charge on the donor. Thus in the pair the donor and acceptor energy levels move apart, so that

$$\mathcal{E}_p = \mathcal{E}_d - \mathcal{E}_a + \frac{q^2}{4\pi\epsilon r_{da}} \tag{3.50}$$

From (3.33) and (3.34) this becomes

$$\mathcal{E}_p = \mathcal{E}_g - (\Delta\mathcal{E}_d + \Delta\mathcal{E}_a) + \frac{q^2}{4\pi\epsilon r_{da}} \tag{3.51}$$

$$\mathcal{E}_p = \mathcal{E}_g + \frac{q^2}{4\pi\epsilon} \left(\frac{1}{r_{da}} - \frac{1}{2r_d} - \frac{1}{2r_a} \right) \tag{3.52}$$

where r_d and r_a are the orbital radii of the electron in the ground state of the isolated donor and the hole in the ground state of the isolated acceptor, respectively. From (3.52) we see that when

$$r_{da} < \frac{2r_d r_a}{r_d + r_a} \tag{3.53}$$

the ionization energy of the pair is greater than the energy gap. That is, the donor and acceptor energy levels are removed from the energy gap as the two ions are brought together. Since the Bohr radii of electrons and holes around isolated hydrogenic donors and acceptors are typically large, the formation of a complex will generally remove states from the forbidden band. This situation is shown in Fig. 3.12(a).

When one of the ions in the pair is doubly charged, the deeper second ionization state can remain in the forbidden gap. In this case the complex will have properties similar to those of a singly charged acceptor. The formation of such a complex is shown in Fig. 3.12(b).

If the ionization energy of one or both of the ions forming the complex is large, the orbital radius of the electron or hole will be small. In this situation it may be difficult for the complex to fulfill the inequality of (3.53), and the

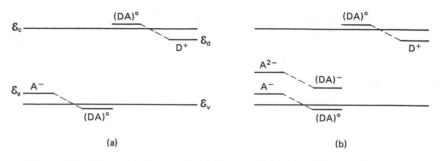

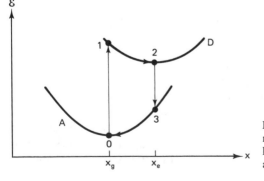

Figure 3.12 Formation of a complex (a) from a singly charged donor and acceptor and (b) from a singly charged donor and doubly charged acceptor. $\Delta\mathscr{E}_d = \mathscr{E}_c - \mathscr{E}_d$, $\Delta\mathscr{E}_a = \mathscr{E}_a - \mathscr{E}_v$.

levels of the complex derived from the isolated donor and acceptor levels will remain in the forbidden gap. Since these levels are deep (nonhydrogenic), an electron or hole in the level is tightly bound to the acceptor or donor. Under these conditions it is necessary to take into account interactions between the electron or hole and the complex in addition to the Coulombic attraction. Specifically, we have to consider the effect of the thermal vibrations of the complex and neighboring host atoms on the energy of the electron or hole in a bound state. This interaction is not important and can be neglected for hydrogenic impurities that weakly bind electrons and holes.

We can examine this interaction qualitatively with the configuration coordinate model shown in Fig. 3.13. The perturbation of the energy levels by vibrations of the complex and neighboring host atoms can be represented as a three-dimensional quantized harmonic oscillator. For simplicity, we consider only a one-dimensional classical oscillator. The ground state of the complex consists of an electron at zero in the level A derived from the isolated acceptor. In the ground state, A is thus negatively charged and D is positively charged. The coupling between the electron and complex causes a distortion of the lattice to minimize the electron energy. Thus the equilib-

Figure 3.13 Configuration coordinate model, taking into account the effect of lattice vibrations on the energy levels of a complex.

rium position of the lattice will be at x_g when the electron is in the ground state. The excited state of the complex consists of an electron at 2 in the level D derived from the isolated donor. In the excited state, A and D are now both neutral. The coupling between the electron and complex is different from the ground-state situation and the equilibrium position of the lattice is at x_e.

When an electron is excited from the ground state to the excited state, it first makes a transition from 0 to 1. This is because an electron transition can occur much faster than the atoms can move (Franck–Condon principle). Once the electron is in the excited state at 1, the lattice must move to its excited-state equilibrium position x_e. The electron in the excited state then goes from 1 to 2, giving the energy difference between 1 and 2 to the lattice. In a similar manner, when an electron makes a transition from the excited state to the ground state, it first makes the transition 2 to 3 and then 3 to 0 as the lattice moves to its new equilibrium position, x_g. Notice that the energy acquired by the electron going from 0 to 1 (photon absorption) is larger than the energy given off going from 2 to 3 (photon emission). The difference between these two energies is called the *Franck – Condon shift*. Since differences in absorption and emission are referred to as Stokes shifts, we see that the Franck–Condon shift is a Stokes shift due to atomic displacement. A similar analysis can be used to explain energy shifts and broadening due to electron or hole transitions between deep levels and conduction or valence bands.

3.5 DISLOCATIONS

The crystal imperfections we have examined so far are classified as point defects. Dislocations are linear defects which can extend throughout large regions of the crystal. Although they play an important role in the strength and fracture of metals, dislocations serve mainly to degrade the properties of semiconductors. If the dislocation has dangling bonds, they can act as a linear array of acceptors that scatter mobile carriers. The dangling bonds can also attract impurities and interstitials or act as an internal source and sink for vacancies. In practical applications, dislocations can degrade the performance of semiconductor devices in a number of ways. As a result, a substantial amount of effort has been expended in the semiconductor industry to minimize the number of dislocations introduced into a material during crystal growth and subsequent device processing.

Figure 3.14 shows an edge dislocation in the sphalerite crystal structure. Dislocations are characterized by their axis, $\mathbf{a}$, Burgers vector, $\mathbf{b}$, and slip plane. The axis indicates the direction along which the deviation from crystal periodicity occurs. This dislocation has a set of two dangling bonds in each atomic plane along the direction of its axis. The Burgers vector

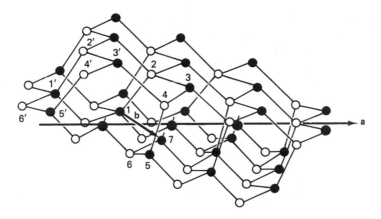

Figure 3.14 An edge dislocation in the sphalerite structure. The axis is along ⟨211⟩ and the Burgers vector is along ⟨110⟩. The slip plane is {111}.

represents the direction and displacement of the dislocated region compared to a perfect crystal. It can be determined by first taking a closed circuit around the bonds in a perfect region of the crystal and then taking the same circuit around a dislocated region. The vector required to close the circuit around the dislocated region is the Burgers vector. For example, in Fig. 3.14, the circuit 1′ to 2′ to 3′ to 4′ to 5′ to 6′ to 1′ represents a circuit in a perfect region. The same circuit around a dislocated region would be 1 to 2 to 3 to 4 to 5 to 6 to 7. The vector required to close the circuit from 1 to 7 is the Burgers vector. Notice that the Burgers vector is a direct lattice vector. The slip plane is the plane along which the dislocation can move under applied stresses, so it must contain both the axis and the Burgers vector. For this dislocation **a** is along ⟨211⟩, **b** is along ⟨110⟩, and the slip

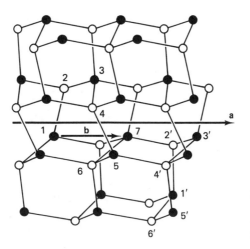

Figure 3.15 A screw dislocation in the sphalerite structure. The axis and Burgers vector are along ⟨110⟩. The slip plane is not defined.

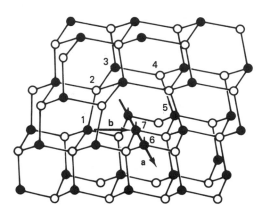

Figure 3.16 A 60° dislocation in the sphalerite structure. The axis and the Burgers vector are along different ⟨110⟩. The slip plane is {111}.

plane is {111}. Since the angle between **a** and **b** is 90°, this dislocation is referred to as an *edge dislocation*.

When the angle between **a** and **b** is zero, the dislocation is called a *screw dislocation*. If we take a circuit similar to that taken in Fig. 3.14 for the dislocation in Fig. 3.15, we find that **b** is in the same direction as **a**. Notice that this screw dislocation has no dangling bonds, but perturbs the lattice by a spiral rearrangement of bonds around its axis. Since **a** and **b** are in the same direction, a unique slip plane is not defined for this dislocation.

Although the dislocations we have seen so far were either pure edge or screw, most dislocations have both edge and screw components. Figure 3.16 shows a dislocation in the sphalerite structure in which the angle between **a** and **b** is 60°, so it has partially edge and partially screw properties. Other dislocations of this nature are possible in the sphalerite structure [J. Hornstra, *J. Phys. Chem. Solids 5*, 129 (1958)]. Dislocations must either extend throughout a crystal with each end terminated at a surface or form a closed loop within the crystal. Generally, the dislocation axis will twist and turn as it winds its way through the crystal, going from edge to screw to edge, and so on.

3.6 PLANAR DEFECTS

Two planar defects often observed in semiconductors can be most easily examined by first looking at the projections of the sphalerite and wurtzite crystal structures on a {110} and a {1100} surface, respectively. As can be seen in Fig. 3.17, there is a distinct difference between the hexagonal arrangement of atoms in the two projections. This is because the six atoms in the hexagonal projection for the sphalerite structure are arranged in a chair-like configuration in three dimensions (Fig. 1.10), whereas these six atoms in the wurtzite structure are arranged in a boatlike configuration (Fig. 1.12).

With these two projections in mind, let us examine the *intrinsic stacking*

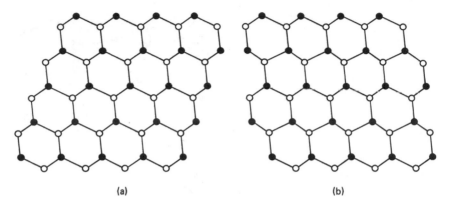

Figure 3.17   (a) Projection of the sphalerite structure on a {110} surface. (b) Projection of the wurtzite structure on a {1100} surface.

fault shown in Fig. 3.18(a) for the sphalerite structure. The top row of six atom sets has the sphalerite structure. The next two rows of six atom sets (shaded region) fits in the structure nicely but have a wurtzite configuration. The bottom two rows have the sphalerite structure. Thus an intrinsic stacking fault in the sphalerite structure is an included wurtzite region. Conversely, an intrinsic stacking fault in the wurtzite structure is an included sphalerite region. Figure 3.18(b) shows an *extrinsic* stacking fault, which has two rows of wurtzite-type bonding (shaded region) separated by a sphalerite region. Notice that for both the intrinsic and extrinsic stacking faults, two rows of included wurtzite region is required to return the crystal to its original struc-

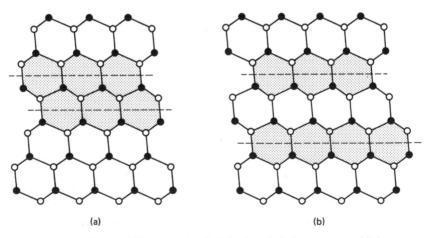

Figure 3.18   (a) An intrinsic stacking fault in the sphalerite structure. (b) An extrinsic stacking fault in the same structure. Twinning planes are indicated by dashed lines in both figures.

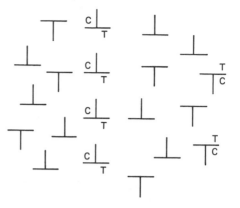

Figure 3.19 Diagram showing the formation of a low angle grain boundary or polygonization wall from a random field of dislocations. The letters "T" and "C" represent tensile and compressive stress fields, respectively.

ture. Stacking faults can terminate on the surface of a crystal or on dislocations referred to as partial dislocations.

The dashed lines in Fig. 3.18 indicate *twin planes*. It can be seen that the crystal structure on one side of a twin plane is a mirror image of the other side. If a sphalerite crystal has only one row of included wurtzite, the entire crystal on one side of the twin plane will be a mirror image of the other side.

Another planar defect that is often observed in semiconductors is the low-angle grain boundary or polygonization wall. This defect is formed as a low-energy configuration for a random field of dislocations. In Fig. 3.19 we represent each dislocation by an upright or an inverted T. The leg of the T represents the extra half-plane of atoms in the dislocation, while the arms of the T indicate the slip plane. The atoms near the extra half-plane above the slip plane are in compression, while the atoms just below the slip plane are in tension. If the dislocations are free to move around on their slip planes, a low-energy configuration is obtained when the compressive stress field of one dislocation reduces the tension from another. When a number of dislocations line up in this manner, a low-angle grain boundary is formed.

3.7 SURFACES

In the discussion of energy bands, we avoided the difficulties of dealing with the surfaces of a crystal by imposing a cyclic boundary condition. This allowed us to obtain some understanding of the bulk properties of the material. Real crystals, of course, have surfaces and these surfaces can influence the results of physical measurement and device fabrication in the material. In the band picture, a surface produces a termination of the periodic potential of the crystal. From this point of view, the deviation from perfect periodicity results in acceptor-like electronic states in the forbidden energy gap at the surface [I. Tamm, *Phys. Z. Sowjetunion 1*, 733 (1932)]. To avoid a tedious

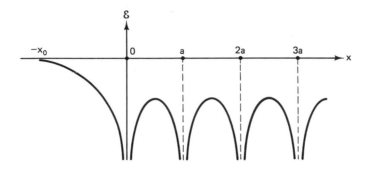

Figure 3.20   Diagram showing the termination of the periodic potential of a crystal at the surface.

calculation, let us look at a plausibility argument for the existence of these states.

Figure 3.20 shows the basic features of the argument. We look at the crystal in a direction, x, perpendicular to the surface and set the last atom at $x = 0$. Outside the crystal, the potential increases to $\mathscr{E} \simeq 0$, the minimum energy an electron must have to escape the crystal, at $x = -x_0$. For $x < -x_0$ Schrödinger's equation is

$$\frac{-\hbar^2}{2m} \frac{d^2\psi_0}{dx^2} = \mathscr{E}\psi_0 \tag{3.54}$$

The solution to (3.54) is

$$\psi_0 = A \exp(k_0 x) \tag{3.55}$$

and

$$\mathscr{E} = \frac{-\hbar^2 k_0^2}{2m} \tag{3.56}$$

If the electron had escaped from the crystal, $\mathscr{E}$ would be greater than zero and k_0 would be imaginary. In this case (3.56) would simply be the free-electron energy as in (2.5). However, we are looking for a bound state. This means that $\mathscr{E}$ is less than zero and k_0 must be real.

Inside the crystal, the wavefunction must have the form of (2.10) for a Bloch electron,

$$\psi = A \exp(ikx)u(x) \tag{3.57}$$

From (2.21) we concluded that k must be real. However, this result came from the Born–von Kármán boundary condition. When the crystal is terminated at a surface, k must be complex to match boundary conditions. That is, the wavefunction and its derivative must be continuous at the surface. Therefore, let

$$k = k_r + ik_i \tag{3.58}$$

Matching boundary conditions at the surface, $x = -x_0$, equations (3.55), (3.57), and (3.58) give us

$$k_0 = ik_r - k_i + \left[\frac{1}{u(x)} \frac{du(x)}{dx} \right]_{-x_0} \tag{3.59}$$

From (2.45) we can see that the last term on the right-hand side of (3.59) is, in general, complex. If we separate this term into its real and imaginary components, (3.59) becomes simply

$$k_0 = ik_r - k_i + \gamma + i\delta \tag{3.60}$$

Since k_0 is real, the allowed energy levels from (3.56) are

$$\mathscr{E} = \frac{-\hbar^2}{2m} (\gamma - k_i)^2 \tag{3.61}$$

When we set k_i equal to zero in (3.61) this puts us back to the Born–von Kármán boundary condition for a perfect crystal. From (2.45) γ is indexed by K, so

$$\mathscr{E} = \frac{-\hbar^2}{2m} \gamma^2 \tag{3.62}$$

represents the infinite energy bands of a perfect crystal. Since k_i has a value different from zero only for a crystal with surfaces, (3.61) shows that a surface drops states corresponding to k_i out of each band indexed by γ or K. To determine whether or not these states lie in the forbidden energy gap would require a lengthy calculation on a given crystal structure with a particular surface. If we look at the wavefunctions generated by the boundary,

$$\psi_0 = A_0 \exp (k_o x)$$
$$\psi = A \exp (ik_r x - k_i x) u(x) \tag{3.63}$$

we see an exponential decay on either side of the surface with a boundary layer characterized by k_o on the outside and by k_i on the inside. This localization of bound states at the surface is shown schematically in Fig. 3.21.

The surface states discussed above are referred to as intrinsic or Tamm states. In the bond picture these would be represented as dangling bonds. In the modern conception of surfaces [J. Bardeen, *Phys. Rev. 71*, 717 (1947)] we know that the unsaturated bonds can react chemically with various contaminants and that these contaminants can also be physically adsorbed on the surface. The result is that real surface states can be donors or acceptors and the net number of surface states can be greater or less than the number of intrinsic states. Some of the more important phenomena associated with these extrinsic states, such as surface band bending and surface recombination, are examined in more detail in later chapters.

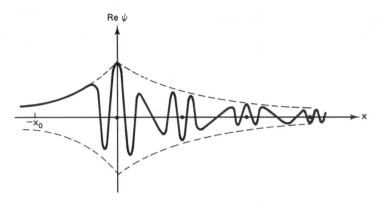

Figure 3.21 Schematic diagram for the localization of surface states.

3.8 LATTICE VIBRATIONS

In a perfect periodic structure the atoms are assumed to be fixed at the basis sites of a Bravais lattice. We know, however, that the atoms exhibit random thermal vibrations around their equilibrium positions, where the amplitude of the vibrations depends on temperature. Because of the bonding forces that produce the crystal structure, the random motion of individual atoms are strongly coupled to one another. This coupling produces random wave propagation in the crystal. When electrons and holes move through this crystal of randomly vibrating atoms, they can exchange energy and momentum with the atoms. In this manner the charge carriers are scattered by the thermal vibrations of the crystal and a thermodynamic equilibrium is established between the crystal and the electrons and holes. Since these scattering events play an important role in the transport properties of semiconductors, it is necessary to obtain a mathematical description of lattice vibrations.

Let us first define a vector $\mathbf{u}_{R,d}$ which indicates the displacement of an atom from its equilibrium position given by basis vector $\mathbf{d}$ and direct lattice vector $\mathbf{R}$. We take $\mathbf{R}$ to represent a particular primitive unit cell and $\mathbf{d}$ to represent the position of each atom in the unit cell. Thus, $\mathbf{R}$ and $\mathbf{d}$ index every atom in the crystal. The mass of each atom is M_d so that $\mathbf{d}$ indexes atoms of different mass in each primitive unit cell. Using these definitions the kinetic energy of all the atoms in the crystal is

$$T = \sum_{R,d} \tfrac{1}{2} M_d \, | \, \dot{\mathbf{u}}_{R,d} \, |^2 \tag{3.64}$$

where $\dot{\mathbf{u}}_{R,d}$ indicates the total derivative with respect to time. In (3.64) the sum in $\mathbf{R}$ is taken over the N unit cells and the sum in $\mathbf{d}$ is taken over the n atoms per unit cell.

We next assume that the potential energy of the atoms in the crystal,

U, depends on the displacements, $\mathbf{u}_{R,d}$, and has a minimum, U_0 when all atoms are in their equilibrium positions $\mathbf{u}_{R,d} = 0$. If we resolve $\mathbf{u}_{R,d}$ into Cartesian components,

$$\mathbf{u}_{R,d} = \sum_j u_{j,R,d}\hat{j}, \qquad \hat{j} = \hat{x}, \hat{y}, \hat{z} \tag{3.65}$$

we can expand the potential energy in a power series around the minimum potential energy as

$$U = U_0 + \sum_{j,R,d} \left[\frac{\partial U}{\partial u_{j,R,d}}\right]_0 u_{j,R,d}$$

$$+ \frac{1}{2} \sum_{j,R,d,j',R',d'} \left[\frac{\partial^2 U}{\partial u_{j,R,d}\partial u_{j',R',d'}}\right]_0 u_{j,R,d}u_{j',R',d'} + \cdots \tag{3.66}$$

Since the expansion is around the minimum potential energy point, the slope $[\partial U/\partial u_{j,R,d}]_0$ is zero. For small displacements, we can neglect higher-order terms and approximate the potential energy as

$$U = U_0 + \frac{1}{2} \sum_{j,R,d,j',R',d'} \left[\frac{\partial^2 U}{\partial u_{j,R,d}\partial u_{j',R',d'}}\right]_0 u_{j,R,d}u_{j',R',d'} \tag{3.67}$$

We can obtain the classical equations of motion for the atoms of the crystal with Lagrange's formulation

$$\frac{d}{dt}\left(\frac{\partial L}{\partial \dot{u}}\right) - \frac{\partial L}{\partial u} = 0 \tag{3.68}$$

where

$$L = T - U \tag{3.69}$$

Using (3.64) and (3.67) in (3.69) and (3.68), we have

$$M_d\ddot{u}_{j,R,d} = -\frac{1}{2} \sum_{j',R',d'} \left[\frac{\partial^2 U}{\partial u_{j,R,d}\partial u_{j',R',d'}}\right]_0 u_{j',R',d'} \tag{3.70}$$

Equation (3.70) can be simplified somewhat by writing it in vector notation as

$$M_d\ddot{\mathbf{u}}_{R,d} = \frac{1}{2} \sum_{R',d'} [\mathbf{A}_{R,d,R',d'}]\cdot\mathbf{u}_{R',d'} \tag{3.71}$$

where $[\mathbf{A}_{R,d,R',d'}]$ is a coupling tensor consisting of the terms $[\partial^2 U/\partial u_{j,R,d}\partial u_{j',R',d'}]_0$.

Notice that (3.71) has the form of Newton's law. Each term on the right-hand side of (3.71) represents the force exerted on $\mathbf{d}$ atom in $\mathbf{R}$ unit cell through the coupling tensor by a displacement of $\mathbf{d}'$ atom in $\mathbf{R}'$ unit cell. We also see that (3.71) represents a large number of coupled differential equations in very compact form. There are three Cartesian equations for each of n atoms in N unit cells, for a total of $3nN$ equations. We know intuitively, however, that the force terms on the right-hand side of (3.71) cannot depend on the absolute position of the atoms but only on the relative spacing between them. If we let $S = R' - R$, where S is another direct lattice vector, (3.71) takes the form

$$M_d \ddot{\mathbf{u}}_{R,d} = -\frac{1}{2} \sum_{S,d'} [\mathbf{A}_{S,d,d'}] \cdot \mathbf{u}_{R+S,d'} \tag{3.72}$$

which, like the static crystal potential in (2.9), is translationally invariant. Because of this, Bloch's theorem can be used to reduce the number of equations of motion.

Let us assume a traveling-wave solution to (3.72) of the form

$$\mathbf{u}_{R,d}(t) = \mathbf{u}_d \exp [i(\mathbf{q}_s \cdot \mathbf{R} - \omega_s t)] \tag{3.73}$$

for any wavevector $\mathbf{q}_s$. The angular frequency of vibration is ω. The wavevector $\mathbf{q}_s$ for these lattice vibrations belongs in the same reciprocal space as the wavevector $\mathbf{k}$ for electron waves. From (3.73)

$$\mathbf{u}_{R,d}(t) = \exp (i\mathbf{q}_s \cdot \mathbf{R})\mathbf{u}_{0,d}(t) \tag{3.74}$$

where

$$\mathbf{u}_{0,d} = \mathbf{u}_d \exp (-i\omega_s t) \tag{3.75}$$

is the displacement of the atoms in the unit cell chosen as the origin. For (3.72) to satisfy Bloch's equation, the motion of the atoms on sites $\mathbf{d}$ differ only in phase. Substituting (3.73) into (3.72), we obtain

$$\omega_s^2 M_d \mathbf{u}_d = \frac{1}{2} \sum_{S,d'} \exp (i\mathbf{q}_s \cdot \mathbf{S})[\mathbf{A}_{S,d,d'}] \cdot \mathbf{u}_{d'} \tag{3.76}$$

However, the Fourier transform of the coupling tensor is

$$\mathbf{A}_{d,d'}(\mathbf{q}) = \sum_S [\mathbf{A}_{S,d,d'}] \exp (i\mathbf{q} \cdot \mathbf{S}) \tag{3.77}$$

so that (3.76) is reduced to

$$\omega_s^2 M_d \mathbf{u}_d = \frac{1}{2} \sum_{d'} [\mathbf{A}_{d,d'}(\mathbf{q}_s)] \cdot \mathbf{u}_{d'} \tag{3.78}$$

In (3.78) the problem of $3nN$ coupled differential equations is reduced to $3n$ equations, which are the equations of motion for the basis atoms in a single unit cell. We notice from (3.77) that the coupling tensor in (3.78) is different for each wavevector $\mathbf{q}_s$. By going through arguments identical to those in Section 2.1 for Bloch electrons, we find that (1) when $\mathbf{q}_s$ is restricted to the first Brillouin zone, each value of $\mathbf{q}_s$ in (3.73) defines a unique vibrational mode, $\mathbf{u}_{R,d}(t)$; and (2) from the Born–von Kármán boundary condition (2.19) there are N allowed values of $\mathbf{q}_s$ in the first Brillouin zone, just as there are N allowed values of $\mathbf{k}$. Therefore, the coupling tensor in (3.78) sums up the interaction among the atoms in one unit cell with the atoms in all other unit cells.

Although we have reduced the lattice vibration problem to one of manageable dimensions, (3.78) is still difficult to solve for a three-dimensional crystal. Equation (3.78) gives us the relationship between ω_s and $\mathbf{q}_s$ for lattice vibrations which is determined by the crystal structure and elastic constants of the material. Qualitatively this relationship is similar to that between $\mathscr{E}$ and $\mathbf{k}$ for electrons. Since the energy of a vibrational mode (3.73) is proportional to its frequency, ω, both of these relationships can be plotted in the first Brillouin zone. Unlike the infinite number of energy bands for electrons, however, (3.78) shows that there are only $3n$ frequency bands for lattice vibrations or three for each atom in the unit cell. These three bands, of course, correspond to the three degrees of freedom of the Cartesian coordinates. When the Cartesian coordinates are selected so that one lies in the direction of propagation of the traveling wave in (3.73), these three bands are classified by their polarization. That is, the component of the random atomic vibrations, $\mathbf{u}_{R,d}$, along the direction of propagation is referred to as a *longitudinal band,* while the two components perpendicular to the direction of propagation are called *transverse bands.*

To understand some of the basic features of vibrational frequency bands, let us examine a one-dimensional Bravais lattice with a basis of two atoms per unit cell. As shown in Fig. 3.22, we assume that the atoms have masses M_1 and M_2, are equally spaced at a distance a, and have a coupling coefficient α. We will consider interactions between nearest neighbors only. From (3.72) the equations of motion are

$$M_1 \ddot{u}_{01} = -\tfrac{1}{2}\alpha[(u_{01} - u_{a2}) + (u_{01} - u_{-a2})]$$
$$M_2 \ddot{u}_{a2} = -\tfrac{1}{2}\alpha[(u_{a2} - u_{2a1}) + (u_{a2} - u_{01})] \tag{3.79}$$

Expressions for the displacement of each atom are given by (3.73) as

$$u_{01} = u_1 \exp[-i\omega_s t]$$
$$u_{a2} = u_2 \exp[i(q_s a - \omega_s t)]$$
$$u_{2a1} = u_1 \exp[i(2q_s a - \omega_s t)] \tag{3.80}$$
$$u_{-a2} = u_2 \exp[i(-q_s a - \omega_s t)]$$

Using (3.80) in (3.79), we have

$$\omega_s^2 M_1 u_1 = \alpha(u_1 - u_2 \cos q_s a)$$
$$\omega_s^2 M_2 u_2 = \alpha(u_2 - u_1 \cos q_s a) \tag{3.81}$$

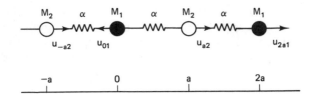

Figure 3.22 One-dimensional Bravais lattice with a basis of two atoms per unit cell. The two atoms of different mass are equally spaced and coupled.

The coupled set of equations (3.81) can be put in matrix form as

$$\begin{bmatrix} \alpha - \omega_s^2 M_1 & -\alpha \cos q_s a \\ -\alpha \cos q_s a & \alpha - \omega_s^2 M_2 \end{bmatrix} \begin{bmatrix} u_1 \\ u_2 \end{bmatrix} = 0 \tag{3.82}$$

Equation (3.82) is similar to (2.63) for the nearly free electron energy bands. To keep the displacements from vanishing identically, the determinant of coefficients for the matrix must be set equal to zero. Solving for the frequency, we obtain

$$\omega_{s\pm}^2 = \frac{\alpha}{2}\left(\frac{1}{M_1} + \frac{1}{M_2}\right) \pm \frac{\alpha}{2}\left[\left(\frac{1}{M_1} + \frac{1}{M_2}\right)^2 - \frac{4\sin^2 q_s a}{M_1 M_2}\right]^{1/2} \tag{3.83}$$

Equation (3.83) tells us that the vibrational frequency spectrum has two allowed bands separated by a forbidden gap. The results are plotted versus wavevector in Fig. 3.23.

For small values of q_s, the low-frequency band, ω_{s-}, is linearly dependent on q_s. Since this relationship between frequency and wavevector is a characteristic of elastic waves in continuous media, ω_{s-} is referred to

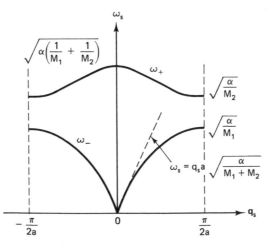

Figure 3.23 Vibrational frequency bands for a one-dimensional Bravais lattice with a basis of two atoms in the first Brillouin zone.

as an *acoustic* band. As a result of this dependence, the group velocity $d\omega_s/dq_s$ and the phase velocity ω_s/q_s are approximately given by the velocity of sound in the material. Also, for this band the two atoms in each unit cell have displacements that are about in phase. For the much higher frequency band, ω_{s+}, near $\mathbf{q}_s = 0$, however, the two atoms in each unit cell have displacements that are in opposite directions at any instant of time. This motion produces the maximum relative displacement between atoms, which, from (3.72), gives the maximum restoring force and thus the highest vibrational frequency. In ionic crystals this opposing motion of positively and negatively charged atoms in a unit cell results in an oscillating dipole moment which interacts strongly with electromagnetic radiation. For this reason ω_{s+} is called an *optical* band. At values of $\mathbf{q}_s$ near the edge of the first Brillouin zone, the differences between the acoustic and optical branches are not as pronounced. As indicated in Fig. 3.23, when the masses of the two atoms in the unit cell are identical, the forbidden gap disappears and the two branches become degenerate at the zone edge.

   We have seen that the random thermal motion of the atoms in a crystal can be described by displacement vectors, $\mathbf{u}_{R,d}(t)$, given by (3.73). From the classical equations of motion, we were able to deduce the relationship between the frequency and wavevector of these displacement vectors. To determine the energy of the lattice vibrations, a Hamiltonian for the crystal can be constructed from the kinetic energy terms of (3.64) and the potential energy terms of (3.67). The result constitutes the Hamiltonian for a set of harmonic oscillators. This Hamiltonian can then be used in Schrödinger's equation to determine the energy of the system. However, the solution for a quantum harmonic oscillator is well known. We can use this result to obtain the total lattice energy as the sum of energies from each vibrational mode,

$$\mathscr{E} = U_0 + \sum_{j,d,qs} \mathscr{E}_{j,d,qs} \tag{3.84}$$

where

$$\mathscr{E}_{j,d,qs} = \hbar\omega_{j,d}(n_{qs} + \tfrac{1}{2}), \qquad n_{qs} = 0, 1, 2, \ldots \tag{3.85}$$

In (3.84) the summation in $\mathbf{q}_s$ is over all N vibrational modes in the first Brillouin zone, in $\mathbf{d}$ over the one acoustic and $(n - 1)$ optical bands, and in j over the one longitudinal and two transverse bands.

   For (3.85) n_{qs} is the excitation state of a lattice vibrational mode indexed by $\mathbf{q}_s$. Since n_{qs} can take on only integer values, we see that each vibrational mode $\mathbf{q}_s$ can only gain or lose energy in quanta of $\hbar\omega_s$. This behavior is exactly analogous to the quanta of electromagnetic radiation in a cavity. Electromagnetic quanta are referred to as photons. For this reason lattice vibrational quanta are called *phonons*. The excitation state of a given mode, n_{qs}, is then taken as the number of phonons in that mode. The advantage of using the phonon–particle concept rather than the vibrational mode–wave concept is that a phonon interacts with other particles as if it had momentum $\hbar\mathbf{q}_s$ and energy $\hbar\omega_{j,d}$. Phonons are classified according to their frequency band. That is, depending on whether j is transverse (T) or longitudinal (L) and d is acoustic (A) or optical (O), a phonon is a transverse acoustic (TA), a transverse optical (TO), a longitudinal acoustic (LA), or a longitudinal optical (LO) phonon.

Let us look at the number of phonons in each mode. From the statistical mechanics for a set of harmonic oscillators, we find the probability that a mode of oscillation frequency ω is excited at temperature T is the same as the probability of finding a photon of frequency ω in a spectrum of radiant energy. Thus the average number of phonons at thermal equilibrium in a vibrational mode of frequency ω_s is

$$\langle n_{qs} \rangle = \frac{1}{\exp\,(\hbar\omega_s/kT) - 1} \tag{3.86}$$

where k is Boltzmann's constant and T is the lattice temperature. From (3.86) we see that at a given temperature the number of lower frequency acoustic phonons is considerably larger than the number of higher frequency optical phonons.

From (3.85) we can also obtain the average energy per vibrational mode as

$$\langle \mathscr{E} \rangle = \hbar\omega_s(\langle n_{qs} \rangle + \tfrac{1}{2}) \tag{3.87}$$

$$\langle \mathscr{E} \rangle = \tfrac{1}{2}\hbar\omega_s + \frac{\hbar\omega_s}{\exp\,(\hbar\omega_s/kT) - 1} \tag{3.88}$$

This is the energy that can be exchanged with an electron or hole in a scattering process. Both acoustic and optical phonons play important roles in

TABLE 3.2   Longitudinal and Transverse Optical Phonon Frequencies
in 10^{12} Hz

	ω_{LO}	ω_{TO}		ω_{LO}	ω_{TO}
C	39.96	39.96	InP	10.35	9.10
SiC	29.15	23.90	InAs	7.30	6.57
Si	15.69	15.69	InSb	6.00	5.37
Ge	9.02	9.02	ZnO	17.4	11.8
BN	40.20	31.95	ZnS	10.47	8.22
BP	25.0	24.6	ZnSe	7.53	6.12
AlN	27.5	20.0	ZnTe	6.18	5.47
AlP	15.03	13.20	CdS	9.1	6.9
AlSb	10.20	9.60	CdSe	6.52	5.13
GaN	24.0	16.5	CdTe	5.13	4.20
GaP	12.08	10.96			
GaAs	8.76	8.06			
GaSb	7.30	6.90			

the transport properties of semiconductors. The optical phonons near $\mathbf{q}_s =$ 0 are more important in optical processes. Longitudinal and transverse optical phonon frequencies near $\mathbf{q}_s = 0$ are listed in Table 3.2 for a number of semiconductors.

PROBLEMS

3.1. The electrons in a sample of n-type GaAs are frozen out on their donor levels by decreasing the temperature. Contacts are applied to the sample and the photoconductivity is then measured by exciting the electrons back into the conduction band with far-infrared radiation. The resulting spectra are shown in Fig. P3.1.
  (a) Explain the spacing between peaks, the temperature dependence, and the merging of the spectra at high energies.
  (b) If the effective mass is $0.0665m$, what is the dielectric constant?

3.2. A sample of chalcopyrite $CdSnP_2$ is annealed in excess P_4 pressure from an external gaseous phase.
  (a) Assuming Schottky disorder, write the quasi-chemical equations that describe this situation.
  (b) Are the resulting defects expected to be donors or acceptors? Explain.

3.3. Assume that the impurities As^+ and In^- form a complex pair in Si. Determine and plot versus spacing the ionization energies for nearest, second, and third nearest-neighbor pairs. The lattice constant is 5.43 Å, the dielectric constant is 11.8, and the energy gap is 1.1 eV for Si. Assume that each level shifts the same amount. Do any of these levels lie in the energy gap?

3.4. **(a)** Show that the optical and acoustic phonon bands for a one-dimensional Bravais lattice with a basis of two atoms reduce to one acoustic band when the atomic masses are identical.
  (b) Are the number of modes in each band the same? Explain.

3.5. Determine and plot in the {100} direction the three-dimensional vibrational bands for a simple cubic Bravais lattice with a basis of one atom. Consider interactions between nearest neighbors only.

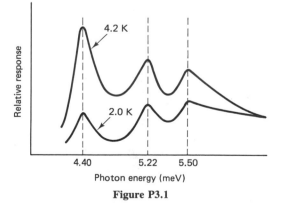

Figure P3.1

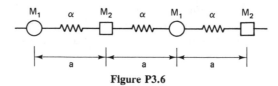

Figure P3.6

3.6. An electromagnetic wave with variable energy $\hbar\omega$ and wave vector ω/c is incident on the one-dimensional lattice shown in Fig. P3.6. The lattice absorbs energy from the electromagnetic wave at a wavelength of 20π μm. If the spacing between atoms of mass 2×10^{-26} and 2.5×10^{-26} kg is 5 Å, what is the force constant in newtons per meter?

3.7. A donor–acceptor complex in a one-dimensional lattice with a force constant of 10 N/m is found to exhibit a Franck–Condon shift of 0.1 eV. What is the difference in the equilibrium donor–acceptor spacing when the electron is in the ground state and in the excited state?

3.8. The point imperfection in a perfect crystal can be studied using a δ-function approximation for the defect potential. Let us approximate the point imperfection potential by a three-dimensional square well potential with well depth $-V_0$ and well radius a.
   (a) Give the explicit solution of the normalized ground-wave function.
   (b) If the bound-state energy is 0.5 V, $m^* = m$, and $a = 1.0$ Å, what is V_0 in eV?

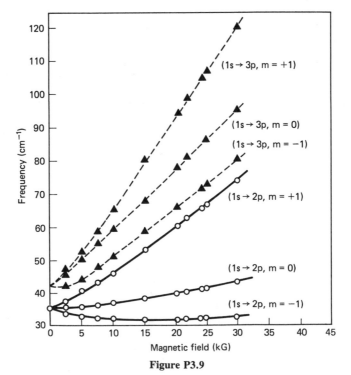

Figure P3.9

(c) What is the radius of the bound electron for the value used in part (b)?

(d) How many atoms for Ge and how many valence electrons are enclosed inside the sphere whose radius is given in part (c)? What is wrong with this square-well potential assumption?

3.9. The data for the energy separation between the ground state ($1s$) and first excited states ($2p$) of shallow donor impurity levels in a direct bandgap semiconductor are shown in Fig. P3.9.

(a) Explain the origin and meaning of these data.

(b) Calculate the electron effective mass for this semiconductor.

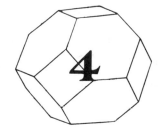

Equilibrium
Distributions

In Section 2.6 the distribution of electrons among conduction and valence bands in a semiconductor was considered briefly. We used the tight-binding approach to follow the banding of atomic states as the atoms were brought together to form a crystal. From the known electronic configuration of the atoms, the total number of electrons available to occupy the energy bands of the crystal was obtained. Qualitatively, we found that all of these available electrons were used to form the bonds of the crystal at 0 K, so that, in the band picture, the valence bands were completely full and the conduction bands were completely empty. When we abandon the concept of a perfect crystal and introduce imperfections in a crystal, such as impurities and lattice vibrations, the distribution of electrons among available states has to be considered in more detail. This is the purpose of the present chapter.

4.1 STATISTICS

Let us consider the number of electronic energy levels and states in a band. For a crystal with N primitive unit cells there are N energy levels per energy band. Equation (2.30) shows that each of these energy levels can be uniquely indexed by the wavevector $\mathbf{k}$ when we use only values of $\mathbf{k}$ in the first Brillouin zone. Thus the appropriate notation for each of these N energy levels in a band is $\mathscr{E}_k$. (In direct space $\mathbf{R}$ can be used to index the energy levels.) Each of these $\mathscr{E}_k$ energy levels can be occupied by two electrons of opposite spin. In addition, states in the same band at different points in the first Brillouin zone can also have the same energy. For example, the multiple conduction band minima in indirect materials are equivalent in energy. We can indicate this degeneracy or multiple states per energy level $\mathscr{E}_k$ as g_k. The number of energy bands is determined by the number of atoms per primitive unit cell, indexed by the basis vector $\mathbf{d}$, and the angular momentum degeneracy of the atomic states. That is, a crystal with one atom per unit cell would have one s-like band, three p-like bands, and so on, while a crystal with two atoms per unit cell would have two s-like bands, six p-like bands, and so on. The problem is to determine how n' electrons occupy the g_k states of the $\mathscr{E}_k$ energy levels for the various energy bands.

For this purpose we assume that each of the g_k states can be occupied by only one electron (Pauli's exclusion principle) and that each state has the same probability of being occupied. Let n_k of the g_k states be occupied with n_k electrons, where n_k is less than or equal to g_k. We then consider the number of ways the g_k states can be filled with n_k electrons. From Fig. 4.1 we can see that there are g_k ways of inserting the first electron, $g_k - 1$ ways of inserting the second, $g_k - 2$ ways of inserting the third, and $g_k - n_k + 1$ ways of inserting the n_k electron. The total number of ways is then

$$\prod_{n_k} (g_k - n_k + 1) = \frac{g_k!}{(g_k - n_k)!} \tag{4.1}$$

In this arrangement of electrons among states, however, we have counted permutations of electrons among themselves. Since the position and momentum of these electrons cannot be precisely defined, it is not possible for us to distinguish one electron from another. Because the electrons are indistinguishable, these permutations do not count as separate arrangements and must be removed from (4.1). To remove these permutations, maintain the arrangement of the n_k occupied levels in (4.1) and see how many ways each of the electrons can be rearranged. In this manner we see there are n_k ways of rearranging the first electron inserted, $n_k - 1$ ways of rearranging the second electron inserted, for a total of

$$\prod_{n_k} n_k = n_k! \tag{4.2}$$

permutations. Therefore, the number of independent ways of arranging n_k electrons among g_k states in the $\mathscr{E}_k$ energy level is

$$W_k = \frac{g_k!}{n_k! (g_k - n_k)!} \tag{4.3}$$

Figure 4.1   Schematic of the occupancy of the g_k states of the $\mathscr{E}_k$ energy level with n_k electrons.

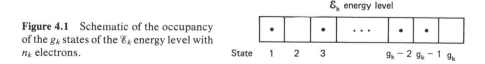

For all N energy levels in a band we have

$$W_b = \prod_k W_k = \prod_k \frac{g_k!}{n_k!\,(g_k - n_k)} \tag{4.4}$$

In most situations of practical interest the occupancy of band states is influenced by localized states due to impurities or other defects. Because of their localized nature the statistics of these states is substantially different from band states [N. F. Mott and R. W. Gurney, *Electronic Processes in Ionic Crystals* (Oxford: Clarendon Press, 1948)]. For example, if we consider the occupancy of a donor atom, it can be empty (D^+) in one way, or occupied in two ways by one unpaired electron of either spin (D^0). Although there is an extra spin state available for occupancy (D^-), the Coulomb repulsion of two electrons in the same localized region raises the energy of the doubly occupied level so high that this possibility must be excluded from the statistics. The situation is similar for an acceptor. It can be occupied in two ways by one unpaired electron of either spin and a bound hole (A^0), or in one way by two paired electrons when the hole is free (A^-). However, for the acceptor to be empty of electrons (A^+) is very unlikely, since this would be equivalent to occupancy by two holes.

Keeping this in mind, let us examine the occupancy of the first n impurity levels as in (3.32). Although there are an infinite number of these levels, let us consider only the first n, since in a practical situation the remainder are merged with the associated band states. As before we look at the $\mathscr{E}_n$ energy level with g_n degenerate states occupied by n_n electrons. However, in the band situation, the degenerate atomic levels interact to produce different $\mathscr{E}_k$ and the degeneracy factor g_k includes only spin and equivalent points in the first Brillouin zone. In the isolated impurity situation, the degenerate atomic levels do not interact and the degeneracy factor g_n has to take into account spin and the number of impurity atoms.

As before, there are g_n ways of inserting the first electron, but now there are only $g_n - 2$ ways of inserting the second electron since the electron that occupies the first state repulses the occupancy of the opposite spin state on the same impurity. There are then $g_n - 4$ ways of inserting the third electron, and $g_n - 2n_n + 2$ ways of inserting the n_n electron. Taking into account the indistinguishability of electrons, the total number of independent ways of arranging n_n electrons among the g_n states in the $\mathscr{E}_n$ impurity level is

$$W_n = \prod_{n_n} \frac{g_n - 2n_n + 2}{n_n}$$

$$= \frac{2^{n_n}(g_n/2)!}{n_n!\,(g_n/2 - n_n)!} \tag{4.5}$$

If we want to determine how n' electrons are distributed among the n impurity levels, we find

$$W = \prod_n W_n = \prod_n \frac{2^{n_n}(g_n/2)!}{n_n!\,(g_n/2 - n_n)!} \tag{4.6}$$

Usually, we simply want to determine what fraction of the impurity *atoms* is occupied. Since an impurity atom is occupied if any of its energy *levels* (not emerged with the associated band) is occupied, the appropriate expression is

$$W_i = \frac{2^{n_i}(g_i/2)!}{n_i!\,(g_i/2 - n_i)!} \tag{4.7}$$

Let us apply (4.7) to a monovalent donor. For $N_d \,(= N_d^0 + N_d^+)$ total donor atoms there are $2N_d$ states, including spin. Notice that the term $g_i/2$ in (4.7) includes only the degeneracy due to the number of atoms. The number of occupied states is N_d^0. The factor 2 in (4.7) accounts for the two different ways the N_d^0 states can be occupied with an electron of either spin. In general, there can be more than two ways in which the N_d^0 states can be occupied. That is, if the donor states are derived from multiple or degenerate conduction band minima, the factor 2 in (4.7) should be replaced by a factor g_d to account for the additional degeneracy. For example, Si has six equivalent conduction band minima. Taking spin into account, donors associated with these minima would have $g_d = 12$. Making the substitutions discussed above in (4.7), we obtain for the number of independent ways donor atoms can be occupied,

$$W_d = \frac{g_d^{N_d^0} N_d!}{N_d^0!\,(N_d - N_d^0)!} \tag{4.8}$$

For $N_a = N_a^0 + N_a^-$ total acceptor atoms, the N_a^- atoms are occupied with two paired electrons. Thus these atoms can be occupied in only one way. The N_a^0 atoms, however, can be occupied in two ways with an unpaired electron. To account for this difference, the factor 2 or, in general, g_a for an acceptor must be taken to the N_a^0 power. The value of g_a for an acceptor associated with the light and heavy hole valence bands of Si would be 4. Otherwise, we proceed as before to obtain for the number of independent ways acceptor atoms can be occupied,

$$W_a = \frac{g_a^{(N_a - N_a^-)} N_a!}{N_a^-!\,(N_a - N_a^-)!} \tag{4.9}$$

In the general case of a semiconductor with donors and acceptors and their associated energy bands, we can multiply (4.4), (4.8), and (4.9) together, with the result

$$W = \frac{g_d^{N_d^0} N_d!}{N_d^0!\,(N_d - N_d^0)!} \, \frac{g_a^{(N_a - N_a^-)} N_a!}{N_a^-!\,(N_a - N_a^-)!} \prod_k \frac{g_k!}{n_k!\,(g_k - n_k)!} \tag{4.10}$$

Equation (4.10) gives the total number of ways all of the donor, acceptor, and band states can be occupied by n' electrons, assuming that each state has an equal probability of being occupied. From this we must now determine the *most probable* distribution of electrons among all the states. In a real crystal the electrons exchange energy and momentum with the random vibrations of the atomic lattice until the electrons and the lattice are at the same temperature (thermal equilibrium). Under this thermal equilibrium condition, the most probable distribution or arrangement of electrons is the one that is most disordered. That is, the distribution of electrons which can occur in the largest number of ways is the most probable one. To determine the most probable distribution, we could therefore proceed to maximize (4.10) with respect to n'. However, that maximization procedure must be subjected to two constraints: (1) the number of electrons must remain constant and (2) the total energy of the electrons must remain constant.

The procedure we use to perform this maximization is that of Lagrange's undetermined multipliers. This method can be summarized as follows. Given an arbitrary function $f(x_i)$ of n variables, $i = 1, 2, \ldots, n$, find the maximum or minimum value of $f(x_i)$ subject to the constraint that some other arbitrary function $g(x_i)$ remains constant. For $f(x_i)$ to be an extremum, $df = 0$, and for $g(x_i)$ to be constant $dg = 0$. Because of this,

$$df + \alpha \, dg = 0 \tag{4.11}$$

independent of the value of α. When we carry out the total differential, (4.11) becomes

$$\sum_i \left(\frac{\partial f}{\partial x_i} + \alpha \frac{\partial g}{\partial x_i} \right) dx_i = 0 \tag{4.12}$$

and (4.11) is also satisfied by

$$\frac{\partial}{\partial x_i} [f(x_i) + \alpha g(x_i)] = 0 \tag{4.13}$$

for $i = 1, 2, \ldots, n$. Equation (4.13) together with

$$g(x_i) = \text{const.} \tag{4.14}$$

constitute $n + 1$ simultaneous equations which can be solved for α and the n values of x_i. If it is necessary to find the extrema of $f(x_i)$ with two constraints, $g(x_i)$ and $h(x_i)$ staying constant, (4.11) becomes

$$df + \alpha \, dg + \beta \, dh = 0 \tag{4.15}$$

and

$$\frac{\partial}{\partial x_i} [f(x_i) + \alpha g(x_i) + \beta h(x_i)] = 0 \tag{4.16}$$

for $i = 1, 2, \ldots, n$.

We can now apply (4.16) to the problem of maximizing (4.10) subject to the constraints

$$n' = N_d^0 + N_a^- + \sum_k n_k \tag{4.17}$$

and

$$\mathscr{E}' = N_d^0 \mathscr{E}_d + N_a^- \mathscr{E}_a + \sum_k n_k \mathscr{E}_k \tag{4.18}$$

where n' is the total number of electrons in the crystal and $\mathscr{E}'$ is the total internal electron energy. Rather than maximizing W, it is more convenient to use $\ln W$. Taking the natural logarithm of (4.10) and using Stirling's approximation,

$$\ln n! = n \ln n - n \tag{4.19}$$

for large numbers, we have

$$\ln W = N_d \ln \frac{N_d}{N_d - N_d^0} - N_d^0 \ln \frac{N_d^0}{g_d(N_d - N_d^0)}$$

$$+ N_a \ln \frac{g_a N_a}{N_a - N_a^-} - N_a^- \ln \frac{g_a N_a^-}{N_a - N_a^-}$$

$$+ \sum_k \left\{ g_k \ln \frac{g_k}{g_k - n_k} - n_k \ln \frac{n_k}{g_k - n_k} \right\} \tag{4.20}$$

The maximization procedure (4.16) is then

$$\frac{\partial}{\partial n_k} [\ln W - \alpha n' - \beta \mathscr{E}'] = 0 \tag{4.21}$$

$$\frac{\partial}{\partial N_d^0} [\ln W - \alpha n' - \beta \mathscr{E}'] = 0 \tag{4.22}$$

$$\frac{\partial}{\partial N_a^-} [\ln W - \alpha n' - \beta \mathscr{E}'] = 0 \tag{4.23}$$

where $\ln W$ is given by (4.20), n' by (4.17), and $\mathscr{E}'$ by (4.18).

Carrying out the operations of (4.21), (4.22), and (4.23), we obtain the most probable distributions,

$$\frac{n_k}{g_k} = \frac{1}{1 + \exp [\beta(\mathscr{E}_k - \mathscr{E}_f)]} \tag{4.24}$$

$$\frac{N_d^0}{N_d} = \frac{1}{1 + \dfrac{1}{g_d} \exp [\beta(\mathscr{E}_d - \mathscr{E}_f)]} \tag{4.25}$$

$$\frac{N_a^-}{N_a} = \frac{1}{1 + g_a \exp\left[\beta(\mathscr{E}_a - \mathscr{E}_f)\right]} \tag{4.26}$$

respectively, where we have defined a *Fermi energy* as

$$\mathscr{E}_f \equiv \frac{-\alpha}{\beta} \tag{4.27}$$

Equations (4.24), (4.25), and (4.26) are the Fermi–Dirac distribution functions for band, donor, and acceptor states, respectively. When the exponentials in the denominators of these equations are much greater than 1, the distribution of electrons can be approximated by the classical Maxwell–Boltzmann functions. The undetermined multiplier $\mathscr{E}_f$ is evaluated from the constraint (4.17) that the total number of electrons in the system remain constant. Thus there is an intimate relationship between the Fermi energy and the number of electrons. The other undetermined multiplier, β, can be evaluated by thermodynamic analysis.

4.2 THERMODYNAMICS

The first and second laws of thermodynamics can be represented by Euler's equation,

$$\mathscr{E}' = TS - PV + \sum_i \mu_i n_i + \psi Q \tag{4.28}$$

which relates the internal energy of a system, $\mathscr{E}'$, to the intensive variables of temperature (T), pressure (P), chemical potential (μ_i), and internal electrostatic potential (ψ); and the extensive variables of entropy (S), volume (V), particle number (n_i), and total electric charge (Q). Extensive variables are those which depend on concentration, while intensive variables are those which do not. Equation (4.28) can be taken as the basic law of thermodynamics. From it we can define the Helmholtz function,

$$F \equiv \mathscr{E}' - TS \tag{4.29}$$

and the Gibbs function as

$$G \equiv \mathscr{E}' - TS + PV \tag{4.30}$$

both of which have minimal properties.

The Helmholtz function, F, is a minimum at thermal equilibrium for systems in which T, V, n_i, and Q are constant. Since any change in F under constant temperature conditions can be completely transformed into work, it is called the free energy of the system. The Gibbs function, G, is a minimum at thermal equilibrium for systems in which T, P, n_i, and Q are constant. With these equations and Boltzmann's definition of entropy, at thermal equilibrium,

$$S = k \ln W_m \tag{4.31}$$

we have most of the thermodynamic concepts we currently need. In (4.31) k is Boltzmann's constant (not to be confused with the electron wavevector **k**) and W_m is the most probable arrangement of particles in a system.

Let us apply these thermodynamic relationships to evaluate the undetermined multiplier, β, in the Fermi–Dirac distribution function. The most probable distribution of electrons is obtained by substituting (4.24), (4.25), and (4.26) into (4.20). The result is

$$\ln W_m = \beta(\mathscr{E}' - \mathscr{E}_f n') + \sum_k g_k \ln \{1 + \exp [\beta(\mathscr{E}_f - \mathscr{E}_k)]\}$$

$$+ N_d \ln \{1 + g_d \exp [\beta(\mathscr{E}_f - \mathscr{E}_d)]\}$$

$$+ N_a \ln \{g_a + \exp [\beta(\mathscr{E}_f - \mathscr{E}_a)]\} \tag{4.32}$$

If we take the partial derivative of (4.32) with respect to $\mathscr{E}'$, keeping in mind that both β and $\mathscr{E}_f$ could depend on $\mathscr{E}'$, we obtain

$$\frac{\partial}{\partial \mathscr{E}'} (\ln W_m)_{n'} = \beta \tag{4.33}$$

From (4.29) and (4.31) we find that

$$\frac{1}{k} \left(\frac{\partial S}{\partial \mathscr{E}'} \right)_{n'} = \frac{\partial}{\partial \mathscr{E}'} (\ln W_m)_{n'} = \frac{1}{kT} \tag{4.34}$$

Therefore, β is just the reciprocal of Boltzmann's constant multiplied by the temperature of the electrons.

The undetermined Fermi energy, $\mathscr{E}_f$, can also be related to thermodynamic quantities. Using (4.31) in (4.29), the Helmholtz function is

$$F = \mathscr{E}' - kT \ln W_m \tag{4.35}$$

Substituting (4.32) into (4.35) and taking the partial derivative of F with respect to n' at constant T and V, keeping in mind that $\mathscr{E}_f$ depends on n', we obtain

$$\left(\frac{\partial F}{\partial n'} \right)_{T,V} = \mathscr{E}_f \tag{4.36}$$

Thus the Fermi energy is the change in free energy of the crystal when an electron is added or taken away. For our purposes (4.28) is

$$\mathscr{E}' = TS - PV + \mu n' + \psi Q \tag{4.37}$$

since we are considering only electrons. In this case the total charge, Q, is equal to $-qn'$. Using this in (4.37) and (4.37) in (4.29), we have the Helmholtz function in the form

$$F = -PV + (\mu - q\psi)n' \tag{4.38}$$

so that

$$\left(\frac{\partial F}{\partial n'}\right)_{T,V} = \mathcal{E}_f = (\mu - q\psi) \tag{4.39}$$

Thus, the Fermi energy is equal to the sum of the chemical potential and the internal electrostatic potential energy. Since the electrochemical potential for electrons is

$$\zeta \equiv \mu - q\psi \tag{4.40}$$

we see that the Fermi energy in general is equal to the electrochemical potential.

Although the equilibrium Fermi energy is often referred to as the chemical potential and the nonequilibrium Fermi energy is often referred to as the electrochemical potential, this nomenclature is not correct. That is, it neglects the contribution of an internal or built-in electrostatic potential to the equilibrium Fermi energy. In both equilibrium cases with excess carriers (space-charge) and nonequilibrium situations with no excess carriers, the Fermi energy is equal to the electrochemical potential. The distinction between the two situations is that in the nonequilibrium case an external electrostatic potential is added to the internal electrostatic potential. The concept of a quasi-Fermi level, which has no apparent thermodynamic significance, is used in nonequilibrium situations with excess carrier distributions. This is discussed in more detail in Chapter 8.

Except for the evaluation of $\mathcal{E}_f$ to maintain constant n', we have completed the analysis of the equilibrium distribution of electrons. For the energy band states (4.24), the equilibrium Fermi–Dirac distribution function has the form

$$f_0(\mathcal{E}, T) = \frac{1}{1 + \exp[(\mathcal{E} - \mathcal{E}_f)/kT]} \tag{4.41}$$

Equation (4.41) gives the probability that a band state of energy $\mathcal{E}$ is occupied by an electron at temperature T. We can see that at 0 K all states with energy below $\mathcal{E}_f$ are occupied while all states above $\mathcal{E}_f$ are empty (Fig. 4.2). For a state at 0 K with energy $\mathcal{E}_f$, the occupation probability is not defined and the derivative with respect to energy is a delta function. For a state at finite temperatures with energy $\mathcal{E}_f$, the occupation probability is 0.5 and the derivative is finite. Equation (4.41) will be used to determine the occupancy of conduction bands with electrons.

The probability that a state is not occupied is obviously given by

$$1 - f_0(\mathcal{E}, T) = \frac{1}{1 + \exp[(\mathcal{E}_f - \mathcal{E})/kT]} \tag{4.42}$$

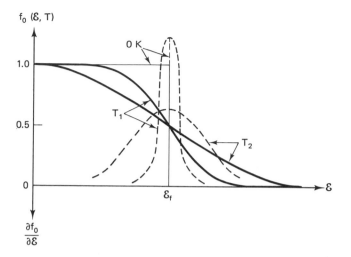

Figure 4.2 The equilibrium Fermi–Dirac distribution function and its energy derivative (dashed lines) for $0 < T_1 < T_2$. The derivative at 0 K is a delta function.

This is the distribution that is appropriate for the occupancy of holes in a valence band. Similar distribution functions can also be obtained from (4.25) and (4.26) for the occupancy of donor and acceptor levels with electrons and holes. When the exponential factor is sufficiently large, the Fermi–Dirac distribution function can be approximated, in each case, by a classical Maxwell–Boltzmann distribution.

4.3 DENSITY OF STATES

To determine the relationship between the total number of electrons in a band and the Fermi energy, we must first evaluate the density of states in the first Brillouin zone. From (4.24) and (4.41) the number of electrons in each state indexed by k is

$$n_k = g_k f_0(\mathscr{E}_k) \tag{4.43}$$

The total number of electrons, n', is found by summing (4.43) over all k. When g_k is sufficiently large, the summation can be replaced by an integral over all energies and

$$n' = \int_{-\infty}^{+\infty} g(\mathscr{E}) f_0(\mathscr{E}) \, d\mathscr{E} \tag{4.44}$$

The factor $g(\mathscr{E})$ contains the spin degeneracy and the total density of energy levels in the crystal. The concentration of electrons in the conduction band can be calculated from

$$n = \frac{1}{V} \int_{\mathscr{E}_C}^{\mathscr{E}_T} N(\mathscr{E}) f_0(\mathscr{E}) \, d\mathscr{E} \tag{4.45}$$

where V is the crystal volume, $N(\mathscr{E}) \, d\mathscr{E}$ is the total number of states that lie between the constant energy surfaces $\mathscr{E}$ and $\mathscr{E} + d\mathscr{E}$, $N(\mathscr{E})$ is the density of states, $\mathscr{E}_C$ is the energy at the bottom of the conduction band, and $\mathscr{E}_T$ is the energy at the top of the conduction band.

Figure 4.3 indicates that the total volume in reciprocal space between $\mathscr{E}$ and $\mathscr{E} + d\mathscr{E}$ is given by $\int_s dS \, dk_n$. From (2.29) the volume occupied by each value of k is $(2\pi)^3/V$. Thus, since each value of k represents a state that can be occupied by two electrons with different spins, the total number of states between $\mathscr{E}$ and $\mathscr{E} + d\mathscr{E}$ is

$$N(\mathscr{E}) \, d\mathscr{E} = \frac{2V}{(2\pi)^3} \int_s dS \, dk_n \tag{4.46}$$

where dk_n is the component of $d\mathbf{k}$ normal to the constant-energy surfaces and dS is the differential surface area for the constant-energy surface, S. The total derivative of energy is

$$d\mathscr{E} = \nabla_k \mathscr{E} \cdot d\mathbf{k} = |\nabla_k \mathscr{E}| \, dk_n \tag{4.47}$$

Using (4.47) in (4.46), the density of states is

$$N(\mathscr{E}) = \frac{2V}{(2\pi)^3} \int_s \frac{dS}{|\nabla_k \mathscr{E}|} \tag{4.48}$$

This equation can be used to determine the density of states for an energy band with any general dispersion relationship, including a nonquadratic one.

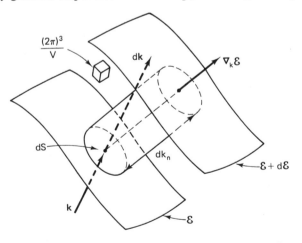

Figure 4.3 Diagram of the vectors and differential volume used to determine the density of states in the first Brillouin zone.

We note in passing that the integrand in (4.48) diverges when $\nabla_k \mathscr{E} = 0$. Although the resulting density of states remains finite at these points, discontinuities in the slope of $N(\mathscr{E})$ versus $\mathscr{E}$ are produced. These commonly occur at band extrema, as well as at other points, and are referred to as van Hove singularities.

Let us determine the density of states for ellipsoidal energy band extrema with a quadratic dispersion relationship as given by (2.123). For (2.123) the coordinate system has been rotated to diagonalize the effective mass tensor. In general, these extrema can occur at any point k_0 in the first Brillouin zone, and there will be equivalent conduction band minima at each equivalent point k_0 in the first Brillouin zone. To simplify the mathematics, consider just one of the equivalent minima and translate coordinates so that $k_0 = 0$. Since the density of states must be independent of the coordinate system, this is easily justified. The relationship between $\mathscr{E}$ and k for the minimum at the origin is then

$$\mathscr{E} - \mathscr{E}_0 = \pm \frac{\hbar^2}{2} \sum_i \frac{k_i^2}{m_i^*} \tag{4.49}$$

where the $+$ and $-$ refer to conduction band and valence band extrema, respectively. Equation (4.49) can be put in the form

$$\sum_i \left(\frac{k_i}{a_i}\right)^2 = 1, \qquad i = 1, 2, 3 \tag{4.50}$$

where the axes of the ellipsoid are

$$a_i = \left(\frac{2m_i^* |\mathscr{E} - \mathscr{E}_0|}{\hbar^2}\right)^{1/2} \tag{4.51}$$

Since the volume of an ellipsoid with axes a_i is

$$\frac{4}{3}\pi \prod_i a_i = \frac{4\pi}{3\hbar^3} (8m_1^* m_2^* m_3^*)^{1/2} |\mathscr{E} - \mathscr{E}_0|^{3/2} \tag{4.52}$$

the differential volume between the constant energy surfaces $\mathscr{E}$ and $\mathscr{E} + d\mathscr{E}$ is

$$\frac{2\pi}{\hbar^3} (8m_1^* m_2^* m_3^*)^{1/2} |\mathscr{E} - \mathscr{E}_0|^{1/2} \, d\mathscr{E} \tag{4.53}$$

Dividing (4.53) by $(2\pi)^3/V$, the volume occupied by each energy level indexed by k, and multiplying by 2 because each energy level can be occupied by two electrons of opposite spin and thus actually represents two electron states, the density of states for electrons is

$$N(\mathscr{E}) = \frac{4\pi V}{h^3} (8m_1^* m_2^* m_3^*)^{1/2} |\mathscr{E} - \mathscr{E}_0|^{1/2} \tag{4.54}$$

where identity $h \equiv \hbar 2\pi$ has been used.

Equation (4.54) gives the density of states for any single energy band extremum which can be represented by a quadratic dispersion relationship. The density of states of a band with multiple equivalent minima can be expressed in the same form as

$$N(\mathscr{E}) = \frac{4\pi V}{h^3} (2m_d)^{3/2} \, | \, \mathscr{E} - \mathscr{E}_0 \, |^{1/2} \tag{4.55}$$

where

$$m_d \equiv g^{2/3}(m_1^* m_2^* m_3^*)^{1/3} = (g^2 m_1^* m_2^* m_3^*)^{1/3} \tag{4.56}$$

is the density-of-states effective mass. For brand extrema with cylindrical symmetry,

$$m_d = (g^2 m_1 m_t^2)^{1/3} \tag{4.57}$$

where m_1 and m_t are the effective masses longitudinal and transverse to the principal axis of revolution. For a single extremum with spherical symmetry,

$$m_d = m^* \tag{4.58}$$

and the density-of-states effective mass is just the electron effective mass.

When nonequivalent extrema occur at the same energy, the total density of states is obtained by adding the contribution of each. In general, if these degenerate extrema occur at the same value of k, they interact to produce warped ellipsoids [B. Lax and J. G. Mavroides, *Phys. Rev. 100,* 1650 (1955)]. However, if we approximate the actual extrema by ellipsoids, the density of states can be written as

$$N(\mathscr{E}) = \frac{4\pi V}{h^3} \sum_i (2m_{di})^{3/2} \, | \, \mathscr{E} - \mathscr{E}_{0i} \, |^{1/2} \tag{4.59}$$

where the sum is now over the number of degenerate nonequivalent extreme valence bands that occur at the same energy in the diamond and sphalerite crystals. Equation (4.59) gives

$$N(\mathscr{E}) = \frac{2\pi V}{h^3} (2m_d)^{3/2} \, | \, \mathscr{E} - \mathscr{E}_v \, |^{1/2} \tag{4.60}$$

where the density-of-states effective mass is

$$m_d = (m_l^{*3/2} + m_h^{*3/2})^{2/3} \tag{4.61}$$

and m_l and m_h are the light- and heavy-hole masses, respectively.

The density of states for localized levels in the energy gap between bands can also be easily determined. If we consider the ground states of hydrogenic impurities, the total number including spin is $2N_i$. However, both spin states cannot be occupied at the same time so the effective number of states is N_i. When the states are discrete, the *density* of states at $\mathscr{E}_i$ is infinite. The integral over all possible states, however, must be N_i. Therefore, the density of states can be represented as

$$N_i(\mathscr{E}) = N_i\delta(\mathscr{E} - \mathscr{E}_i) \tag{4.62}$$

where $\delta(\mathscr{E} - \mathscr{E}_i)$ is the Dirac delta function.

4.4 ELECTRON AND HOLE DISTRIBUTIONS

4.4.1 Effective Density of States

Having determined the density of states in various situations, we are now in a position to examine the distribution of the total number of electrons, n', among the different bands and levels. Let us first look at the concentration of electrons in the lowest-lying conduction bands. From (4.45) the concentration or number of electrons per unit volume in these conduction band minima is

$$n = \frac{1}{V} \int_{\mathscr{E}_c}^{\mathscr{E}_t} f_0(\mathscr{E}) N_c(\mathscr{E}) \, d\mathscr{E} \tag{4.63}$$

where $\mathscr{E}_t$ is the energy at the top of the bands. From Fig. 4.2 we see that $f_0(\mathscr{E})$ approaches zero at high energies, so that we can, to a good approximation, replace $\mathscr{E}_t$ with ∞ in (4.63). Using (4.41) for $f_0(\mathscr{E})$ and (4.55) for $N_c(\mathscr{E})$, we have

$$n = 4\pi \left(\frac{2m_{de}}{h^2}\right)^{3/2} \int_{\mathscr{E}_c}^{\infty} \frac{(\mathscr{E} - \mathscr{E}_c)^{1/2} \, d\mathscr{E}}{1 + \exp\left[(\mathscr{E} - \mathscr{E}_f)/kT\right]} \tag{4.64}$$

for ellipsoidal minima. When we introduce the dimensionless variables,

$$x = \frac{\mathscr{E} - \mathscr{E}_c}{kT} \quad \text{and} \quad \eta = \frac{\mathscr{E}_f - \mathscr{E}_c}{kT} \tag{4.65}$$

(4.64) becomes

$$n = 4\pi \left(\frac{2kTm_{de}}{h^2}\right)^{3/2} \int_0^{\infty} \frac{x^{1/2} \, dx}{1 + \exp(x - \eta)} \tag{4.66}$$

The Fermi–Dirac integral of order $\frac{1}{2}$ is defined as

$$F_{1/2}(\eta) \equiv \frac{2}{\sqrt{\pi}} \int_0^{\infty} \frac{x^{1/2} \, dx}{1 + \exp(x - \eta)} \tag{4.67}$$

With this expression the electron concentration in the conduction band minima is

$$n = 2 \left(\frac{2\pi kTm_{de}}{h^2}\right)^{3/2} F_{1/2}(\eta) \tag{4.68}$$

Values for $F_{1/2}(\eta)$ are tabulated in Appendix B.

When $\mathscr{E}_f$ is less than about $\mathscr{E}_c - 4kT$, the Fermi–Dirac distribution can be approximated by the Maxwell–Boltzmann distribution. In this case (4.66) becomes

$$n = 4\pi \left(\frac{2kTm_{de}}{h^2}\right)^{3/2} \exp(\eta) \int_0^\infty \exp(-x)x^{1/2}\, dx \tag{4.69}$$

But

$$\int_0^\infty \exp(-x)x^{1/2}\, dx = \frac{\sqrt{\pi}}{2} \tag{4.70}$$

and (4.69) is therefore

$$n = 2 \left(\frac{2\pi kTm_{de}}{h^2}\right)^{3/2} \exp\left(\frac{\mathscr{E}_f - \mathscr{E}_c}{kT}\right) \tag{4.71}$$

Notice that the preexponential factor in (4.71) is the same as in the more general case of (4.68). This factor,

$$N_c = 2 \left(\frac{2\pi kTm_{de}}{h^2}\right)^{3/2} \tag{4.72}$$

is referred to as the *effective* conduction band density of states. This terminology reflects the fact that, in this classical approximation, the conduction band can be regarded as a single level with degeneracy N_c at the energy level $\mathscr{E}_c$. Using N_c from (4.72), (4.71) has the simple form

$$n = N_c \exp\left(\frac{\mathscr{E}_f - \mathscr{E}_c}{kT}\right) \tag{4.73}$$

where $(\mathscr{E}_f - \mathscr{E}_c)/kT$ is negative.

A similar approach can be taken for the concentration of holes in the highest-lying valence bands. In this situation we have to use the probability that a state is not occupied, so that the hole concentration is given by

$$p = \frac{1}{V} \int_{\mathscr{E}_b}^{\mathscr{E}_v} [1 - f_0(\mathscr{E})]N_v(\mathscr{E})\, d\mathscr{E} \tag{4.74}$$

where $\mathscr{E}_b$ is the bottom of the valence bands. Using (4.42) for $[1 - f_0(\mathscr{E})]$ and (4.55) for $N_v(\mathscr{E})$, we have

$$p = 4\pi \left(\frac{2m_{dh}}{h^2}\right)^{3/2} \int_{-\infty}^{\mathscr{E}_v} \frac{(\mathscr{E}_v - \mathscr{E})^{1/2}\, d\mathscr{E}}{1 + \exp[(\mathscr{E}_f - \mathscr{E})/kT]} \tag{4.75}$$

With the dimensionless variables,

$$x = \frac{\mathcal{E}_v - \mathcal{E}}{kT} \quad \text{and} \quad \eta = \frac{\mathcal{E}_v - \mathcal{E}_f}{kT} \tag{4.76}$$

(4.75) takes the form

$$p = 4\pi \left(\frac{2kTm_{dh}}{h^2}\right)^{3/2} \int_0^\infty \frac{x^{1/2}\, dx}{1 + \exp(x - \eta)} \tag{4.77}$$

which is the same as (4.66) for electrons in the conduction bands. We can thus write

$$p = N_v F_{1/2}(\eta) \tag{4.78}$$

in the general case and

$$p = N_v \exp\left(\frac{\mathcal{E}_v - \mathcal{E}_f}{kT}\right) \tag{4.79}$$

when $\mathcal{E}_f$ is greater than about $\mathcal{E}_v + 4kT$. The effective valence band density of states is

$$N_v = 2\left(\frac{2\pi kTm_{dh}}{h^2}\right)^{3/2} \tag{4.80}$$

It is interesting to note that the *three*-dimensional effective density of states for electrons and holes described by (4.72) and (4.80) are inversely proportional to the *cube* of the respective de Broglie wavelengths. This can be seen by rearranging (1.35) so that

$$\frac{1}{\lambda^2} = \frac{2\mathcal{E}m}{h^2} \tag{1.35}$$

Thus a *two*-dimensional effective density of states should be inversely proportional to the *square* of the de Broglie wavelength, and so on.

Let us determine the value of the Fermi energy in the situation where the number of electrons in the conduction band is equal to the number of holes in the valence band. Since this equality would apply when there were no impurity or defect levels in the energy gap or when thermal excitation would produce a much larger number of electrons and holes than the impurities, it is called the *intrinsic* case. Here we have

$$n = p = n_i \tag{4.81}$$

where n_i is the intrinsic carrier concentration. From the classical approximation (4.73) and (4.79),

$$\mathcal{E}_f = \frac{\mathcal{E}_c + \mathcal{E}_v}{2} + \frac{kT}{2} \ln \frac{N_v}{N_c} \tag{4.82}$$

Equation (4.82) indicates that in the intrinsic case and at finite temperatures, the Fermi level is displaced from the center of the energy gap, toward the conduction band if N_v is greater than N_c, or toward the valence band if N_c is greater than N_v.

4.4.2 Mass-Action Laws

When we multiply n times p in the classical approximation,

$$np = N_c N_v \exp \left(\frac{\mathscr{E}_v - \mathscr{E}_c}{kT} \right)$$

$$np = N_c N_v \exp \left(\frac{-\mathscr{E}_g}{kT} \right)$$

(4.83)

we find that the product is independent of $\mathscr{E}_f$ and depends only on $\mathscr{E}_g$ and the temperature T. N_c and N_v are also functions of temperature. Since the product is independent of $\mathscr{E}_f$, this means that in the classical approximation (4.83) is valid for *any pair of* values of n and p, including the intrinsic case. We thus have

$$np = n_i^2(T)$$

(4.84)

where for a given material n_i^2 is a function of temperature only. If we recall the quasi-chemical equation (3.38) for the formation of an electron and a hole,

$$0 = e^- + h^+ : \mathscr{E}_g$$

(4.85)

we see that (4.83) or (4.84) indicates a relationship between the concentrations of the products of the quasi-chemical reaction for electrons and holes. This relationship between concentrations is exponentially dependent on the energy (per particle) required to produce the reaction, $\mathscr{E}_g$, and is called a *mass-action law*.

Considering the limitations on the classical equations (4.73) and (4.79), the mass-action law for electrons and holes given in (4.84) is valid for

$$\mathscr{E}_v + 4kT < \mathscr{E}_f < \mathscr{E}_c - 4kT$$

(4.86)

Under these conditions the semiconductor is said to be nondegenerate. When $\mathscr{E}_f$ is greater than $\mathscr{E}_c - 4kT$ or less than $\mathscr{E}_v + 4kT$, (4.84) can no longer be used. Under these conditions the semiconductor is said to be degenerate, and (4.68) and (4.78) must be used to obtain the mass-action law.

$$np = N_c N_v F_{1/2} \left(\frac{\mathscr{E}_f - \mathscr{E}_c}{kT} \right) F_{1/2} \left(\frac{\mathscr{E}_v - \mathscr{E}_f}{kT} \right)$$

$$np \neq n_i^2$$

(4.87)

Thus, under these conditions the np product depends on the Fermi energy.

Let us next look at the relationship between $\mathscr{E}_f$ and the electron occupancy of discrete donor levels. From (4.25) we obtain

$$\frac{N_d^0}{N_d^0 + N_d^+} = \frac{1}{1 + (1/g_d) \exp\left[(\mathscr{E}_d - \mathscr{E}_f)/kT\right]} \tag{4.88}$$

which can be put in the form

$$\frac{N_d^+}{N_d^0} = \frac{1}{g_d} \exp\left(\frac{\mathscr{E}_d - \mathscr{E}_f}{kT}\right) \tag{4.89}$$

Thus the ratio of the concentration of ionized to neutral donors depends exponentially on the separation of the Fermi energy from the donor energy level. If we multiply (4.89) by (4.73), we obtain

$$\frac{nN_d^+}{N_d^0} = \frac{N_c}{g_d} \exp\left(\frac{\mathscr{E}_d - \mathscr{E}_c}{kT}\right) \tag{4.90}$$

which, using (3.33), is simply

$$\frac{nN_d^+}{N_d^0} = \frac{N_c}{g_d} \exp\left(\frac{-\Delta\mathscr{E}_d}{kT}\right) \tag{4.91}$$

This equation applies only for a nondegenerate semiconductor. Notice that (4.91) is independent of the concentrations (or Fermi energy) and depends only on temperature and the donor ionization energy, $\Delta\mathscr{E}_d$. Equation (4.91) is the mass-action law for the donor ionization reaction (3.37),

$$D^0 = D^+ + e^- : \Delta\mathscr{E}_d \tag{4.92}$$

For acceptors (4.26) gives

$$\frac{N_a^-}{N_a^- + N_a^0} = \frac{1}{1 + g_a \exp\left[(\mathscr{E}_a - \mathscr{E}_f)/kT\right]} \tag{4.93}$$

which can be easily solved for the ratio

$$\frac{N_a^-}{N_a^0} = \frac{1}{g_a} \exp\left(\frac{\mathscr{E}_f - \mathscr{E}_a}{kT}\right) \tag{4.94}$$

of ionized to neutral acceptors. When (4.94) is multiplied by (4.79), the mass-action law for the ionization of an acceptor is obtained:

$$\frac{pN_a^-}{N_a^0} = \frac{N_v}{g_a} \exp\left(-\Delta\mathscr{E}_a\right) \tag{4.95}$$

$$A^0 = A^- + e^+ : \Delta\mathscr{E}_a \tag{4.96}$$

Just as with the donor mass-action law, (4.95) is valid only for a nondegenerate semiconductor. In the degenerate case (4.94) must be multiplied by (4.78).

Let us now consider the electron and hole concentrations for an extrinsic semiconductor. In general, we have a concentration of N_d donors, of which N_d^+ are ionized, and N_a acceptors per unit volume, of which N_a^- are ionized. For a nondegenerate semiconductor the electron and hole concentrations are given by (4.73) and (4.79), respectively. For a degenerate semiconductor n and p are given by (4.68) and (4.78). From (4.88) and (4.89) the ratio of ionized to total donor concentrations is

$$\frac{N_d^+}{N_d} = \frac{1}{1 + g_d \exp\left[(\mathscr{E}_f - \mathscr{E}_d)/kT\right]} \tag{4.97}$$

and from (4.93) the ratio of ionized to total acceptor concentrations is

$$\frac{N_a^-}{N_a} = \frac{1}{1 + g_a \exp\left[(\mathscr{E}_a - \mathscr{E}_f)/kT\right]} \tag{4.98}$$

A relationship among these four charge concentrations n, p, N_d^+, and N_a^- can be obtained by invoking *charge neutrality*.

This concept can be understood with the following thought experiment. Consider an assembly of fixed and mobile positive and negative charges. Then let their spatial distribution be such that one region has a net positive charge relative to another region, which has a net negative charge. From Poisson's equation,

$$\mathbf{\nabla \cdot D(r)} = \rho(\mathbf{r})$$
$$\mathbf{\nabla \cdot E(r)} = \frac{\rho(\mathbf{r})}{\epsilon} \tag{4.99}$$

this net distribution of charge, $\rho(\mathbf{r})$, produces an electric field, $\mathbf{E}$. This electric field, in turn, produces a force such that the negative mobile charges in the negatively charged region tend to move toward the positively charged region, and the positive mobile charges in the positively charged region tend to move toward the negatively charged region. Therefore, *as long as there are mobile charge carriers in either region,* they will move in a direction to maintain charge neutrality. The time constant that characterizes this process is quite fast and is referred to as the *dielectric relaxation time*. Although this is examined in more detail in Chapter 8, for our present purposes we can see that charge neutrality will be maintained for times that are long in comparison to this dielectric relaxation time. We thus have, from (4.99),

$$\mathbf{\nabla \cdot E} = \frac{\rho(\mathbf{r})}{\epsilon} = 0$$
$$= \frac{q}{\epsilon}(N_d^+ - N_a^- + p - n) = 0 \tag{4.100}$$

or

$$N_d^+ - N_a^- + p - n = 0 \tag{4.101}$$

4.4.3 Fermi Energy Variations

With (4.73) or (4.68), (4.79) or (4.78), (4.97), (4.98), and (4.101), we can in principle solve for the five unknown variables N_d^+, N_a^-, p, n, and $\mathscr{E}_f$. In general, however, this is quite difficult. For a nondegenerate semiconductor the equation to be solved for $\mathscr{E}_f$ is

$$\frac{N_d}{1 + g_d \exp [(\mathscr{E}_f - \mathscr{E}_d)/kT]} - \frac{N_a}{1 + g_a \exp [(\mathscr{E}_a - \mathscr{E}_f)/kT]}$$

$$= N_c \exp \left(\frac{\mathscr{E}_f - \mathscr{E}_c}{kT}\right) - N_v \exp \left(\frac{\mathscr{E}_v - \mathscr{E}_f}{kT}\right) \quad (4.102)$$

To see how $\mathscr{E}_f$ varies with $N_d - N_a$, it is necessary to make certain simplifying assumptions. Let us first assume that essentially all the donors and acceptors are ionized. This is equivalent to the condition $\mathscr{E}_a + 4kT < \mathscr{E}_f < \mathscr{E}_d - 4kT$, so that (4.102) reduces to

$$N_d - N_a \simeq N_c \exp \left(\frac{\mathscr{E}_f - \mathscr{E}_c}{kT}\right) - N_v \exp \left(\frac{\mathscr{E}_v - \mathscr{E}_f}{kT}\right) \quad (4.103)$$

When $\mathscr{E}_f$ is in the upper half of the energy gap, $\mathscr{E}_v + \mathscr{E}_g/2 < \mathscr{E}_f < \mathscr{E}_d - 4kT$, the hole concentration will be small, and from (4.103) we obtain

$$\mathscr{E}_f \simeq \mathscr{E}_c - kT \ln \frac{N_c}{N_d - N_a} \quad (4.104)$$

When $\mathscr{E}_f$ is in the lower half of the energy gap $\mathscr{E}_a + 4kT < \mathscr{E}_f < \mathscr{E}_c - \mathscr{E}_g/2$, the electron concentration will be small and (4.103) gives us

$$\mathscr{E}_f \simeq \mathscr{E}_v + kT \ln \frac{N_v}{N_a - N_d} \quad (4.105)$$

From (4.104), (4.105), and (4.82) the variation of the Fermi energy with $N_d - N_a$ appears as shown in Fig. 4.4.

For sufficiently low temperatures the assumption that the Fermi energy is $4kT$ removed from the donor or acceptor energies, $\mathscr{E}_d$ or $\mathscr{E}_a$, is no longer valid. In n-type material, $\mathscr{E}_f$ moves toward $\mathscr{E}_d$ as the temperature is lowered and electrons in the conduction band begin "freezing out" on the donor levels. Under these conditions the concentration of neutral donors can no longer be neglected. The acceptors, however, are still all ionized and the hole concentration can be neglected. Assuming that the electron concentration in the conduction band is small, (4.102) gives us, for $\mathscr{E}_v + \mathscr{E}_g/2 < \mathscr{E}_f$,

$$\mathscr{E}_f = \mathscr{E}_d + kT \ln \frac{N_d - N_a}{g_d N_a} \quad (4.106)$$

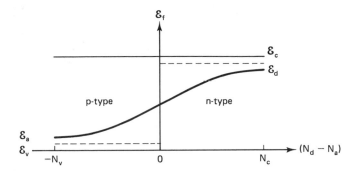

Figure 4.4 Variation of the Fermi energy with donor and acceptor concentrations.

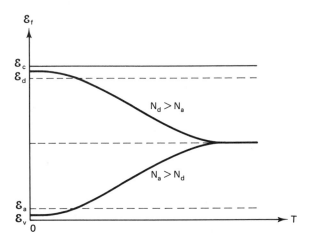

Figure 4.5 Variation of the Fermi energy with temperature.

Thus the Fermi energy lies above the donor energy. In p-type material under similar conditions,

$$\mathscr{E}_f = \mathscr{E}_a - kT \ln \frac{N_a - N_d}{g_a N_d} \qquad (4.107)$$

From (4.102) the temperature dependence of the Fermi energy under conditions such that either N_d is much greater than N_a or N_a is much greater than N_d is shown in Fig. 4.5.

4.5 IMPURITY AND DEFECT DISTRIBUTIONS

In most practical applications the properties of semiconductors are strongly influenced by impurities and stoichiometric defects. Unlike other crystal imperfections, such as dislocations, the concentrations of impurities and defects depends largely on the thermal history of the material during crystal growth and device fabrication. For this reason, we examine the thermal equilibrium concentrations of these species and the laws that govern their behavior.

Let us consider a general quasi-chemical reaction among impurities and defects of the form

$$\sum_i \nu_i R_i = \sum_i \nu_i P_i \tag{4.108}$$

where R_i are the reactants, P_i the products, and ν_i the constants needed to balance the equation. Equation (4.28) for the energy of the reaction (4.108) is

$$\mathscr{E}' = TS - PV + \sum_i \nu_i \zeta_i n_i \tag{4.109}$$

where the electrochemical potential is

$$\zeta_i = \mu_i + Z_i q \psi \tag{4.110}$$

and Z_i is the net charge of each species. From (4.30) the Gibbs function is

$$G = \sum_i \nu_i \zeta_i n_i \tag{4.111}$$

and in thermal equilibrium

$$\left[\frac{dG}{dn_i}\right]_{T,P} = \sum_i \nu_i \zeta_i = 0 \tag{4.112}$$

In (4.112) ν_i is positive for the products on the right side of (4.108) and negative for the reactants on the left-hand side.

We first examine the situation where the concentrations of impurities and defects are sufficiently dilute that they have a negligible effect on the total concentration of possible sites. The concentration of a given species under these conditions can be determined by maximizing the number of possible configurations. The result is similar to (4.73) for conduction band electrons in a nondegenerate material.

$$n_i = N_i \exp\left(\frac{\zeta_i - \mathscr{E}_i}{kT}\right) \tag{4.113}$$

In this equation N_i is the total concentration of available sites for the species, ζ_i the electrochemical potential, and $\mathscr{E}_i$ the energy of formation. The concentration of available sites can be assimilated in the exponential to obtain

$$\zeta_i = \zeta_i^0 + kT \ln n_i \tag{4.114}$$

Using (4.114) in (4.112), the equilibrium condition becomes

$$-\sum_i \nu_i \zeta_i^0 = kT \sum_i \nu_i \ln n_i \tag{4.115}$$

or in an equivalent form,

$$\prod_i n_i^{\nu_i} = \exp\left(\frac{-\sum_i \nu_i \zeta_i^0}{kT}\right) \qquad (4.116)$$

We can see that the exponential in (4.116) depends only on temperature and the energy required to form the reaction (4.108). Therefore, (4.116) can be written as

$$\prod_i n_i^{\nu_i} = K(T) \qquad (4.117)$$

where $K(T)$ is a constant, independent of concentration, at a given temperature.

Equation (4.117) is the mass-action law for the general quasi-chemical reaction (4.108). With it we can obtain the mass-action law (4.84) for the formation of an electron and a hole (4.85), (4.91) for the ionization of a donor (4.92), and (4.95) for the ionization of an acceptor (4.96). When the quasi-chemical reaction describes an interaction with a gaseous species such as in (3.41) to (3.47), the partial pressure, p_i, of the gaseous species is used in the mass-action law instead of the concentration n_i. Under conditions where the assumption of a dilute solution of impurities or defects is not valid, it is customary to use *activity* in place of concentration or *fugacity* in place of partial pressure in (4.117). Activity and fugacity are represented as

$$a_i = \gamma_i(n_i)n_i \qquad (4.118)$$

and

$$f_i = \gamma_i(p_i)p_i \qquad (4.119)$$

respectively, where γ_i is the activity or fugacity coefficient. The use of activity and fugacity in the mass-action law is equivalent to the use of Fermi–Dirac statistics for electrons and holes in degenerate semiconductors.

As an example of the use of quasi-chemical reactions and mass-action laws to describe the behavior of impurities and defects, let us examine the amphoteric impurity Si in GaAs. Assuming Schottky disorder and a negligible gallium partial pressure, the equations governing the intrinsic condition are

$$0 = e^- + h^+ \qquad\qquad np = n_i^2 = K_1 \qquad (4.120)$$

$$0 = V_{Ga}^0 + V_{As}^0 \qquad\qquad \{V_{Ga}^0\}\{V_{As}^0\} = K_2 \qquad (4.121)$$

$$\tfrac{1}{4}As_4(g) + V_{As}^0 = As_{As} \qquad \{V_{As}^0\} = \frac{1}{K_3 p_{As_4}^{1/4}} \qquad (4.122)$$

where the braces indicate the concentration of the species enclosed. If we assume that Si_{Ga}^+ is a donor and Si_{As}^- is an acceptor, the equations governing the extrinsic conditions are

$$\text{Si(s)} + V^0_{\text{Ga}} = \text{Si}^0_{\text{Ga}} \qquad \{\text{Si}^0_{\text{Ga}}\} = \{V^0_{\text{Ga}}\}N_{\text{Si}}K_4 \tag{4.123}$$

$$\text{Si}^0_{\text{Ga}} = \text{Si}^+_{\text{Ga}} + e^- \qquad \{\text{Si}^+_{\text{Ga}}\}n = \{\text{Si}^0_{\text{Ga}}\}K_5 \tag{4.124}$$

$$\text{Si(s)} + V^0_{\text{As}} = \text{Si}^0_{\text{As}} \qquad \{\text{Si}^0_{\text{As}}\} = \{V^0_{\text{As}}\}N_{\text{Si}}K_6 \tag{4.125}$$

$$\text{Si}^0_{\text{As}} = \text{Si}^-_{\text{As}} + h^+ \qquad \{\text{Si}^-_{\text{As}}\}p = \{\text{Si}^0_{\text{As}}\}K_7 \tag{4.126}$$

where N_{Si} is the total concentration of Si atoms. Taking $N^+_d = \{\text{Si}^+_{\text{Ga}}\}$ and $N^-_a = \{\text{Si}^-_{\text{As}}\}$, the charge neutrality condition is

$$n + N^-_a = N^+_d + \frac{n_i^2}{n} \tag{4.127}$$

Equations (4.120) to (4.127) can be solved in a straightforward manner for N^+_d, N^-_a, and n under the assumption that all the impurities are ionized. We then have

$$N^+_d = K_2K_3K_4K_5 \frac{N_{\text{Si}}}{n} p^{1/4}_{\text{As}_4} \tag{4.128}$$

$$N^-_a = \frac{K_6K_7}{K_3} \frac{nN_{\text{Si}}}{n_i^2} \frac{1}{p^{1/4}_{\text{As}_4}} \tag{4.129}$$

$$n^2 = \frac{n_i^2 + K_2K_3K_4K_5N_{\text{Si}}p^{1/4}_{\text{As}_4}}{n_i^2 + (K_6K_7/K_3)(N_{\text{Si}}/p^{1/4}_{\text{As}_4})} \tag{4.130}$$

These equations give us the expected dependence of the donor and acceptor concentrations on arsenic pressure, total impurity concentration, and temperature. Notice that the electron concentration depends on the impurity concentration only when N_{Si} is appropriately larger than n_i.

Similar models can be constructed for any assumed impurity and defect situation. The resulting equilibrium concentrations of species, however, depend strongly on the initial assumptions regarding the nature of the species. Also, in general, the equilibrium constants are not well known, so that there is usually considerable uncertainty in the temperature dependence of the concentrations. For these reasons, models for the concentrations of impurities and defects in a semiconductor must usually be based on experimental results.

PROBLEMS

4.1. Use the Lagrange undetermined multiplier method to find the maximum and minimum distances from the origin to the curve $5x^2 + 6xy + 5y^2 - 8 = 0$.

4.2. An epitaxial layer of GaAs growing at 1000 K is doped with zinc under conditions such that the concentration of zinc atoms is much less than the intrinsic carrier concentration. If the ionization energy of the zinc impurity is 0.030 eV with a degeneracy factor of 2, what is the ratio of the concentration of ionized zinc impurities to neutral zinc impurities at the growth temperature? Assume that the energy gap is 1.32 eV with an electron effective mass of $0.0665m_0$ and a hole effective mass of $0.340m_0$.

4.3. Assuming that the amphoteric impurity Si in GaAs forms Si^+ donors and $(Si_{Ga}V_{Ga})^-$ acceptors, obtain expressions for N_d^+ and N_a^- and plot versus N_{Si} for n-type material. Indicate the position of n_i on the graph. What experiment could be done to distinguish this from the example discussed in Section 4.5?

4.4. A sample of GaAs at 300 K has a total acceptor concentration of 4.3×10^{16} cm^{-3} and a total donor concentration of 8.6×10^{16} cm^{-3}. If the conduction band density of states is 4.3×10^{17} cm^{-3} and the donor ionization energy is 5.87 meV with a degeneracy factor of 2, what is the separation of the Fermi level from the conduction band in units of kT?

4.5. For the band structure shown in Fig. P4.5 at 77 K, what value of $\Delta\mathscr{E}$ gives the same number of electrons in the Γ (direct) conduction band minimum as in the X (indirect) conduction band minimum? $\mathscr{E}_g = 2.3$ eV. Assume nondegenerate material.

4.6. At room temperature, intrinsic GaAs appears as shown in Fig. P4.6 with $m_e(\Gamma)$ $= 0.065m$, $m_e(X) \simeq 0.30m$ (for each of six minima) and $m_h \simeq 0.47m$.
(a) What is the ratio of the carrier density in X to that in Γ at 300 K?
(b) At what temperature are the two carrier densities equal?

4.7. Calculate and plot the density of states for the first band of a simple cubic lattice in the same approximations used for the constant-energy contours in Problem 2.9.
(a) The tight-binding approximation
(b) The nearly free electron approximation
(c) The free-electron approximation
Discuss your results in terms of the constant-energy contours of Problem 2.9.

4.8. If a magnetic field B in the z direction is applied to a semiconductor, the

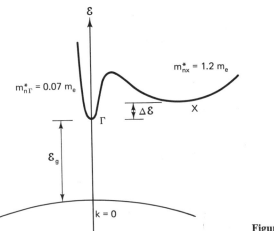

Figure P4.5

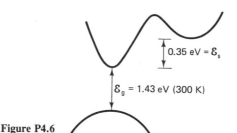

Figure P4.6

electron energy levels are given by

$$\mathscr{E}(k_z, N, B) = \mathscr{E}_0 + \frac{\hbar^2 k_z^2}{2M^*} + \omega_c \left(N + \frac{1}{2}\right)$$

where $\omega_c = eB/M^*$, and $N = 0, 1, 2, \ldots$ designate the particular Landau level. For the limit as $B \to 0$, this energy must reduce to the usual free-electron approximation (parabolic energy bands). Calculate and plot the density of states in a magnetic field B, and discuss the significance of your results.

4.9. Derive the following expression for the electron concentration in a compensated n-type semiconductor:

$$\frac{n(N_a + n - n_i^2/n)}{(N_d - N_a - n + n_i^2/n)} = \frac{1}{\beta} N_c e^{-\Delta\mathscr{E}_d/kT} \quad \begin{array}{c} \mathscr{E}_c \; - \\ \updownarrow \; \Delta\mathscr{E}_d \\ \mathscr{E}_d \; - \end{array}$$

where β is the ground-state degeneracy of the donor impurity level. Sketch the variation of n with temperature, showing the distinct regions that occur in different temperature ranges.

4.10. Near a parabolic band edge with the dispersion relation $\mathscr{E} = p^2/2m^*$, the density of states is

$$n_{\mathscr{E}}(\mathscr{E}) \, d\mathscr{E} = 2\pi \left(\frac{2m^*}{\hbar^2}\right)^{3/2} \mathscr{E}^{1/2} \, d\mathscr{E}$$

Other dispersion relations give other density functions. Find $n_{\mathscr{E}}(\mathscr{E}) \, d\mathscr{E}$ for phonons having the dispersion relation

$$\mathscr{E} = \hbar v k$$

in a cubic sample of length L. What is the group velocity, and what is it physically?

4.11. A particle random walks a step Δx in time Δt with equal probability in either direction on a line (one dimension). After a *large* number of steps N, what is:
(a) The probability $P(O, N)$ that the particle remains at the origin?
(b) The probability $P(2n, N)$ that the particle is $2n$ steps beyond the origin?
(*Hint:* Remember that N is *large* and that $k! \simeq \sqrt{2\pi k} \, (k/e)^k$, for k large, and $\ln(1 + \alpha) \simeq \alpha - \alpha^2/2$, for α small.)

4.12. A sample of GaAs at 300 K ($\mathscr{E}_g = 1.43$ eV, $m_e = 0.067m$, $m_{lh} = 0.12m$, $m_{hh} = 0.5m$) exhibits a hole concentration of 2×10^{17} cm^{-3} due to acceptors having $g_a = 4$ and $\Delta\mathscr{E}_a = 120$ meV.

(a) What is the concentration of acceptors? (Assume that $N_d = 0$.)

(b) How many donors ($g_d = 2$, $\Delta \mathscr{E}_d = 6$ meV) must be introduced to reduce the hole concentration to 1×10^{17} cm^{-3}?

(c) Consider a sample having the impurity concentrations calculated in parts (a) and (b). If the temperature increases slightly, will the hole concentration increase or decrease? Why?

4.13. A semiconductor sample is made quite thin in its Z dimension so that k_z may assume very few discrete values. What is the concentration of electrons in the conduction band at equilibrium (electrons per unit area)? Assume that the conduction band is spherically symmetric, the material is nondegenerate, and the effective mass approximation is valid. If the Fermi energy is $4kT$ below the conduction band energy, what is the concentration of conduction band electrons per unit area in GaAs at 300 K? ($m_e = 0.067m$.)

Transport Properties

When electrons are in thermal equilibrium with the lattice, they are distributed among possible energy levels in a manner given by the Fermi–Dirac function of (4.41). Under these conditions no net transport of charge or energy occurs since the probability that a state with wavevector $\mathbf{k}$ is occupied is the same as that for a state with wavevector $-\mathbf{k}$. That is, the equilibrium distribution function, f_0, is symmetrical about the origin in $\mathbf{k}$-space.

When external forces or temperature gradients are applied to the material, however, this is no longer true. Under these conditions we can, in principle, determine the nonequilibrium distribution function, f, in a manner similar to that used for f_0 in Section 4.1. In the nonequilibrium case, however, we would have to maximize (4.10) for the most probable distribution subject to the additional constraints that a steady flow of charge and energy be maintained. That is,

$$\mathbf{J} = -q \sum_k n_k \mathbf{v}_k \tag{5.1}$$

$$\mathbf{W} = \sum_k e_k n_k \mathbf{v}_k \tag{5.2}$$

where $\mathbf{J}$ is the electrical current density, $\mathbf{W}$ is the heat flow density, and e_k is the heat content per electron. These additional constraints produce an asymmetry in the nonequilibrium distribution function which shifts its center away from the origin in k-space. In this chapter we examine this nonequilibrium distribution of electrons and use it to determine the transport of charge and energy in semiconductors.

5.1 BOLTZMANN'S EQUATION

The approach we take in determining the transport properties of semiconductors is to first construct an electron wave packet from plane wave solutions to the time-dependent Schrödinger equation. This is the same method as that used in Section 2.8. Then, from the correspondence principle, we can use a semiclassical approach.

On this basis, let $f(\mathbf{k}, \mathbf{r}, t)$ be the probability that a state with wavevector $\mathbf{k}$ is occupied by an electron with position $\mathbf{r}$ at time t. The electrons are continually changing their position according to (2.109) and, under the influence of forces $\mathbf{F}_t$ (other than the periodic crystal forces), are continually changing their wavevector according to (2.111). $\mathbf{F}_t$ includes applied forces $\mathbf{F}$ and forces $\mathbf{F}_c$ due to electron collisions with lattice vibrations and other imperfections in the crystal. Therefore, at time $t + dt$ the probability that a state with wavevector $\mathbf{k} + d\mathbf{k}$ is occupied by an electron with position $\mathbf{r} + d\mathbf{r}$ is given by

$$f\left(\mathbf{k} + \frac{1}{\hbar} \mathbf{F}_t \, dt, \mathbf{r} + \mathbf{v} \, dt, t + dt\right)$$

The total rate of change in the distribution function in the region of the point $\mathbf{r}$ is then

$$\frac{df}{dt} = \frac{1}{\hbar} \mathbf{F}_t \cdot \nabla_k f + \mathbf{v} \cdot \nabla_r f + \frac{\partial f}{\partial t} \tag{5.3}$$

On the right side of (5.3), the first term takes into account changes in the distribution due to forces, the second term accounts for changes due to concentration gradients, and the last term is the local change in the distribution at the point $\mathbf{r}$. Equation (5.3) is referred to as Boltzmann's transport equation.

Since the total number of states in the crystal is constant, the total rate of change of the distribution function must be zero (Liouville's theorem), and

$$\frac{\partial f}{\partial t} = -\frac{1}{\hbar} \mathbf{F}_t \cdot \nabla_k f - \mathbf{v} \cdot \nabla_r f = \frac{\partial f}{\partial t}\bigg|_c - \frac{1}{\hbar} \mathbf{F} \cdot \nabla_k f - \mathbf{v} \cdot \nabla_r f \tag{5.4}$$

Because of the difficulty of finding a value for $\mathbf{F}_c$, we separate the collision forces from the applied forces by defining a local change in the distribution due to collisions only as

$$\frac{\partial f}{\partial t}\bigg|_c \equiv -\frac{1}{\hbar} \mathbf{F}_c \cdot \nabla_k f \tag{5.5}$$

Let us examine this collision term. The action of applied forces and gradients tends to disturb the distribution function f from its equilibrium value f_0. If this disturbance is removed, the scattering processes will tend to restore equilibrium. When the change in the distribution is not large compared to its initial value, it is reasonable to assume that

$$\frac{\partial f}{\partial t} = \frac{\partial f}{\partial t}\bigg|_c = \frac{-(f - f_0)}{\tau_m} \tag{5.6}$$

where τ_m is a constant of proportionality called the *momentum relaxation time*. In general, τ_m depends on the electron energy and is different for different scattering mechanisms. We examine τ_m for various scattering processes in some detail in Chapter 6. In the meantime, integrating (5.6), we find that

$$f(t) - f_0 = [f(0) - f_0] \exp\left(-\frac{t}{\tau_m}\right) \tag{5.7}$$

That is, the momentum relaxation time τ_m characterizes an exponential relaxation of the distribution function f to its equilibrium value f_0.

In the steady state $\partial f/\partial t = 0$ and using this and (5.6) in (5.4), we obtain the steady-state Boltzmann equation in the relaxation time approximation,

$$f = f_0 - \frac{\tau_m}{\hbar} \mathbf{F} \cdot \boldsymbol{\nabla}_k f - \tau_m \mathbf{v} \cdot \boldsymbol{\nabla}_r f \tag{5.8}$$

Since

$$\boldsymbol{\nabla}_k f = \frac{\partial f}{\partial \mathscr{E}} \boldsymbol{\nabla}_k \mathscr{E} \tag{5.9}$$

the Boltzmann equation can be put in the form

$$f = f_0 - \frac{\tau_m}{\hbar} \frac{\partial f}{\partial \mathscr{E}} \mathbf{F} \cdot \boldsymbol{\nabla}_k \mathscr{E} - \tau_m \mathbf{v} \cdot \boldsymbol{\nabla}_r f \tag{5.10}$$

From (2.109)

$$\mathbf{v} = \frac{1}{\hbar} \boldsymbol{\nabla}_k \mathscr{E}$$

so that (5.10) is finally

$$f = f_0 - \tau_m \mathbf{v} \cdot \left(\frac{\partial f}{\partial \mathscr{E}} \mathbf{F} + \boldsymbol{\nabla}_r f\right) \tag{5.11}$$

or

$$f = f_0 - \frac{\tau_m}{\hbar} \boldsymbol{\nabla}_k \mathscr{E} \cdot \left(\frac{\partial f}{\partial \mathscr{E}} \mathbf{F} + \boldsymbol{\nabla}_r f\right) \tag{5.12}$$

The Boltzmann equation in the form of (5.12) tells us that the nonequilibrium distribution of electrons depends on the scattering processes through the term τ_m, on the band structure through $\nabla_k \mathscr{E}$, on applied forces through $(\partial f / \partial \mathscr{E})\mathbf{F}$, and on concentration gradients through $\nabla_r f$. Therefore, we have, in general, a rather difficult partial differential equation to solve for f, the nonequilibrium distribution function.

5.2 DISTRIBUTION FUNCTION

Before looking at a more general solution for f, let us look at the simplest possible case. We will assume that the only applied force is a small electric field $\mathbf{E}$ and that there are no concentration or temperature gradients. Under these conditions (5.11) becomes

$$f = f_0 + q\tau_m \frac{\partial f}{\partial \mathscr{E}} \mathbf{v} \cdot \mathbf{E} \qquad (5.13)$$

Equation (5.13) can be integrated to obtain an analytical expression for f provided that the energy dependence of τ_m is known. The solution, however, is nonlinear in $\mathbf{E}$. Under the relaxation time assumption that the change in distribution function is not large, we can also make the approximation that

$$\frac{\partial f}{\partial \mathscr{E}} \simeq \frac{\partial f_0}{\partial \mathscr{E}} \qquad (5.14)$$

so that

$$f = f_0 + q\tau_m \frac{\partial f_0}{\partial \mathscr{E}} \mathbf{v} \cdot \mathbf{E} \qquad (5.15)$$

This retains only a linear term in E, which is consistent with our initial assumption of a small electric field.

Let us now look at a more general situation, where we include a small electric field $\mathbf{E}$ and an arbitrary magnetic field $\mathbf{B}$ in the force term and retain the term for concentration and temperature gradients. Under these conditions (5.8) is

$$\frac{f - f_0}{\tau_m} = \frac{+q}{\hbar} (\mathbf{E} + \mathbf{v} \times \mathbf{B}) \cdot \nabla_k f - \mathbf{v} \cdot \nabla_r f \qquad (5.16)$$

We will assume that the solution for (5.16) has the form of (5.15),

$$f = f_0 + \frac{\partial f_0}{\partial \mathscr{E}} \mathbf{v} \cdot \mathbf{G} \qquad (5.17)$$

and then solve for the unknown vector $\mathbf{G}$. Inserting (5.17) into (5.16) the term on the left-hand side of (5.16) is simply

$$\frac{f - f_0}{\tau_m} = \frac{1}{\tau_m} \frac{\partial f_0}{\partial \mathscr{E}} \mathbf{v} \cdot \mathbf{G} \tag{5.18}$$

The terms on the right-hand side of (5.16) are more difficult to evaluate.
Ignoring $+q/\hbar$ for the moment, the first term on the right-side of (5.16) is

$$(\mathbf{E} + \mathbf{v} \times \mathbf{B}) \cdot \nabla_k f = \mathbf{E} \cdot \nabla_k f_0 + (\mathbf{v} \times \mathbf{B}) \cdot \nabla_k f_0$$

$$+ \mathbf{E} \cdot \nabla_k \left(\frac{\partial f_0}{\partial \mathscr{E}} \mathbf{v} \cdot \mathbf{G} \right) + (\mathbf{v} \times \mathbf{B}) \cdot \nabla_k \left(\frac{\partial f_0}{\partial \mathscr{E}} \mathbf{v} \cdot \mathbf{G} \right) \tag{5.19}$$

In (5.19) the third term on the right has both $\mathbf{E}$ and $\mathbf{G}$ and is thus a second-order term in $\mathbf{E}$. Neglecting this third term and making the substitution

$$\nabla_k f_0 = \frac{\partial f_0}{\partial \mathscr{E}} \hbar \mathbf{v} \tag{5.20}$$

in the first and second term on the right, (5.19) becomes

$$(\mathbf{E} + \mathbf{v} \times \mathbf{B}) \cdot \nabla_k f = \hbar \frac{\partial f_0}{\partial \mathscr{E}} \mathbf{v} \cdot \mathbf{E} + \hbar \frac{\partial f_0}{\partial \mathscr{E}} (\mathbf{v} \times \mathbf{B}) \cdot \mathbf{v}$$

$$+ (\mathbf{v} \times \mathbf{B}) \cdot \nabla_k \left(\frac{\partial f_0}{\partial \mathscr{E}} \mathbf{v} \cdot \mathbf{G} \right) \tag{5.21}$$

Since $(\mathbf{v} \times \mathbf{B}) \cdot \mathbf{v}$ is identically zero, the second term on the right in (5.21) is zero. When we perform the gradient operation in the third term, (5.21) is

$$(\mathbf{E} + \mathbf{v} \times \mathbf{B}) \cdot \nabla_k f = \hbar \frac{\partial f_0}{\partial \mathscr{E}} \mathbf{v} \cdot \mathbf{E} + \frac{\partial f_0}{\partial \mathscr{E}} (\mathbf{v} \times \mathbf{B}) \cdot \nabla_k (\mathbf{v} \cdot \mathbf{G})$$

$$+ (\mathbf{v} \cdot \mathbf{G})(\mathbf{v} \times \mathbf{B}) \cdot \nabla_k \frac{\partial f_0}{\partial \mathscr{E}} \tag{5.22}$$

From (5.20) we see that

$$\nabla_k \frac{\partial f_0}{\partial \mathscr{E}} = \frac{\partial^2 f_0}{\partial \mathscr{E}^2} \hbar \mathbf{v}$$

and the third term on the right in (5.22) vanishes because of $(\mathbf{v} \times \mathbf{B}) \cdot \mathbf{v}$. The second term on the right in (5.22) can be resolved into Cartesian components and rearranged to obtain the result,

$$\frac{-q}{\hbar} (\mathbf{E} + \mathbf{v} \times \mathbf{B}) \cdot \nabla_k f = -q \frac{\partial f_0}{\partial \mathscr{E}} \mathbf{v} \cdot \mathbf{E} - \frac{q}{\hbar^2} \frac{\partial f_0}{\partial \mathscr{E}} \mathbf{v} \cdot [\mathbf{B} \times (\mathbf{G} \cdot \nabla_k) \nabla_k \mathscr{E}] \tag{5.23}$$

This is the desired form for the first term on the right in (5.16).

Let us now examine the second term on the right in (5.16),

$$\mathbf{v}\cdot\mathbf{\nabla}_r f = \mathbf{v}\cdot\mathbf{\nabla}_r f_0 + \mathbf{v}\cdot\mathbf{\nabla}_r \left(\frac{\partial f_0}{\partial \mathscr{E}}\,\mathbf{v}\cdot\mathbf{G}\right) \tag{5.24}$$

For our purposes we can assume that the spatial dependence of $\mathbf{G}$ is small and consider only the first term in (5.24). We then have

$$\mathbf{v}\cdot\mathbf{\nabla}_r f = \frac{\partial f_0}{\partial((\mathscr{E}-\mu)/kT)}\,\mathbf{v}\cdot\mathbf{\nabla}_r \left(\frac{\mathscr{E}-\mu}{kT}\right)$$

$$= kT\frac{\partial f_0}{\partial \mathscr{E}}\,\mathbf{v}\cdot\mathbf{\nabla}_r \left(\frac{\mathscr{E}-\mu}{kT}\right) \tag{5.25}$$

where μ is the chemical potential.

Using (5.18), (5.23), and (5.25), Boltzmann's equation is now

$$\frac{1}{\tau_m}\frac{\partial f_0}{\partial \mathscr{E}}\,\mathbf{v}\cdot\mathbf{G} = +q\frac{\partial f_0}{\partial \mathscr{E}}\,\mathbf{v}\cdot\mathbf{E} + \frac{q}{\hbar^2}\frac{\partial f_0}{\partial \mathscr{E}}\,\mathbf{v}\cdot[\mathbf{B}\times(\mathbf{G}\cdot\mathbf{\nabla}_k)\mathbf{\nabla}_k\mathscr{E}]$$

$$- kT\frac{\partial f_0}{\partial \mathscr{E}}\,\mathbf{v}\cdot\mathbf{\nabla}_r \left(\frac{\mathscr{E}-\mu}{kT}\right) \tag{5.26}$$

Since each term in (5.26) has a common factor $(\partial f_0/\partial \mathscr{E})\mathbf{v}$ on the left, it can be eliminated to obtain

$$\frac{1}{\tau_m}G = +q\mathbf{E} - kT\,\mathbf{\nabla}_r \left(\frac{\mathscr{E}-\mu}{kT}\right) + \frac{q}{\hbar^2}[\mathbf{B}\times(\mathbf{G}\cdot\mathbf{\nabla}_k)\mathbf{\nabla}_k\mathscr{E}] \tag{5.27}$$

Defining an electrothermal field for electrons, $\mathscr{F}$, by

$$q\mathscr{F} = +q\mathbf{E} - T\mathbf{\nabla}_r \left(\frac{\mathscr{E}-\mu}{T}\right) \tag{5.28}$$

(5.27) has the form

$$\mathbf{G} = q\tau_m\mathscr{F} + \frac{q\tau_m}{\hbar^2}[\mathbf{B}\times(\mathbf{G}\cdot\mathbf{\nabla}_k)\mathbf{\nabla}_k\mathscr{E}] \tag{5.29}$$

Equation (5.29) can be solved for $\mathbf{G}$ by using an explicit expression for the conduction band minima. For this purpose we will assume ellipsoidal minima with a quadratic dispersion relationship as given by (2.117). In vector notation we have

$$\mathscr{E} = \mathscr{E}_c + \tfrac{1}{2}\hbar^2\mathbf{k}\cdot\mathbf{M}\cdot\mathbf{k} \tag{5.30}$$

where

$$
\mathbf{M} = \begin{bmatrix} \dfrac{1}{m_1^*} & 0 & 0 \\[2mm] 0 & \dfrac{1}{m_2^*} & 0 \\[2mm] 0 & 0 & \dfrac{1}{m_3^*} \end{bmatrix} \tag{5.31}
$$

is the effective mass tensor, the $\mathbf{k}$ on its left is a row vector, and the $\mathbf{k}$ on its right is a column vector. From (5.30),

$$(\mathbf{G} \cdot \nabla_k) \nabla_k \mathscr{E} = \hbar^2 \mathbf{M} \cdot \mathbf{G} \tag{5.32}$$

and (5.29) takes the form

$$\mathbf{G} = q\tau_m \mathscr{F} + q\tau_m \mathbf{B} \times (\mathbf{M} \cdot \mathbf{G}) \tag{5.33}$$

Reducing this equation to its components and solving for $\mathbf{G}$, we have, finally,

$$\mathbf{G} = q\tau_m \left[\frac{\mathscr{F} - q\tau_m \mathbf{M} \cdot (\mathscr{F} \times \mathbf{B}) + (q\tau_m)^2 (\det \mathbf{M})(\mathscr{F} \cdot \mathbf{B})(\mathbf{M}^{-1} \cdot \mathbf{B})}{1 + (q\tau_m)^2 (\det \mathbf{M})(\mathbf{M}^{-1} \cdot \mathbf{B}) \cdot \mathbf{B}} \right] \tag{5.34}$$

The nonequilibrium distribution function for electrons in ellipsoidal conduction band minima is obtained by using $\mathbf{G}$ from (5.34) in (5.17). For spherical minima the distribution function is

$$f = f_0 + \frac{\partial f_0}{\partial \mathscr{E}} \, q\tau_m \mathbf{v} \cdot \left[\frac{\mathscr{F} - (q\tau_m/m^*)(\mathscr{F} \times \mathbf{B}) + (q\tau_m/m^*)^2 (\mathscr{F} \cdot \mathbf{B})\mathbf{B}}{1 + (q\tau_m/m^*)^2 \mathbf{B} \cdot \mathbf{B}} \right]$$

$$\tag{5.35}$$

We can see that there are four components to the distribution function. The first is simply the equilibrium function f_0 given by (4.41), which does not contribute to the transport of charge and energy. The term that involves $\mathscr{F}$ is the ohmic contribution to the transport properties. This term accounts for electrical and thermal conductivity as well as the Seebeck, Peltier, and Thomson effects. The term with $\mathscr{F} \times \mathbf{B}$ is the Hall contribution to transport and accounts for the Hall, Ettinghausen, Nernst, and Righi–Leduc effects. The $\mathbf{B}^2$ terms in the numerator and denominator of (5.35) account for magnetoresistive effects. We discuss these various effects in more detail later.

The distribution function derived above for electrons can also be used for holes when the appropriate parameters of q, m^*, and τ_m are substituted in the equations.

From (5.1) and (5.2) we can determine the current density $\mathbf{J}$ and heat flow density $\mathbf{W}$ by summing (or integrating) $n_k \mathbf{v}_k$ and $e_k n_k \mathbf{v}_k$, respectively, over the first Brillouin zone. In (4.68), however, we have already obtained an expression for n, the number of electrons in the conduction band minima. For this reason we can approach the problem from a different point of view. That is, the current density can be determined by

$$\mathbf{J} = -qn\langle\mathbf{v}\rangle \tag{5.36}$$

where $\langle\mathbf{v}\rangle$ is the average velocity of the n electrons in the nonequilibrium distribution. In a similar manner the heat flow density can be obtained from

$$\mathbf{W} = n\langle e\mathbf{v}\rangle \tag{5.37}$$

where the heat content per electron and the velocity are averaged over the distribution. The problem is to determine how this averaging should be performed.

5.3 CHARGE TRANSPORT

For this purpose let us examine the current density for spherical conduction band minima in a small electric field. The average velocity is obtained by summing the velocities of all the electrons in the distribution and normalizing the result. That is,

$$\langle\mathbf{v}\rangle = \frac{\int_{-\infty}^{\infty} \mathbf{v}f \, d\mathbf{v}}{\int_{-\infty}^{\infty} f \, d\mathbf{v}} \tag{5.38}$$

where f is given by (5.15). Inserting (5.15) in (5.38), we have

$$\langle\mathbf{v}\rangle = \frac{\int_{-\infty}^{\infty} \mathbf{v}f_0 \, d\mathbf{v} + q\int_{-\infty}^{\infty} \tau_m(\partial f_0/\partial\mathscr{E})\mathbf{v}(\mathbf{v}\cdot\mathbf{E}) \, d\mathbf{v}}{\int_{-\infty}^{\infty} f_0 \, d\mathbf{v} + q\int_{-\infty}^{\infty} \tau_m(\partial f_0/\partial\mathscr{E})(\mathbf{v}\cdot\mathbf{E}) \, d\mathbf{v}} \tag{5.39}$$

The term on the left in the numerator of this equation is an average over the equilibrium distribution. Since there is no transport of charge in equilibrium, this term is zero. The term on the right in the denominator provides for additional nonequilibrium carriers over the equilibrium concentration. We will take this term to be zero as well. Equation (5.39) is therefore

$$\langle\mathbf{v}\rangle = \frac{+q\int_{-\infty}^{\infty} \tau_m(\partial f_0/\partial\mathscr{E})\mathbf{v}(\mathbf{v}\cdot\mathbf{E}) \, d\mathbf{v}}{\int_{-\infty}^{\infty} f_0 \, d\mathbf{v}} \tag{5.40}$$

For spherical conduction band minima we can replace the integrals over three-dimensional velocity space by integrals over energy with relative ease. From (2.109) and (2.118) the relationship between $\mathcal{E}$ and $\mathbf{v}$ is

$$\mathcal{E} - \mathcal{E}_c = \tfrac{1}{2}m^*\mathbf{v}^2 = \tfrac{1}{2}m^*v^2 \tag{5.41}$$

and the differential volume in velocity space is

$$d\mathbf{v} = 4\pi v^2\, dv \tag{5.42}$$

With (5.41) and (5.42), (5.40) becomes

$$\langle \mathbf{v} \rangle = \frac{+q \displaystyle\int_{\mathcal{E}_c}^{\infty} \tau_m(\partial f_0/\partial \mathcal{E})\mathbf{v}(\mathbf{v}\cdot\mathbf{E})(\mathcal{E} - \mathcal{E}_c)^{1/2}\, d\mathcal{E}}{\displaystyle\int_{\mathcal{E}_c}^{\infty} f_0(\mathcal{E} - \mathcal{E}_c)^{1/2}\, d\mathcal{E}} \tag{5.43}$$

If we consider an electric field in the x direction, the term

$$\mathbf{v}(\mathbf{v}\cdot\mathbf{E}) = +v_x^2 E_x \tag{5.44}$$

Assuming equipartition of energy, each degree of freedom has the same average kinetic energy and

$$v^2 = v_x^2 + v_y^2 + v_z^2 = 3v_x^2 \tag{5.45}$$

Using (5.44) and (5.45), (5.43) is then

$$\langle v_x \rangle = \frac{2qE_x}{3m^*} \frac{\displaystyle\int_{\mathcal{E}_c}^{\infty} \tau_m(\partial f_0/\partial \mathcal{E})(\mathcal{E} - \mathcal{E}_c)^{3/2}\, d\mathcal{E}}{\displaystyle\int_{\mathcal{E}_c}^{\infty} f_0(\mathcal{E} - \mathcal{E}_c)^{1/2}\, d\mathcal{E}} \tag{5.46}$$

Defining a drift velocity $\mathbf{v}_d$ as the average velocity of the carriers over the distribution, and introducing the dimensionless variables of (4.65), (5.46) becomes

$$v_{dx} = \frac{-qE_x}{m^*}\langle \tau_m \rangle \tag{5.47}$$

where

$$\langle \tau_m \rangle = \frac{2}{3}\frac{\displaystyle\int_0^{\infty} \tau_m(-\partial f_0/\partial x)x^{3/2}\, dx}{\displaystyle\int_0^{\infty} f_0 x^{1/2}\, dx} \tag{5.48}$$

Equation (5.48) gives the proper form for the averaging procedure over the distribution of electrons. By evaluating the average momentum relaxation time in the manner proscribed, we can determine the drift velocity from (5.47). Equation (5.47) tells us that for small electric fields, the drift velocity is directly proportional to the field. The constant of proportionality is called the *conductivity mobility*, μ_c. Thus (5.47) can be put in the form

$$v_{dx} = -\mu_c E_x \tag{5.49}$$

where

$$\mu_c = \frac{q\langle\tau_m\rangle}{m^*} \tag{5.50}$$

From (5.36) and (5.47) the current density in the x direction is

$$J_x = \frac{q^2 n\langle\tau_m\rangle}{m^*} E_x \tag{5.51}$$

Since the constant of proportionality between current density and electric field is referred to as the conductivity, we have

$$\sigma = \frac{q^2 n\langle\tau_m\rangle}{m^*} \tag{5.52}$$

and

$$\sigma = qn\mu_c \tag{5.53}$$

Thus, for the simple case of a small applied electric field, we can define all the transport parameters in terms of the average momentum relaxation time, $\langle\tau_m\rangle$. Once $\langle\tau_m\rangle$ has been obtained, the transport problem is solved.

In general, however, the quantity to be averaged is more complex. For example, to determine the energy transport from (5.37), an extra energy term is included in the average. Also, from (5.34) the vector **G** depends on multiple powers of τ_m and depends on energy through both τ_m and $\mathcal{F}$. Thus the quantity that must be averaged over the electron distribution in more complex transport problems has the form $\tau_m^s x^t$, where s and t are to be determined. Equation (5.48) shows that the averaging procedure for this quantity is

$$\langle\tau_m^s x^t\rangle = \frac{2}{3} \frac{\int_0^\infty \tau_m^s(-\partial f_0/\partial x)x^{t+3/2}\, dx}{\int_0^\infty f_0 x^{1/2}\, dx} \tag{5.54}$$

Equation (5.54) can be evaluated if we know the dependence of τ_m on electron energy. In Chapter 6 we will find that τ_m can be represented as having a simple power dependence on energy for most scattering mechanisms. Therefore, let us take the momentum relaxation time as having the form

$$\tau_m = \tau_0 x^r \tag{5.55}$$

where τ_0 is independent of energy. Equation (5.54) is then

$$\langle \tau_m^s x^t \rangle = \frac{2}{3} \tau_0^s \frac{\int_0^\infty (-\partial f_0/\partial x) x^{sr+t+3/2} \, dx}{\int_0^\infty f_0 x^{1/2} \, dx} \tag{5.56}$$

This equation can be solved by integrating by parts.

Let $u = x^{sr+t+3/2}$ and $dv = (-\partial f_0/\partial x) \, dx$. Then $du = (sr + t + \frac{3}{2})x^{sr+t+1/2} \, dx$ and $v = -f_0$. Using these expressions in the numerator of (5.56) gives us

$$\langle \tau_m^s x^t \rangle = \frac{2}{3} \tau_0^s \frac{-[f_0 x^{sr+t+3/2}]_0^\infty + (sr + t + \frac{3}{2}) \int_0^\infty f_0 x^{sr+t+1/2} \, dx}{\int_0^\infty f_0 x^{1/2} \, dx} \tag{5.57}$$

or

$$\langle \tau_m^s x^t \rangle = \frac{2}{3} \left(sr + t + \frac{3}{2} \right) \tau_0^s \frac{\int_0^\infty f_0 x^{sr+t+1/2} \, dx}{\int_0^\infty f_0 x^{1/2} \, dx} \tag{5.58}$$

The integral in the numerator is a Fermi–Dirac integral of order j given by

$$F_j(\eta) = \frac{1}{j!} \int_0^\infty f_0 x^j \, dx \tag{5.59}$$

The integral in the denominator is simply a Fermi–Dirac integral of order $\frac{1}{2}$ which we saw before in (4.67). Values for these integrals are tabulated in Appendix B. Using (5.59) and (4.67), we obtain

$$\langle \tau_m^s x^t \rangle = \frac{4}{3\sqrt{\pi}} \left(sr + t + \frac{3}{2} \right)! \, \tau_0^s \frac{F_{sr+t+1/2}(\eta)}{F_{1/2}(\eta)} \tag{5.60}$$

which is the final form for the average.

To determine transport parameters, we will use expressions of the form $\langle \tau_m^s x^t \rangle$, which can then be evaluated with (5.60). As an example, the conductivity mobility in (5.50) and the conductivity in (5.52) can be obtained from

$$\langle \tau_m \rangle = \frac{4(r + \frac{3}{2})! \, \tau_o}{3\sqrt{\pi}} \frac{F_{r+1/2}(\eta)}{F_{1/2}(\eta)} \tag{5.61}$$

when the value of r for the appropriate scattering mechanism is known.

Let us now look at the transport of electrons in both electric and magnetic fields. Assuming no concentration or temperature gradients, (5.34) reduces to

$$G = +q\tau_m \left[\frac{E - (q\tau_m/m^*)(E \times B) + (q\tau_m/m^*)^2 B(E \cdot B)}{1 + (q\tau_m/m^*)^2 B \cdot B} \right] \quad (5.62)$$

From (5.47) and (5.48) the *drift velocity* is

$$v_d = \frac{2}{3m^*} \frac{\int_0^\infty G(-\partial f_0/\partial x) x^{3/2} \, dx}{\int_0^\infty f_0 x^{1/2} \, dx} \quad (5.63)$$

and the current density is given by (5.36). Because of (5.58), it is not necessary to average each term of **G** to obtain the current density. We simply have, by inspection,

$$J = \frac{q^2 n}{m^*} \left\langle \frac{\tau_m}{1 + (\omega_c \tau_m)^2} \right\rangle E - \frac{q^3 n}{m^{*2}} \left\langle \frac{\tau_m^2}{1 + (\omega_c \tau_m)^2} \right\rangle (E \times B)$$

$$+ \frac{q^4 n}{m^{*3}} \left\langle \frac{\tau_m^3}{1 + (\omega_c \tau_m)^2} \right\rangle B(E \cdot B) \quad (5.64)$$

where we have introduced the cyclotron frequency

$$\omega_c \equiv \frac{q |B|}{m^*} \quad (5.65)$$

The first term in (5.64) is the ohmic term. The factor $1 + (\omega_c \tau_m)^2$ in the denominator of the average in this term reflects the magnetoresistance or reduction in conductivity due to the magnetic field. The second term reflects the Hall effect; it also has magnetoresistance associated with it. The third term is an additional magnetoresistance term.

Let us look at (5.64) for small magnetic fields. Under this condition the second-order terms in **B**, which produce the magnetoresistance, are small and

$$J = \frac{q^2 n}{m^*} \langle \tau_m \rangle E - \frac{q^3 n}{m^{*2}} \langle \tau_m^2 \rangle (E \times B) \quad (5.66)$$

If we take $B = \hat{z} B_z$, (5.66) becomes

$$J_x = \frac{q^2 n}{m^*} \langle \tau_m \rangle E_x - \frac{q^3 n}{m^{*2}} \langle \tau_m^2 \rangle E_y B_z \quad (5.67)$$

$$J_y = \frac{q^2 n}{m^*} \langle \tau_m \rangle E_y + \frac{q^3 n}{m^{*2}} \langle \tau_m^2 \rangle E_x B_z \quad (5.68)$$

$$J_z = \frac{q^2 n}{m^*} \langle \tau_m \rangle E_z \quad (5.69)$$

When $J_y = 0$, (5.68) gives us

$$E_x = - \frac{m^*}{qB_z} \frac{\langle \tau_m \rangle}{\langle \tau_m^2 \rangle} E_y \tag{5.70}$$

Using (5.70) in (5.67) and neglecting a second-order term in B_z, we have

$$J_x = \frac{-qn}{B_z} \frac{\langle \tau_m \rangle^2}{\langle \tau_m^2 \rangle} E_y \tag{5.71}$$

That is, J_x and B_z induce a field E_y. This is the Hall effect. These geometric constraints are obtained experimentally by applying a magnetic field in the z direction, a current in the x direction, and measuring the voltage in the y direction with a high-impedance voltmeter, so that the current in the y direction is negligible.

The *Hall constant* is defined as

$$R_H \equiv \frac{E_y}{J_x B_z} = - \frac{1}{qn} \frac{\langle \tau_m^2 \rangle}{\langle \tau_m \rangle^2} \tag{5.72}$$

From (5.72) we can see that the concentration of electrons in the conduction band can be obtained from an experimental determination of the Hall constant. If the charge carriers are holes in the valence bands, the negative q's in (5.28), (5.34), and (5.36) are replaced by positive q's. The resulting Hall constant is

$$R_H = \frac{1}{qp} \frac{\langle \tau_m^2 \rangle}{\langle \tau_m \rangle^2} \tag{5.73}$$

Thus the sign of the Hall constant (and Hall field) indicates the sign of the charge carriers and Hall measurements can be used to distinguish between n- and p-type material. With (5.53) for the conductivity, we can define a *Hall mobility* as

$$\mu_H = R_H \sigma = \mu_c \frac{\langle \tau_m^2 \rangle}{\langle \tau_m \rangle^2} \tag{5.74}$$

This mobility differs from the conductivity mobility μ_c by the factor

$$r_H \equiv \frac{\langle \tau_m^2 \rangle}{\langle \tau_m \rangle^2} \tag{5.75}$$

which is referred to as the *Hall factor*. For a nondegenerate semiconductor, we find, from (5.60),

$$r_H = \frac{3\sqrt{\pi}}{4} \frac{(2r + \frac{3}{2})!}{[(r + \frac{3}{2})!]^2} \tag{5.76}$$

(In the analysis of experimental data, the Hall factor is often assumed to be

1. Depending on the relevant scattering mechanisms and temperature, this can produce about an 80% error in the carrier concentration.)

Let us next examine the flow of charge for small electric fields in the presence of electron concentration gradients. We will assume there are no magnetic fields or temperature gradients. Under these conditions the electrothermal field (5.28) for electrons is

$$\mathscr{F} = +\mathbf{E} + \frac{1}{q} \nabla_r \mu = +\frac{1}{q} \nabla_r \zeta \tag{5.77}$$

and from (5.34),

$$\mathbf{G} = q\tau_m \mathbf{E} + \tau_m \nabla_r \mu = +\tau_m \nabla_r \zeta \tag{5.78}$$

Equations (5.36), (5.47), and (5.34) tell us that the electron current density is given by

$$\mathbf{J} = \frac{+qn}{m^*} \langle \mathbf{G} \rangle \tag{5.79}$$

or

$$\mathbf{J} = \frac{q^2 n}{m^*} \langle \tau_m \rangle \mathbf{E} + \frac{qn}{m^*} \langle \tau_m \rangle \nabla_r \mu \tag{5.80}$$

Using (5.50) for the conductivity mobility, (5.80) becomes

$$\mathbf{J} = qn\mu_n \mathbf{E} + n\mu_n \nabla_r \mu \tag{5.81}$$

where μ_n indicates conductivity mobility for electrons.

The gradient of the chemical potential can be written in terms of a concentration gradient as

$$\nabla_r \mu = \frac{\partial \eta}{\partial n} \frac{\partial \mu}{\partial \eta} \nabla_r n \tag{5.82}$$

Since

$$\frac{d}{d\eta} F_j(\eta) = F_{j-1}(\eta) \tag{5.83}$$

[J. McDougall and E. C. Stoner, *Philos. Trans. R. Soc. London 237*, 67 (1938)], (4.68) gives us

$$\frac{\partial n}{\partial \eta} = N_c F_{-1/2}(\eta) \tag{5.84}$$

Also, from (4.65),

$$\frac{\partial \eta}{\partial \mu} = \frac{1}{kT} \tag{5.85}$$

Using (5.84), (5.85), and (4.68) the chemical potential gradient is

$$\nabla_r \mu = \frac{kT}{N_c F_{1/2}(\eta)} \nabla_r n \tag{5.86}$$

and the electron current density is

$$\mathbf{J} = q n \mu_n \mathbf{E} + kT \mu_n \frac{F_{1/2}(\eta)}{F_{-1/2}(\eta)} \nabla_r n \tag{5.87}$$

Equation (5.87) shows that in the presence of an electric field and a concentration gradient, the electron current density consists of two components: The first component is proportional to the electric field and is called the *drift* term. The second component is directly related to the concentration gradient and is referred to as the *diffusion* term. Notice that the electron current density is in the same direction as the concentration gradient, which is in the direction of increasing concentration.

The diffusion component of current is usually obtained from Fick's first law as $q D_n \nabla_r n$, where D_n is the diffusion constant of the electrons. In comparison with (5.87), we find that

$$D_n = \frac{kT}{q} \mu_n \frac{F_{1/2}(\eta)}{F_{-1/2}(\eta)} \tag{5.88}$$

Equation (5.88) is the Einstein relationship between the diffusion coefficient and the conductivity mobility. Although this relationship is easily derived for an equilibrium condition where the total current density is zero, the approach we have taken shows that Einstein's relation is also valid under nonequilibrium conditions.

Following similar arguments for valence band holes, the current density is

$$\mathbf{J} = q p \mu_p \mathbf{E} - kT \mu_p \frac{F_{1/2}(\eta)}{F_{-1/2}(\eta)} \nabla_r p \tag{5.89}$$

where μ_p is the conductivity mobility for holes. Notice that the diffusion component of hole current is opposite to the direction of the hole gradient. From Fick's law the diffusion coefficient for holes is also of the form of (5.88). When the Fermi energy is in the energy gap at least $4kT$ removed from either band edge, (5.88) reduces to

$$D_n = \frac{kT}{q} \mu_n \tag{5.90}$$

The equations we have derived for the mobility (5.50), conductivity (5.52), and Hall constant (5.72) are applicable for electrons in spherical conduction band minima. When the electrons transport charge in an ellipsoidal minimum, the situation is somewhat more complicated. Consider one ellipsoidal conduction band minimum at Γ given by (5.30). If the x direction is taken as one of the axes of the constant energy ellipsoids, an electric field in the x direction will produce a current in the x direction,

$$J_x = \frac{q^2 n \langle \tau_m \rangle}{m_1^*} E_x \tag{5.91}$$

Similar expressions containing m_2^* and m_3^* are obtained in the y and z directions. The total current density is therefore

$$\mathbf{J} = q^2 n \langle \tau_m \rangle \mathbf{M \cdot E} \tag{5.92}$$

where the effective mass tensor $\mathbf{M}$ is given by (5.31). This can also be put in the form

$$\mathbf{J} = \boldsymbol{\sigma \cdot E} \tag{5.93}$$

where $\boldsymbol{\sigma}$ is a conductivity tensor and $\mathbf{E}$ is a column vector. Thus, for an ellipsoidal minimum at Γ, the current density is not necessarily in the same direction as the applied electric field.

When there are g_c equivalent ellipsoidal conduction band minima, it is necessary to account for the fact that the concentration of electrons in each minimum is n/g_c. In this case the current density is obtained by summing the concentration of electrons in each minimum, while allowing for the effective mass each minimum has in the direction of the current. For a semiconductor with conduction band minima in the direction of X, such as Si, this is relatively easy. As shown in Fig. 5.1, when the current is in the x direction, the two minima along the k_x axis each contribute $n/6$ electrons with effective mass m_1^*, while the two minima along the k_y axis contribute $n/6$ electrons each with effective mass m_2^*. In the third dimension, the two minima along the k_z axis also contribute $n/6$ electrons of mass m_3^*. The total

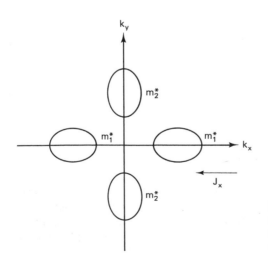

Figure 5.1 Diagram showing how equivalent ellipsoidal minima contribute to conduction along one of their principal axes.

current density in the x direction is therefore

$$J_x = \frac{q^2 n \langle \tau_m \rangle}{6} \left(\frac{2}{m_1^*} + \frac{2}{m_2^*} + \frac{2}{m_3^*} \right) E_x \tag{5.94}$$

Similar expressions are obtained for the components of current in the y and z directions. Thus the total current can be put in the form

$$\mathbf{J} = \frac{q^2 n \langle \tau_m \rangle}{m_c} \mathbf{E} \tag{5.95}$$

where

$$\frac{1}{m_c} \equiv \frac{1}{3} \left(\frac{1}{m_1^*} + \frac{1}{m_2^*} + \frac{1}{m_3^*} \right) \tag{5.96}$$

defines the *conductivity effective mass*.

Equation (5.96) is also valid for semiconductors, such as Ge, which have equivalent minima in directions other than X. Notice that the conductivity effective mass defined by (5.96) is a scalar, so that the current density and the electric field are always in the same direction. Comparing (5.95) with (5.92), we see that equivalent ellipsoidal minima result in isotropic conductivity, while an ellipsoidal minimum at Γ produces anisotropic conductivity. This difference due to the position of ellipsoidal minima is reflected in the conductivity of sphalerite crystals with indirect bandgaps as compared to wurtzite crystals with direct bandgaps.

When the Hall effect in ellipsoidal minima is examined, similar results are obtained. That is, an ellipsoidal conduction band minimum at Γ produces an anisotropic Hall effect, while equivalent minima produce an isotropic Hall effect. In the latter case, for simplicity, a *Hall effective mass* can be defined to reduce the expression for the Hall mobility to the form of (5.74).

5.4 CHARGE AND ENERGY TRANSPORT

To determine the heat flow density, (5.37) tells that we must find the average

$$\langle e\mathbf{v} \rangle = \frac{\displaystyle \int_{-\infty}^{\infty} e\mathbf{v} f \, d\mathbf{v}}{\displaystyle \int_{-\infty}^{\infty} f \, d\mathbf{v}} \tag{5.97}$$

where e is heat content per electron. This equation has the same form as (5.38) except that, in this case, we must include e in the average since e depends on the electron energy. We have already solved this problem with (5.60), so in the same way we obtained (5.79) we can simply write

$$\mathbf{W} = \frac{-n}{m^*} \langle e\mathbf{G} \rangle \tag{5.98}$$

where the minus sign indicates that **W** is opposite in direction to **G** and thus **J** for electrons. It is now necessary to obtain e in terms of $\mathscr{E}$, so that the average in (5.98) can be determined.

Since heat is that portion of the total electron energy which can be added or removed in disordered form, the heat content of all the electrons in the system is given by the entropy term in Euler's equation (4.28). Thus the heat content per electron is

$$e = \left[\frac{\partial}{\partial n'}(TS)\right]_{T,V} \tag{5.99}$$

From (4.29) for the Helmholtz function,

$$TS = \mathscr{E}' - F \tag{5.100}$$

where $\mathscr{E}'$ is the total energy of all the electrons. Using (5.100) in (5.99), we have

$$e = \left(\frac{\partial \mathscr{E}'}{\partial n'}\right)_{T,V} - \left(\frac{\partial F}{\partial n'}\right)_{T,V} \tag{5.101}$$

The first term in (5.101) is simply the energy per electron, $\mathscr{E}$. From (4.39) the second term is simply the chemical potential, μ. The heat content per electron is therefore

$$e = \mathscr{E} - \mu \tag{5.102}$$

Let us now obtain the equations that determine the transport of charge and energy under the following conditions. We will assume that the electric and magnetic fields are small and allow for temperature and concentration gradients. Also, for simplicity we will assume spherical energy band extrema. From (5.79) and (5.34) the equation for the electron current density is

$$\mathbf{J} = \frac{+q^2 n}{m^*}\left[\langle \tau_m \mathscr{F} \rangle - \frac{q}{m^*}\langle \tau_m^2(\mathscr{F} \times \mathbf{B})\rangle\right] \tag{5.103}$$

Using (5.98), (5.102), and (5.34), the electron heat current density is

$$\mathbf{W} = \frac{-qn}{m^*}\left[\langle \tau_m e\mathscr{F} \rangle - \frac{q}{m^*}\langle \tau_m^2 e(\mathscr{F} \times \mathbf{B})\rangle\right] \tag{5.104}$$

where from (5.28),

$$\mathscr{F} = +\mathbf{E} - \frac{1}{q}\nabla_r e + \frac{e}{qT}\nabla_r T \tag{5.105}$$

We will use (5.103), (5.104), and (5.105) to examine various effects that involve the transport of charge and energy in semiconductors.

5.4.1 Thermal Conductivity

One of the more important thermal transport parameters is the thermal conductivity. Although in lightly doped semiconductors most of the heat is carried by lattice vibrations or phonons, in heavily doped semiconductors a substantial proportion is carried by electrons. The thermal conductivity, κ, is defined as the proportionality factor between the heat current density and the temperature gradient,

$$\mathbf{W} = -\kappa \, \nabla_r T \tag{5.106}$$

The minus sign is required because the heat flows from higher to lower temperatures. To determine the thermal conductivity, we examine a sample under open-circuit conditions ($\mathbf{J} = 0$) with no magnetic field. For small temperature gradients, (5.103) is

$$0 = +\langle \tau_m \rangle \left(\mathbf{E} + \frac{1}{q} \nabla_r \mu \right) + \frac{1}{qT} \langle \tau_m e \rangle \nabla_r T \tag{5.107}$$

and (5.104) is

$$\mathbf{W} = \frac{-qn}{m^*} \left[\langle \tau_m e \rangle \left(\mathbf{E} + \frac{1}{q} \nabla_r \mu \right) + \frac{1}{qT} \langle \tau_m e^2 \rangle \nabla_r T \right] \tag{5.108}$$

Using (5.107) in (5.108) yields

$$\mathbf{W} = \frac{-n}{m^* T} \left[\langle \tau_m e^2 \rangle - \frac{\langle \tau_m e \rangle^2}{\langle \tau_m \rangle} \right] \nabla_r T \tag{5.109}$$

and the thermal conductivity due to electrons is

$$\kappa = \frac{n}{m^* T} \left[\langle \tau_m e^2 \rangle - \frac{\langle \tau_m e \rangle^2}{\langle \tau_m \rangle} \right] \tag{5.110}$$

Numerical values for the averages in (5.110) can be obtained with (5.60).

5.4.2 Thermoelectric Effects

We can see from (5.107) that under open-circuit conditions, the electrons diffuse down the temperature gradient and set up an electric field that opposes the motion of electrons due to the gradient. The production of an electric field by a temperature gradient is referred to as the *Seebeck* or *thermoelectric* effect. In the steady state the electric field is given by (5.107) as

$$\mathbf{E} = -\frac{1}{q} \nabla_r \mu - \frac{1}{q \langle \tau_m \rangle T} \langle \tau_m e \rangle \nabla_r T \tag{5.111}$$

Using

$$\nabla_r \mu = \frac{\partial \mu}{\partial T} \nabla_r T \qquad (5.112)$$

(5.111) gives us

$$\mathbf{E} = -\frac{1}{qT}\left[T\frac{\partial \mu}{\partial T} + \frac{\langle \tau_m e \rangle}{\langle \tau_m \rangle}\right] \nabla_r T \qquad (5.113)$$

or

$$\mathbf{E} = T\frac{d}{dT}\left[\frac{\langle \tau_m e \rangle}{qT\langle \tau_m \rangle}\right] \nabla_r T \qquad (5.114)$$

Thus the electric field is related to the temperature gradient by means of the equation

$$\mathbf{E} = \mathcal{T} \nabla_r T \qquad (5.115)$$

where

$$\mathcal{T} \equiv T\frac{d}{dT}\left[\frac{\langle \tau_m e \rangle}{qT\langle \tau_m \rangle}\right] \qquad (5.116)$$

is the *Thomson coefficient*.

As indicated in Fig. 5.2, the Seebeck effect can be examined by measuring the voltage across a semiconductor in a temperature gradient. The voltage is given by

$$V = -\oint \mathbf{E}\cdot d\mathbf{r} = \oint \mathcal{T} \nabla_r T \cdot d\mathbf{r}$$

$$V = \int_{T_0}^{T_1} \mathcal{T}_m \, dT + \int_{T_1}^{T_2} \mathcal{T}_s \, dT + \int_{T_2}^{T_0} \mathcal{T}_m \, dT \qquad (5.117)$$

$$V = \int_{T_1}^{T_2} (\mathcal{T}_s - \mathcal{T}_m) \, dT$$

where $\mathcal{T}_s$ and $\mathcal{T}_m$ are the Thomson coefficients for the semiconductor and

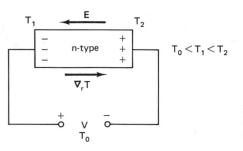

Figure 5.2 Determination of Seebeck effect for *n*-type semiconductor. *V* is negative.

metal, respectively. If the metal is chosen to have $\mathcal{T}_m \simeq 0$, the Seebeck voltage is directly related to the Thomson coefficient of the semiconductor and the temperature difference across the sample. Notice that the Thomson coefficient for electrons is negative, as indicated by (5.115). For holes under the same conditions, the Thomson coefficient is positive, so that the field and temperature gradients are in the same direction.

A transport parameter closely related to the Thomson coefficient is the absolute *thermoelectric power,* $\mathcal{P}$. The relationship is

$$\mathcal{T} = -T \frac{d}{dT} \mathcal{P} \tag{5.118}$$

or from (5.116),

$$\mathcal{P} \equiv -\frac{\langle \tau_m e \rangle}{qT \langle \tau_m \rangle} \tag{5.119}$$

Notice that the thermoelectric power for electrons and holes have opposing signs, due to the dependence on q. Because of this the sign of the thermoelectric power can be used to determine whether a material exhibits n- or p-type conductivity.

When the electric current density is not constrained to be zero, it adds an additional component to the heat current density. Under these conditions (5.103) can be written in the form

$$\mathbf{E} + \frac{1}{q} \nabla_r \mu = \frac{m^*}{q^2 n \langle \tau_m \rangle} \mathbf{J} - \frac{1}{qT} \frac{\langle \tau_m e \rangle}{\langle \tau_m \rangle} \nabla_r T \tag{5.120}$$

Substituting this into (5.108), the heat current density is

$$\mathbf{W} = \frac{-\langle \tau_m e \rangle}{q \langle \tau_m \rangle} \mathbf{J} - \frac{n}{m^* T} \left[\langle \tau_m e^2 \rangle - \frac{\langle \tau_m e \rangle^2}{\langle \tau_m \rangle} \right] \nabla_r T \tag{5.121}$$

Using (5.110) and (5.119), this is simply

$$\mathbf{W} = T\mathcal{P}\mathbf{J} - \kappa \nabla_r T \tag{5.122}$$

Thus the electric current density carries heat in addition to that transported by the temperature gradient. This is referred to as the *Peltier effect.* The constant of proportionality between heat current density and electric current density is the Peltier coefficient, Π, where

$$\Pi \equiv T\mathcal{P} \tag{5.123}$$

$$\Pi = -\frac{\langle \tau_m e \rangle}{q \langle \tau_m \rangle} \tag{5.124}$$

Because of the dependence on q, the Peltier coefficient is negative for electrons and positive for holes.

In addition to transporting heat, the electric current density also generates heat. The net rate at which heat is generated per unit volume is equal to the rate generated per unit volume minus the rate at which it is transported away, or

$$P = \mathbf{J} \cdot \mathbf{E} - \nabla_r \cdot \mathbf{W} \tag{5.125}$$

From (5.120), (5.52), and (5.115),

$$\mathbf{E} = \frac{1}{\sigma} \mathbf{J} + \mathcal{T} \nabla_r T \tag{5.126}$$

Also, from (5.122) and (5.123),

$$\mathbf{W} = \Pi \mathbf{J} - \kappa \nabla_r T \tag{5.127}$$

Using (5.126) and (5.127) in (5.125), we have

$$P = \frac{\mathbf{J} \cdot \mathbf{J}}{\sigma} + \mathcal{T} \mathbf{J} \cdot \nabla_r T - \Pi \nabla_r \cdot \mathbf{J} + \kappa \nabla_r \cdot \nabla_r T \tag{5.128}$$

Thus the net rate at which heat is generated per unit volume, P, has several components. The first term in (5.128) is simply the Joule heat. The second term is referred to as the *Thomson heat*. The third term, which involves the divergence of the electric current, allows for the generation or recombination of electrons in the unit volume and will not be considered further. Finally, the last term provides for the transport of heat out of the volume by thermal conduction.

Notice that the Thomson heat term in (5.128) changes sign when either the current density or the temperature gradient is reversed. Since the Thomson coefficient is negative for electrons, in *n*-type material heating is produced when $\mathbf{J}$ and $\nabla_r T$ are in the same direction. That is, the electrons going from a higher to a lower temperature have to give heat to the lattice. When $\mathbf{J}$ and $\nabla_r T$ are in opposite directions, the electrons produce cooling since they take heat from the lattice in going from a lower to a higher temperature. These effects are indicated schematically in Fig. 5.3. Since the Thomson coefficient for holes is positive, cooling is produced when $\mathbf{J}$ and $\nabla_r T$ are in

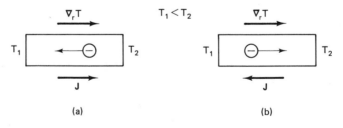

(a)   (b)

Figure 5.3 The Thomson term in an *n*-type semiconductor produces (a) heating when $\mathbf{J}$ and $\nabla_r T$ are in the same direction and (b) cooling when they are in opposite directions.

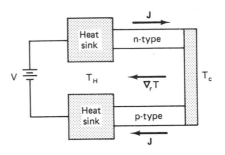

Figure 5.4 Schematic of a thermoelectric cooler. The heat sinks and cold junctions are metals that form ohmic contacts.

the same direction. These results can be used to construct a thermoelectric cooler in the manner shown in Fig. 5.4.

Similar effects are obtained for electrons in concentration gradients. In the absence of a temperature gradient, the electric field for n-type material is given from (5.87) as

$$\mathbf{E} = \frac{\mathbf{J}}{\sigma} - \frac{kT}{qN_cF_{-1/2}(\eta)}\,\mathbf{\nabla}_r n \tag{5.129}$$

In this case, (5.125) is

$$P = \frac{\mathbf{J}\cdot\mathbf{J}}{\sigma} - \frac{kT}{qN_cF_{-1/2}(\eta)}\,\mathbf{J}\cdot\mathbf{\nabla}_r n - \Pi\,\mathbf{\nabla}_r\cdot\mathbf{J} \tag{5.130}$$

Thus, when $\mathbf{J}$ and $\mathbf{\nabla}_r n$ are in the same direction, the electrons take heat from the lattice as they go from higher to lower concentrations and cooling is produced.

5.4.3 Thermomagnetic Effects

When we allow for a small magnetic field, in addition to small electric fields and temperature gradients, the transport of heat produces several thermomagnetic effects. We examine these in the Hall configuration shown in Fig. 5.5. In Section 5.3 we examined the Hall effect assuming that no tem-

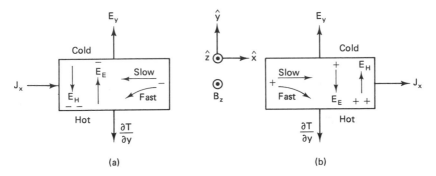

Figure 5.5 Hall and Ettinghausen effects $(0 \leq r)$ for (a) electrons and (b) holes.

perature gradients were present (isothermal conditions) and that there was no electric current in the $\hat{y}$ direction. That is, $J_y = \partial T/\partial x = \partial T/\partial y = 0$. Under these conditions, the electrons were deflected by the magnetic field in the direction shown in Fig. 5.5(a) and an electric field, E_H, was induced in the negative $\hat{y}$ direction to balance the Lorentz force. However, this Hall field can only exactly balance the Lorentz force on electrons with average velocity. If we assume that the momentum relaxation time increases with energy ($\tau_m = \tau_0 x^r$, where $0 \leq r$), the faster or hotter electrons are deflected more and the slower ones less by the magnetic field. As a result, the side of the sample where the faster carriers are deflected becomes warmer and the opposite side cooler, inducing a temperature gradient. In a manner similar to the thermoelectric effect, the warmer electrons tend to diffuse to the cooler surface, where they set up an electric field, as in (5.115), to oppose the diffusion. The mechanism that produces this electric field is referred to as the *Ettinghausen effect*.

The Ettinghausen coefficient is defined under conditions such that no heat current is transferred to the surroundings (adiabatic conditions). For $J_y = \partial T/\partial x = W_y = 0$, this coefficient is

$$P_E \equiv \frac{\partial T/\partial y}{J_x B_z} \tag{5.131}$$

Applying these conditions to (5.103) and (5.104), we can eliminate the electric field and the chemical potential gradient,

$$P_E = \frac{\mu_c}{q\kappa} \left[\frac{\langle \tau_m^2 \mathscr{E} \rangle}{\langle \tau_m \rangle^2} - \frac{\langle \tau_m^2 \rangle \langle \tau_m \mathscr{E} \rangle}{\langle \tau_m \rangle^3} \right] \tag{5.132}$$

From this equation we expect the Ettinghausen field to change sign when the sign of the charge carrier is reversed. Figure 5.5(b) for holes shows that this is, indeed, what occurs. In either case, however, the direction of the Ettinghausen field depends on the energy dependence of the momentum relaxation time. It can be verified by (5.132) that when r in (5.55) is less than zero, the slower carriers are deflected more than the faster ones and the direction of the Ettinghausen field in Fig. 5.5 are reversed. Thus the direction of the Ettinghausen field depends on the sign of the carrier and the scattering mechanism.

Two other thermomagnetic effects we will mention are the *Nernst* and *Righi–Leduc effects*. As indicated in Fig. 5.6, these effects are the thermal analogues of the Hall and Ettinghausen effects, respectively. The Nernst coefficient is defined under isothermal conditions ($J_x = J_y = \partial T/\partial y = 0$) as

$$Q_N \equiv \frac{E_y}{(\partial T/\partial x)B_z} \tag{5.133}$$

Thus we see that the Nernst effect is a process whereby a transverse electric

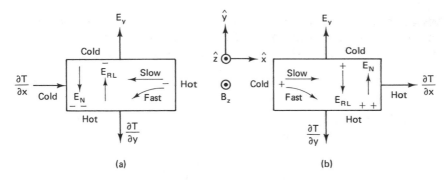

Figure 5.6   Nernst and Righi–Leduc effects ($0 \le r$) for (a) electrons and (b) holes.

field is produced by a temperature gradient. This equation can be compared to (5.72) for the Hall constant. Under these conditions (5.103) and (5.104) give us

$$Q_N = \frac{\mu_c}{qT}\left[\frac{\langle \tau_m^2 \mathscr{E}\rangle}{\langle \tau_m\rangle^2} - \frac{\langle \tau_m^2\rangle\langle \tau_m\mathscr{E}\rangle}{\langle \tau_m\rangle^3}\right] \tag{5.134}$$

or

$$Q_N = \frac{\kappa}{T}P_E \tag{5.135}$$

The Righi–Leduc coefficient is defined under adiabatic conditions ($J_x = J_y = W_y = 0$) as

$$S_{RL} \equiv \frac{\partial T/\partial y}{(\partial T/\partial x)B_z} \tag{5.136}$$

which gives

$$S_{RL} = \frac{n\mu_c^2}{qT\kappa}\left[\frac{\langle \tau_m^2 \mathscr{E}^2\rangle}{\langle \tau_m^2\rangle} + \frac{\langle \tau_m^2\rangle\langle \tau_m\mathscr{E}\rangle^2}{\langle \tau_m\rangle^4} - \frac{2\langle \tau_m^2\mathscr{E}\rangle\langle \tau_m\mathscr{E}\rangle}{\langle \tau_m\rangle^3}\right] \tag{5.137}$$

Arguments regarding the sign of these coefficients are similar to those made for the sign of the Ettinghausen coefficient.

5.5 HIGH-FREQUENCY TRANSPORT

The dc theory of charge and energy transport developed in this chapter can be applied to transport at high frequencies with only slight modification. Let us apply a small sinusoidal electric field,

$$\mathbf{E} = \mathbf{E}_0 \exp\left(-i\omega t\right) \tag{5.138}$$

to a semiconductor sample where ω is the angular frequency of the field.

Assuming no other applied forces and no temperature or concentration gradients, we can examine the form of the nonequilibrium distribution function. In the same manner in which (5.15) was obtained, we have

$$f = f_0 + q\tau_m \frac{\partial f_0}{\partial \mathcal{E}} \mathbf{v} \cdot \mathbf{E}_0 \exp(-i\omega t) \tag{5.139}$$

The time-dependent Boltzmann equation in the relaxation time approximation from (5.4), (5.5), and (5.9) is

$$\frac{\partial f}{\partial t} = -\mathbf{v} \cdot \left(\frac{\partial f}{\partial \mathcal{E}} \mathbf{F} + \nabla_r f \right) - \frac{f - f_0}{\tau_m} \tag{5.140}$$

Taking the time derivative of (5.139), we have

$$\frac{\partial f}{\partial t} = +q\tau_m \frac{\partial f_0}{\partial \mathcal{E}} \left(-i\omega \mathbf{v} + \frac{d\mathbf{v}}{dt} \right) \cdot \mathbf{E}_0 \exp(-i\omega t) \tag{5.141}$$

Since

$$\frac{d\mathbf{v}}{dt} = -q\mathbf{M} \cdot \mathbf{E}_0 \exp(-i\omega t) \tag{5.142}$$

the last term in (5.141) is second order in $\mathbf{E}$ and can be neglected. We now have

$$\frac{\partial f}{\partial t} = -i\omega(f - f_0) \tag{5.143}$$

and the Boltzmann equation is

$$f - f_0 = \frac{-\tau_m}{(1 - i\omega\tau_m)} \mathbf{v} \cdot \left(\frac{\partial f}{\partial \mathcal{E}} \mathbf{F} + \nabla_r f \right) \tag{5.144}$$

Except for the term $(1 - i\omega\tau_m)$, (5.144) has the same form as (5.11). Because of this, all of the dc transport equations can be used at high frequency if we replace τ_m by τ_m^* where

$$\tau_m^* \equiv \frac{\tau_m}{1 - i\omega\tau_m} \tag{5.145}$$

The general solution of the Boltzmann equation (5.144) for ellipsoidal minima is

$$f = f_0$$
$$+ \frac{\partial f_0}{\partial \mathcal{E}} q\tau_m^* \mathbf{v} \cdot \left[\frac{\mathcal{F} - q\tau_m^* \mathbf{M} \cdot (\mathcal{F} \times \mathbf{B}) + (q\tau_m^*)^2 (\det \mathbf{M})(\mathcal{F} \cdot \mathbf{B})(\mathbf{M}^{-1} \cdot \mathbf{B})}{1 + (q\tau_m^*)^2 (\det \mathbf{M})(\mathbf{M}^{-1} \cdot \mathbf{B}) \cdot \mathbf{B}} \right] \tag{5.146}$$

and this is used to determine the transport parameters. Since τ_m^* is a complex number, all the terms in the distribution function (except for f_0) are complex. Thus, in general, the high-frequency components of electric current density and heat current density are not in phase with the applied forces.

A useful effect can be observed when we examine the denominator of the last term in (5.146). If we define a cyclotron frequency as in (5.65) by

$$\omega_c^2 \equiv q^2(\det \mathbf{M})(\mathbf{M}^{-1} \cdot \mathbf{B}) \cdot \mathbf{B} \tag{5.147}$$

the denominator becomes $1 + (\omega_c \tau_m^*)^2$. Using (5.145) gives

$$1 + (\omega_c \tau_m^*)^2 = 1 + \frac{(\omega_c \tau_m)^2}{(1 - i\omega \tau_m)^2} \tag{5.148}$$

which can be put in the form

$$1 + (\omega_c \tau_m^*)^2 = \frac{(\omega_c^2 - \omega^2)\tau_m^2 - 2i\omega\tau_m + 1}{(1 - i\omega\tau_m)^2} \tag{5.149}$$

Equation (5.149) tells us that for conditions such that $\omega\tau_m$ is much greater than 1, $1 + (\omega_c\tau_m^*)^2$ exhibits a sharp minimum when $\omega = \omega_c$. Thus the transport properties, such as charge current density, $\mathbf{J}$, will exhibit a resonant peak when the applied frequency is equal to the cyclotron frequency. This effect is referred to as *cyclotron resonance*. From (5.65) or (5.147) the frequency at which cyclotron resonance occurs can be used to determine effective mass.

5.6 HIGH ELECTRIC FIELD EFFECTS

Up to now we have limited our analysis of transport properties to small electric fields. Under these conditions the energy the carrier distribution gains from the electric field is lost to the lattice through collisions with low-energy acoustic phonons or impurities. The average energy of the electrons, therefore, remains close to the thermal equilibrium value, $\frac{3}{2}kT$, and the drift velocity of the distribution is linearly related to the electric field. However, because the average electron energy in semiconductors is so small, it is relatively easy to obtain significant deviations from this ohmic behavior. For moderate electric fields the collisions with acoustic phonons and impurities, which serve to maintain the electron distribution and the lattice at the same temperature, become less effective and the electrons gain energy from the field faster than they can lose it to the lattice. In this situation, the electron distribution can be characterized by an effective temperature, T_e, which is "hotter" than the lattice temperature, T. The relationship between the drift velocity and electric field is no longer linear and nonohmic electrical behavior is observed.

When the electrons have gained sufficient energy from the field, they can transfer energy to the lattice by the generation of high energy optical phonons. Since this process is an efficient energy loss mechanism for the electrons, the drift velocity of the distribution reaches a limiting value where it no longer increases with the electric field. This value is referred to as the *saturated drift velocity* and is obtained for electron energies of the order of the optical phonon frequencies given in Table 3.2. Figure 5.7(a) and (b) show experimental drift velocity–electric field characteristics of electrons and holes, respectively, for Ge, Si, and GaAs up to the saturated regions. As can be seen, all of these characteristics, except for electrons in GaAs, exhibit, qualitatively, the nonohmic behavior described above. The negative resistance region in the GaAs v_d versus **E** curve is due to the transfer of electrons from the Γ to L conduction band minima. This forms the basis of the Gunn effect.

Since these *hot electron effects* play an important role in the operation of several semiconductor devices, let us examine them in more detail. Although there are several approaches to the problem of high electric field transport in semiconductors [Esther M. Conwell, *High Field Transport in Semiconductors* (New York: Academic Press, 1967)], we will use an approach with which analytical results can readily be obtained. Let us assume that the electron distribution can be described by

$$f = \frac{1}{1 + \exp\left[(\mathcal{E} - \zeta)/kT_e\right]} \tag{5.150}$$

where T_e is the effective temperature of the electron distribution. Notice that this distribution function has the form of the equilibrium distribution

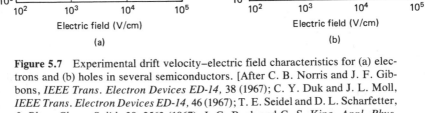

Figure 5.7 Experimental drift velocity–electric field characteristics for (a) electrons and (b) holes in several semiconductors. [After C. B. Norris and J. F. Gibbons, *IEEE Trans. Electron Devices ED-14,* 38 (1967); C. Y. Duk and J. L. Moll, *IEEE Trans. Electron Devices ED-14,* 46 (1967); T. E. Seidel and D. L. Scharfetter, *J. Phys. Chem. Solids 28,* 2563 (1967); J. G. Ruch and G. S. Kino, *Appl. Phys. Lett. 10,* 40 (1967); V. L. Dalal, *Appl. Phys. Lett. 16,* 489 (1970).]

function, f_0, derived in (4.41), except for T_e, which is a function of electric field, **E**. We will also maintain the form of the low-field relationship between the drift velocity and the electric field (5.49).

$$\mathbf{v}_d = -\mu(T_e)\mathbf{E} \tag{5.151}$$

where the mobility, μ, is now a function of T_e or electric field. In a similar manner,

$$\mu(T_e) = \frac{q\langle\tau_m(T_e)\rangle}{m^*} \tag{5.152}$$

where

$$\langle\tau_m(T_e)\rangle = \frac{\displaystyle\int_{-\infty}^{\infty} \tau_m(T_e)f \, d\mathbf{v}}{\displaystyle\int_{-\infty}^{\infty} f \, d\mathbf{v}} \tag{5.153}$$

and f is given by (5.150). Under these assumptions, we can proceed with our examination of high-field transport.

The average over velocity space of the momentum relaxation time in (5.153) can be simplified to an average over energy

$$\langle\tau_m(T_e)\rangle = \tau_0(T_e)\frac{\displaystyle\int_0^{\infty} fx_e^{r+1/2} \, dx_e}{\displaystyle\int_0^{\infty} fx_e^{1/2} \, dx_e} \tag{5.154}$$

or, using (5.59),

$$\langle\tau_m(T_e)\rangle = \frac{2}{\sqrt{\pi}}\left(r + \frac{1}{2}\right)! \, \tau_0(T_e)\frac{F_{r+1/2}(\eta_e)}{F_{1/2}(\eta_e)} \tag{5.155}$$

where

$$x_e \equiv \frac{\mathscr{E} - \mathscr{E}_c}{kT_e} \quad\text{and}\quad \eta_e \equiv \frac{\zeta - \mathscr{E}_c}{kT_e} \tag{5.156}$$

Equation (5.155) tells us that, in general, the appropriate average is

$$\langle\tau_m^s(T_e)x_e^t\rangle = \frac{2}{\sqrt{\pi}}\left(sr + t + \frac{1}{2}\right)! \, \tau_0^s(T_e)\frac{F_{sr+t+1/2}(\eta_e)}{F_{1/2}(\eta_e)} \tag{5.157}$$

To proceed further we need to know the dependence of τ_0 on T_e. This dependence, of course, depends on the particular scattering mechanism. For several scattering mechanisms, the low-field momentum relaxation time can be put in the form

$$\tau_m = \tau_0 x^r = CT^u x^r \tag{5.158}$$

where C is independent of temperature. If we assume that the high-field momentum relaxation time has the same form,

$$\tau_m(T_e) = \tau_0(T_e)x_e^r = CT_e^u x_e^r \tag{5.159}$$

then

$$\tau_m(T_e) = \tau_0 \left(\frac{T_e}{T}\right)^u x_e^r \tag{5.160}$$

From (5.152), (5.155), and (5.160) the field-dependent mobility is

$$\mu(T_e) = \frac{q\tau_0}{m^*} \left(\frac{T_e}{T}\right)^u \frac{2}{\sqrt{\pi}} \left(r + \frac{1}{2}\right)! \frac{F_{r+1/2}(\eta_e)}{F_{1/2}(\eta_e)} \tag{5.161}$$

or simply

$$\mu(T_e) = \mu_0 \left(\frac{T_e}{T}\right)^u \tag{5.162}$$

Thus if we can determine the dependence of T_e on $\mathbf{E}$, we can obtain the dependence of μ and, from (5.151), the dependence of $\mathbf{v}_d$ on $\mathbf{E}$.

The effective electron temperature can be determined from conservation of energy. From (5.4) and (5.6) the time-dependent Boltzmann equation is

$$\frac{\partial f}{\partial t} = \frac{+q}{\hbar} \mathbf{E} \cdot \nabla_k f - \mathbf{v} \cdot \nabla_r f - \frac{f - f_0}{\tau_m} \tag{5.163}$$

If we multiply each term of this equation by the electron energy, $\mathscr{E}$, and average it over the electron distribution, we obtain

$$\frac{\int_{-\infty}^{\infty} \mathscr{E} \frac{\partial f}{\partial t} d\mathbf{v}}{\int_{-\infty}^{\infty} f \, d\mathbf{v}} = \frac{\frac{+q}{\hbar} \mathbf{E} \cdot \int_{-\infty}^{\infty} \mathscr{E} \nabla_k f \, d\mathbf{v}}{\int_{-\infty}^{\infty} f \, d\mathbf{v}} - \frac{\int_{-\infty}^{\infty} \mathscr{E} \mathbf{v} \cdot \nabla_r f \, d\mathbf{v}}{\int_{-\infty}^{\infty} f \, d\mathbf{v}}$$

$$- \frac{\int_{-\infty}^{\infty} \mathscr{E} \frac{f - f_0}{\tau_m} d\mathbf{v}}{\int_{-\infty}^{\infty} f \, d\mathbf{v}} \tag{5.164}$$

With (2.109) and (5.38), the first term on the right-hand side can be reduced to $q\mathbf{v}_d \cdot \mathbf{E}$. Neglecting the diffusion term, (5.164) then reduces to the form

$$\frac{d}{dt} \langle \mathscr{E} \rangle = q\mathbf{v}_d \cdot \mathbf{E} - \frac{\langle \mathscr{E} \rangle - \langle \mathscr{E}_0 \rangle}{\tau_e} \tag{5.165}$$

where we have defined an *energy relaxation time*, τ_e, by

$$\frac{\langle \mathcal{E} \rangle - \langle \mathcal{E}_0 \rangle}{\tau_e} \equiv \frac{\int_{-\infty}^{\infty} \mathcal{E}\left(\dfrac{f - f_0}{\tau_m}\right) d\mathbf{v}}{\int_{-\infty}^{\infty} f \, d\mathbf{v}} \tag{5.166}$$

Equation (5.165) is the energy balance equation for the hot electron distribution. It tells us that the net energy gained per unit time is equal to the power supplied by the electric field minus the energy lost to collisions. Under steady-state conditions the electron distribution attains a temperature T_e and

$$\frac{\langle \mathcal{E} \rangle - \langle \mathcal{E}_0 \rangle}{\tau_e} = q\mathbf{v}_d \cdot \mathbf{E} \tag{5.167}$$

Notice from (5.165) that τ_e characterizes the relaxation of the hot electron distribution to its average thermal equilibrium energy $\langle \mathcal{E}_0 \rangle$ when the electric field is turned off. We see from (5.166) that, since τ_m is in general a function of $\mathcal{E}$, τ_m and τ_e are not equal. That is, the times that characterize the relaxation of energy and momentum are different.

From (5.156) and (5.157) the average energy of the hot electron distribution is

$$\langle \mathcal{E} \rangle = kT_e \langle x_e \rangle = \frac{3}{2} kT_e \frac{F_{3/2}(\eta_e)}{F_{1/2}(\eta_e)} \tag{5.168}$$

In a similar manner the average energy of the equilibrium distribution is

$$\langle \mathcal{E}_0 \rangle = kT \langle x \rangle = \frac{3}{2} kT \frac{F_{3/2}(\eta)}{F_{1/2}(\eta)} \tag{5.169}$$

Using (5.156) and (5.166) yields

$$\frac{\langle \mathcal{E} \rangle}{\tau_e} \equiv \left\langle \frac{\mathcal{E}}{\tau_m} \right\rangle = kT_e \left\langle \frac{x_e}{\tau_m} \right\rangle \tag{5.170}$$

The average in this equation can be evaluated with (5.157) and (5.160) as

$$\frac{\langle \mathcal{E} \rangle}{\tau_e} = \frac{2}{\sqrt{\pi}} \left(\frac{3}{2} - r\right)! \frac{kT_e}{\tau_0} \left(\frac{T}{T_e}\right)^u \frac{F_{3/2-r}(\eta_e)}{F_{1/2}(\eta_e)} \tag{5.171}$$

and

$$\tau_e = \frac{3\sqrt{\pi}}{4} \frac{\tau_0}{(3/2 - r)!} \left(\frac{T_e}{T}\right)^u \frac{F_{3/2}(\eta_e)}{F_{3/2-r}(\eta_e)} \tag{5.172}$$

Under nondegenerate conditions, we can readily obtain an expression relating T_e to E. Using (5.151), (5.161), and (5.171) in (5.167), we have

$$\left(\frac{3}{2} - r\right)! \frac{k}{\tau_0}\left(\frac{T}{T_e}\right)^u (T_e - T) = \left(r + \frac{1}{2}\right)! \frac{q^2\tau_0}{m^*}\left(\frac{T_e}{T}\right)^u E^2 \quad (5.173)$$

or

$$\left(\frac{T_e}{T} - 1\right)\left(\frac{T}{T_e}\right)^{2u} = \frac{q^2\tau_0^2}{m^*kT}\frac{(r + \frac{1}{2})!}{(\frac{3}{2} - r)!} E^2$$

$$= (\beta E)^2 \quad (5.174)$$

For hot electrons T_e is much greater than T and (5.174) can be solved for T_e, with the result

$$\frac{T_e}{T} = (\beta E)^{-2/(2u-1)} \quad (5.175)$$

From (5.162) the mobility is

$$\mu(T_e) = \mu_0 \left(\frac{T_e}{T}\right)^u$$

or

$$\mu(E) = \mu_0(\beta E)^{-2u/(2u-1)} \quad (5.176)$$

With (5.176) we can examine the dependence of the mobility on electric field for scattering processes in which the relaxation time approximation can be used.

The temperature dependence of the momentum relaxation time for acoustic phonon scattering is $u = -\frac{3}{2}$. Using this value in (5.176), the mobility is

$$\mu(T_e) = \mu_0 \left(\frac{T}{T_e}\right)^{3/2}$$

$$\mu(E) = \mu_0 \left(\frac{1}{\beta E}\right)^{3/4} \quad (5.177)$$

Thus the mobility for this scattering mechanism decreases as the electric field increases. From (5.151) and (5.177) the drift velocity,

$$\mathbf{v}_d = -\mu_0\beta^{-3/4}\mathbf{E}^{1/4} \quad (5.178)$$

does not saturate at high electric fields. For ionized impurity scattering $u = +\frac{3}{2}$ and the mobility is

$$\mu(T_e) = \mu_0 \left(\frac{T_e}{T}\right)^{3/2}$$

$$\mu(E) = \mu_0 \left(\frac{1}{\beta E}\right)^{3/2} \quad (5.179)$$

Thus, for this scattering mechanism the mobility also decreases with electric field, and the drift velocity does not saturate. From (5.176) we see that u must be infinite to obtain a completely saturated drift velocity. This corresponds to scattering by optical phonons, where the energy changes are so large that the use of a momentum relaxation time is a poor approximation.

Under these conditions a reasonable approximation to the saturated drift velocity can be obtained by assuming that the energy of the hot electron distribution is dissipated in the generation of longitudinal optical phonons

$$\frac{\langle \mathscr{E} \rangle - \langle \mathscr{E}_0 \rangle}{\tau_e} \simeq \frac{\hbar \omega_{LO}}{\tau_e} \tag{5.180}$$

and that the proportionality factor between drift velocity and electric field is

$$\mu = \frac{q \tau_e}{m^*} \tag{5.181}$$

Using (5.180) and (5.181) in the energy balance equation (5.167), we find that

$$\tau_e = \frac{(\hbar \omega_{LO} m^*)^{1/2}}{qE} \tag{5.182}$$

and the drift velocity is saturated at the value

$$v_d = \left(\frac{\hbar \omega_{LO}}{m^*} \right)^{1/2} \tag{5.183}$$

Values for the longitudinal optical phonon frequencies from Table 3.2 can be used in (5.183) to obtain saturated drift velocities which are in reasonable agreement with experimental results.

If the electric field is increased further in the saturated region, at some point the charge carriers will have sufficient energy to generate an electron–hole pair in a collision with the lattice. This process is known as *impact ionization*. When each of these electron–hole pairs creates an additional pair by impact ionization, an unstable situation is obtained where, in principle, the number of charge carriers increases without limit. This situation is referred to as *avalanche breakdown* and is readily observed in the reverse-bias current–voltage characteristics of *p-n* junctions. These phenomena are examined in more detail in Chapter 9.

PROBLEMS

5.1. Find the nonequilibrium distribution function to second order in $\mathbf{E} = \hat{x} E_x$ for a parabolic band. Assume a Maxwellian equilibrium distribution and that $\tau_m = \tau_0 (\mathscr{E}/kT)^{-1/2}$.

5.2. Derive an expression for the Hall factor assuming Fermi statistics and $\tau_m = \tau_0 x^r$. Plot r_H versus $\eta (-4 \leq \eta \leq 10)$ for $r = -\frac{1}{2}$ and $r = \frac{3}{2}$. Explain the dependence on η.

5.3. Determine the diffusion coefficient (D_n) for electrons in a fully *degenerate n*-type semiconductor, specifically for Ge with $\mu_n = 300 \text{ cm}^2/\text{V·s}$, $n = 10^{19} \text{ cm}^{-3}$, and $m_n = 0.2m_0$.

5.4. Show how all but one of the following secondary effects can be canceled in a Hall experiment by taking four measurements with B and J reversals: Ettinghausen, Nernst, Righi–Leduc, thermoelectric, and IR drop (probe misalignment).

5.5. Assume that a semiconductor has spherical energy surfaces in both the conduction and the valence bands, but with two species of holes. Find an expression for the Hall coefficient in terms of appropriate averages of τ_{p1}, τ_{p2}, τ_n, μ_{p1}, μ_{p2}, μ_n; p_1, p_2, n.

5.6. Consider an *n*-type semiconductor material as shown in Fig. P5.6 of mobility $10^4 \text{ cm}^2/\text{V·s}$ which is to be used as a fast microwave switch. The device, a conductive bar, is switched from the more conductive to the less conductive state, with an applied electric field, by "heating" electrons from the direct to the indirect conduction band minima. Estimate the applied field needed.

$$m^*(\Gamma) = 0.07m, \qquad \frac{N_c(X)}{N_c(\Gamma)} \simeq 60$$

5.7. Consider the Righi–Leduc effect in a uniformly doped nondegenerate *n*-type semiconductor with spherical energy surfaces. A thermal gradient is maintained in the x direction and a magnetic field in the z direction, with $J_x = J_y = W_y = 0$. If $\nabla_y T = S_{RL} B_z \nabla_x T$:
(a) Find S_{RL} in terms of appropriate averages of τ^s.
(b) For spherical energy surfaces and $\tau = \tau_0 \mathscr{E}^p$, express S_{RL} in terms of gamma functions.
(c) For $p = -\frac{1}{2}$, show that $S_{RL} = -(21\pi/32)(k^2 T/q\kappa)n\mu_n^2$, where κ is the thermal conductivity.

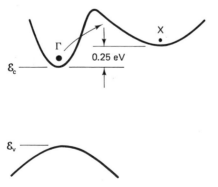

Figure P5.6

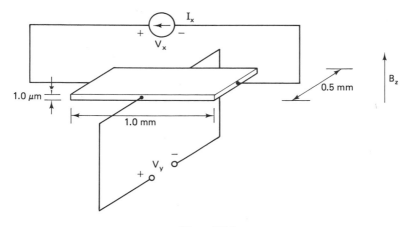

Figure P5.8

5.8. Hall meaurements are performed on a sample in the configuration shown in Fig. P5.8 with $I_x = 5$ mA, $B_z = 0.1$ T, and $T = 77$ K. Measured values are $V_x = 10$ V and $V_y = -5$ V. Assuming that $m_e = 0.1m$, $m_h = 0.5m$, and $\tau_m = \tau_0 x^{3/2}$ (ionized impurities), determine the following:
  (a) Type of material
  (b) Hall constant
  (c) Carrier concentration
  (d) Conductivity
  (e) Conductivity mobility
  (f) Hall mobility

5.9. Calculate the root-mean-square z-directed velocity of conduction band electrons in an isotropic material at equilibrium. Assume that nondegenerate statistics are valid and that the material exhibits spherically symmetric constant energy surfaces about Γ.

5.10. Consider a collection of electrons in an isotropic material which are constrained to possess a nonzero z-directed net velocity (e.g., through the application of an electric field $\mathbf{E} = -E_z\hat{z}$, $\mathbf{J} = \sigma\mathbf{E}$). Show that the distribution function has the form

$$f = \frac{1}{1 + \exp\,[\beta(\mathscr{E} - \alpha - \gamma v_z)]}$$

where v_z is the z-directed velocity.

5.11. The steady-state distribution function for a nondegenerate material under certain conditions is

$$f = \exp\,[-\beta(\mathscr{E} - \zeta)]\,\exp\left(\frac{\hbar^2\beta}{m^*}\gamma k_z\right)$$

where $\beta = 1/kT$, ζ and γ are constants. The conduction band states near Γ

are described by

$$\mathcal{E} = \mathcal{E}_c + \frac{\hbar^2 k^2}{2m^*}$$

Determine the concentration of electrons in the conduction band and their mean z-directed velocity. For the conditions $\gamma = 10^7 \text{ m}^{-1}$, $T = 300 \text{ K}$, $m^* = 0.0665m$, and $n = 10^{16} \text{ cm}^{-3}$, what are the z-directed velocity and current density?

Scattering
Processes

In Chapter 5 we used the relaxation time approximation to examine the transport of charge and energy in electric, magnetic, and thermal fields. There it was assumed that a momentum relaxation time, τ_m, could be defined for various carrier scattering processes, such that

$$\tau_m = \tau_0 x^r \tag{5.55}$$

where x is the electron kinetic energy in units of kT,

$$x \equiv \frac{\mathscr{E} - \mathscr{E}_c}{kT} \tag{4.65}$$

and τ_0 and the exponent r are independent of energy. In this chapter we examine this assumption and discuss the physics of the more important scattering processes. Rather than being all-inclusive, we will derive momentum relaxation times for ionized and neutral impurity scattering, as examples, and then show how these can be combined with values for phonon scattering to model and predict experimental mobility.

6.1 SCATTERING POTENTIALS

As discussed in Chapter 2, an electron moving in a perfect periodic crystal potential with no applied force has a constant velocity and is not scattered by the atoms of the crystal. When a force is applied to an electron, its acceleration can be described by a modified Newton's law where the perfect periodic crystal potential is taken into account by an effective mass. There-fore, to describe the deceleration or scattering of an electron by a crystal defect, it is convenient to examine the perturbation that the defect produces on the perfect crystal potential. This perturbation is referred to as a *scattering potential,* $\Delta U(\mathbf{r})$, and has units of energy. In the following section we examine the scattering processes associated with impurities and phonons and derive their scattering potentials.

6.1.1 Impurities

For an ionized impurity the scattering process is dominated by the electrical interaction between its charge and the charge of the free carrier. For an ion with charge Zq the perturbation on the perfect crystal potential is simply the Coulomb energy,

$$\Delta U(\mathbf{r}) = \frac{\pm Zq^2}{4\pi\epsilon(0)\mathbf{r}} \tag{6.1}$$

where $\mathbf{r}$ is the distance between the ion and the charge carrier. The plus sign is valid when the charges on the ion and the carrier have the same polarity, and the minus sign is for charges of opposite polarity. As indicated in Fig. 6.1, the scattering trajectories of the free carriers are described by a hyperbola with the ion at a focal point.

In (6.1) the screening of the Coulomb potential by atomic and ionic polarization of the constituent atoms is described by the use of the static permittivity of the material, $\epsilon(0)$. Because of the long-range nature of the Coulomb potential, it is also necessary to consider the screening of (6.1) by other free carriers and ionized impurities. We look at this in Section 6.2.

A nonionized or neutral impurity has a scattering potential which is much weaker but more complex than for an ionized impurity. For a hydrogen-like neutral impurity, Coulombic scattering with the ground-state ($1s$) electron cloud occurs. The free carriers, however, can also interact with a neutral impurity by polarizing it or by changing places with a bound electron

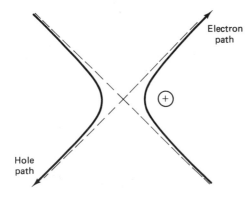

Electron path

Hole path

Figure 6.1 Trajectories of electrons and holes in ionized impurity scattering.

or hole. This raises the possibility of a combination of Coulombic, dipole, and exchange scattering which is not easy to analyze.

An empirical analysis [C. Erginsoy, *Phys. Rev. 79,* 1013 (1950)] of electron scattering by hydrogen atoms indicates that neutral impurity scattering can be approximated by a *differential scattering cross section,*

$$\sigma = \frac{20 r_B}{k} \tag{6.2}$$

where r_B is the ground-state Bohr radius of the impurity, and k is the wavevector of the free electron. It can be shown (see Sections 6.3 and 6.4) that this scattering cross section is approximately equivalent to a scattering potential,

$$\Delta U(\mathbf{r}) \simeq \frac{\hbar^2}{m^*} \left(\frac{r_B}{r^5} \right)^{1/2} \tag{6.3}$$

where $\mathbf{r}$ is the distance between the neutral impurity and the free carrier. It is interesting to note that this $r^{-5/2}$ dependence is longer ranged than the r^{-4} dependence expected for dipole scattering.

6.1.2 Acoustic Phonons

The acoustic phonons in a crystal can scatter carriers by two different and independent processes. These are called *deformation potential scattering* and *piezoelectric scattering*. These scattering mechanisms can be examined, qualitatively, by means of Fig. 6.2, where the displacements, $u(\mathbf{r})$, of a chain of atoms from their Bravais lattice sites are shown for the longitudinal (LA) and transverse (TA) components of (a) zone center and (b) zone edge acoustic phonons.

As can be seen, the distance between adjacent atoms (the size of the unit cell) is strongly affected by the LA phonons, and little affected by the TA phonons. From the tight-binding model of the energy gap variation with lattice constant (see Fig. 2.14), we see that these LA phonons will produce a modulation of the conduction and valence band edges, $\mathscr{E}_c$ and $\mathscr{E}_v$. This modulation in space and time disturbs the periodicity of the crystal potential and produces the so-called deformation potential scattering of the electrons and holes.

For the long-wavelength acoustic phonons, it is convenient to treat the material as an elastic continuum. Then we can see in Fig. 6.2 that the maximum expansion and contraction of the unit cell produced by LA phonons occurs in regions where the divergence of the displacement vector (gradient of the displacement amplitude), or the *strain,* is maximum. Therefore, the scattering potential for deformation potential scattering must be proportional to the strain. Consider the displacement produced by an acoustic phonon

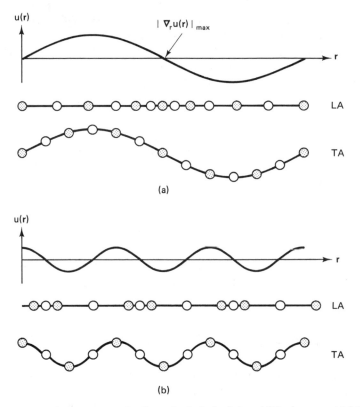

Figure 6.2 Displacements of a diatomic chain for LA and TA phonons at (a) the center and (b) the edge of the Brillouin zone. The lighter mass atoms are indicated by open circles. For zone edge acoustic phonons only the heavier atoms are displaced.

of frequency, ω_s, and wavevector, $\mathbf{q}_s$,

$$\mathbf{u}(\mathbf{r}, t) = \mathbf{a}u(\mathbf{r}, t) \tag{6.4}$$

where

$$u(\mathbf{r}, t) = u \exp{[i(\mathbf{q}_s\cdot\mathbf{r} - \omega_s t)]} \tag{6.5}$$

In these equations $\mathbf{a}$ is the displacement direction, and u is the amplitude. The strain associated with the displacement is

$$\nabla\cdot\mathbf{u}(\mathbf{r}, t) = \mathbf{a}\cdot\nabla u(\mathbf{r}, t) \tag{6.6}$$

$$\nabla\cdot\mathbf{u}(\mathbf{r}, t) = i\mathbf{q}_s\cdot\mathbf{a}u(\mathbf{r}, t) \tag{6.7}$$

Equation (6.7) indicates that for the transverse components of a phonon where the displacement and the wavevector are orthogonal, $\mathbf{q}_s\cdot\mathbf{a} = 0$, and no strain is produced. The scattering potential for the longitudinal component is, therefore,

$$\Delta U(\mathbf{r},\ t) = \mathcal{E}_A \nabla \cdot \mathbf{u}(\mathbf{r},\ t) \tag{6.8}$$

where the *deformation potential, $\mathcal{E}_A$,* in units of energy, is defined as the proportionality constant between the scattering potential (units of energy) and the strain.

For some semiconductors with two or more atoms per unit cell, there is no crystal inversion symmetry. In these crystals the strain, caused predominantly by the LA phonons, polarizes the ions and produces internal electric fields that vary with time and space. The carrier scattering caused by these electric fields is called *piezoelectric scattering.* The scattering potential for electrons is simply

$$\Delta U(\mathbf{r},\ t) = -q\psi(\mathbf{r},\ t) \tag{6.9}$$

where $\psi(\mathbf{r},\ t)$ is the electrostatic potential associated with the internal fields,

$$\psi(\mathbf{r},\ t) = -\int \mathbf{E}(\mathbf{r},\ t) \cdot d\mathbf{r} \tag{6.10}$$

To evaluate (6.9), it is thus necessary to determine the fields produced by the piezoelectric interaction.

At a given frequency, ω, the relationships among the electric flux density, **D**, the electric field, **E**, and the polarization, **P**, are

$$\mathbf{D}(\omega) = \epsilon(\omega)\mathbf{E} = \epsilon_0 \mathbf{E} + \mathbf{P}(\omega) \tag{6.11}$$

where ϵ_0 is the free-space permittivity. In the low-frequency limit (6.11) become

$$\mathbf{D}(0) = \epsilon(0)\mathbf{E} = \epsilon_0 \mathbf{E} + \mathbf{P}(0) \tag{6.12}$$

where $\epsilon(0) = \epsilon_r(0)\epsilon_0$ is the static permittivity. Physically, the source of **D**(0) is the "true" charge (i.e., space and surface charges), while the source of **P**(0) is the "polarization" charge, usually atomic core and ionic dipoles. Since measurements of the so-called "static" dielectric constant do not include piezoelectric polarization, it is necessary to add an additional term to (6.12) to include this effect. [In principle, **P**(0) should include piezoelectric polarization.]

From Fig. 6.2 we see that this polarization must be proportional to the strain induced by the phonons. Ignoring the tensor nature of this interaction, we have

$$\mathbf{D}(0) = \epsilon(0)\mathbf{E}(\mathbf{r},\ t) + e_{\text{pz}} \nabla u(\mathbf{r},\ t) \tag{6.13}$$

where the piezoelectric constant, e_{pz}, has units of coulomb per square meter. With no true charge, the sources of the electric field are piezoelectric, ionic, and atomic polarization,

$$E(\mathbf{r}, t) = -\frac{e_{pz}}{\epsilon(0)} \nabla u(\mathbf{r}, t) \qquad (6.14)$$

Using (6.9), (6.10), and (6.14), the scattering potential in terms of the displacement is

$$\Delta U(\mathbf{r}, t) = \frac{-qe_{pz}}{\epsilon(0)} u(\mathbf{r}, t) \qquad (6.15)$$

With (6.5) and (6.6), equation (6.15) can be expressed in terms of the strain as

$$\Delta U(\mathbf{r}, t) = \frac{iqe_{pz}}{\epsilon(0)q_s} \nabla \cdot \mathbf{u}(\mathbf{r}, t) \qquad (6.16)$$

Comparing (6.8) and (6.16), we see that the scattering potentials for the deformation potential and piezoelectric interactions are separated in phase by 90°. The two acoustic phonon scattering mechanisms therefore operate independently.

6.1.3 Optical Phonons

The optical phonons also scatter carriers by two independent processes. These are referred to as *deformation potential scattering* (the same as for acoustic phonons) and *polar mode scattering*. The deformation potential scattering by optical phonons is similar to that for acoustic phonons and the polar mode scattering is due to the polarization of atoms within the unit cell. The displacements, $u(\mathbf{r})$, of a chain of atoms from their Bravais lattice sites for longitudinal optical (LO) and transverse optical (TO) phonons are shown in Fig. 6.3.

In a manner similar to acoustic phonons it can be seen that the expansion and contraction of the unit cell is dominated by the longitudinal optical phonons. The main difference in Fig. 6.3 at the zone center is that the atoms in the unit cell vibrate against one another. Because of this, for optical phonon deformation potential scattering, it is necessary to consider the *relative* displacement between atoms in the unit cell,

$$\delta\mathbf{u}(\mathbf{r}, t) \equiv \mathbf{u}_1(\mathbf{r}, t) - \mathbf{u}_2(\mathbf{r}, t) \qquad (6.17)$$

where $u_1(\mathbf{r}, t)$ and $\mathbf{u}_2(\mathbf{r}, t)$ have the form given by (6.4) and (6.5). The scattering potential due to modulation of the conduction and valence edges must then be proportional to this relative displacement and

$$\Delta U(\mathbf{r}, t) = D\, \delta u(\mathbf{r}, t) \qquad (6.18)$$

where

$$\delta\mathbf{u}(\mathbf{r}, t) = \mathbf{a}\, \delta u(\mathbf{r}, t) \qquad (6.19)$$

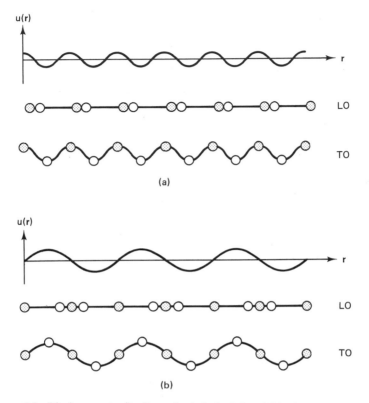

Figure 6.3 Displacements of a diatomic chain for LO and TO phonons at (a) the center and (b) the edge of the Brillouin zone. The lighter mass atoms are indicated by open circles. For zone edge optical phonons only the lighter atoms are displaced.

for optical phonon deformation potential scattering. In (6.18) the deformation potential constant, D, has units of energy per unit length. A similar treatment can be used for intervalley phonon scattering.

The optical phonon polar mode scattering is due to the electric field caused by the polarization of the ions in the unit cell. This polarization is caused mainly by the longitudinal component and is equivalent to the ionic polarization, P_i, which is discussed in Chapter 7. The scattering potential is obtained from (6.9) and (6.10), where the internal electric field is deduced from the low- and high-frequency limits of (6.11),

$$D(0) = \epsilon(0)E = \epsilon_0 E + P(0) \tag{6.20}$$

and

$$D(\infty) = \epsilon(\infty)E = \epsilon_0 E + P(\infty) \tag{6.21}$$

Note that in (6.20) the total low-frequency polarization is due to atomic and ionic polarization,

$$\mathbf{P}(0) = \mathbf{P}(\infty) + \mathbf{P}_i \tag{6.22}$$

Using (6.22) in (6.20) and subtracting (6.21), we obtain

$$\epsilon(0)\mathbf{E} = \epsilon(\infty)\mathbf{E} + \mathbf{P}_i \tag{6.23}$$

or

$$\mathbf{D}(0) = \epsilon(\infty)\mathbf{E} + \mathbf{P}_i \tag{6.24}$$

From (6.24) we can determine the internal fields induced by the optical phonon polarization of the unit cell.

The polarization of a unit cell, $\mathbf{P}_i(\mathbf{r}, t)$, is determined by the relative displacement of the ions in a unit cell, $\delta\mathbf{u}(\mathbf{r}, t)$, and the effective ionic charge, e^*, such that

$$\mathbf{P}_i(\mathbf{r}, t) = \frac{e^*}{\Omega} \delta\mathbf{u}(\mathbf{r}, t) \tag{6.25}$$

In this equation $\Omega = V/N$ is the volume of the N primitive or Wigner–Seitz unit cells and e^* is the Born effective charge given by

$$e^* = \Omega\omega_{LO}\epsilon(\infty)\rho^{1/2} \left[\frac{1}{\epsilon(\infty)} - \frac{1}{\epsilon(0)} \right]^{1/2} \tag{7.174}$$

where ρ is the mass density. This equation is derived in Chapter 7. Assuming no space or surface charges, (6.24) and (6.25) give an internal field,

$$\mathbf{E}(\mathbf{r}, t) = - \frac{e^*}{\Omega\epsilon(\infty)} \delta\mathbf{u}(\mathbf{r}, t) \tag{6.26}$$

Using (6.9), (6.10), and (6.26), the scattering potential for polar mode scattering is

$$\Delta U(\mathbf{r}, t) = \frac{-qe^*}{\Omega\epsilon(\infty)} \int \delta\mathbf{u}(\mathbf{r}, t)\cdot d\mathbf{r} \tag{6.27}$$

or with (6.5) and (6.19),

$$\Delta U(\mathbf{r}, t) = \frac{iqe^*}{\Omega\epsilon(\infty)q_s} \delta u(\mathbf{r}, t) \tag{6.28}$$

A comparison of (6.18) and (6.28) shows that the scattering potentials for deformation potential and polar mode scattering by optical phonons are out of phase by 90° and are thus independent.

Equation (6.28) is sometimes written in the form

$$\Delta U(\mathbf{r}, t) = \frac{iqe_c^*}{\Omega\epsilon_0 q_s} \delta u(\mathbf{r}, t) \tag{6.29}$$

where a Callen effective charge, e_c^* is used. The relationship to the Born effective charge is given by

$$e^* = \epsilon_r(\infty)e_c^* \tag{6.30}$$

6.2 SCREENING

In the derivations of the scattering potentials we assumed, in all cases, that there were no space charges (also, no surface charges). That is, we assumed that the charge carriers were uniformly distributed in the material such that

$$-\rho = q(n - p + N_a^- - N_d^+) = 0 \tag{6.31}$$

In the vicinity of a crystal potential perturbation caused by an impurity or phonon, however, charge carriers can be accumulated or depleted by the scattering potential. This space charge produces an additional potential given by

$$\nabla^2\psi(\mathbf{r}) = -\frac{\rho(\mathbf{r})}{\epsilon(0)} \tag{4.99}$$

where

$$-\rho(\mathbf{r}) = q[n(\mathbf{r}) - p(\mathbf{r}) + N_a^-(\mathbf{r}) - N_d^+(\mathbf{r})] \tag{6.32}$$

which screens the effects of the scattering potential.

In (6.32), $n(\mathbf{r})$, $p(\mathbf{r})$, $N_a^-(\mathbf{r})$, and $N_d^+(\mathbf{r})$ are the *total* electron, hole, ionized acceptor, and ionized donor concentrations as a function of distance, $\mathbf{r}$, from the center of the perturbing potential. These total concentrations can be split into two components,

$$n(\mathbf{r}) = n + \delta n(\mathbf{r})$$
$$N_d^+(\mathbf{r}) = N_d^+ + \delta N_d^+(\mathbf{r}), \quad \text{etc.} \tag{6.33}$$

a uniform concentration and an excess (or deficit) concentration which varies with $\mathbf{r}$ according to the variation of $\psi(\mathbf{r})$. Excess carriers due to built-in potentials in nonuniform materials are discussed in some detail in Chapter 8. For our purposes here, we assume that $q\psi(\mathbf{r}) \ll kT$, so that the excess concentrations are, approximately,

$$\delta n(\mathbf{r}) \simeq \frac{qn}{kT}\psi(\mathbf{r})$$
$$\delta N_d^+(\mathbf{r}) \simeq \frac{-qN_d^+}{kT}\psi(\mathbf{r}), \quad \text{etc.} \tag{6.34}$$

With this approximation we can define an *effective total* electron concentration,

$$n^*(\mathbf{r}) \equiv n(\mathbf{r}) - p(\mathbf{r}) + N_a^-(\mathbf{r}) - N_d^+(\mathbf{r}) \qquad (6.35)$$

which, using (6.33), (6.34), and (6.31), takes the form

$$n^*(\mathbf{r}) = \frac{qn^*}{kT} \psi(\mathbf{r}) \qquad (6.36)$$

From (4.99), (6.32), (6.35), and (6.36), Poisson's equation for the potential is

$$\nabla^2 \psi(\mathbf{r}) = \frac{q^2 n^*}{\epsilon kT} \psi(\mathbf{r}) \qquad (6.37)$$

where the *effective,* uniform electron concentration, n^*, is to be determined later. If we define an effective Debye length,

$$\lambda^2 \equiv \frac{\epsilon kT}{q^2 n^*} \qquad (6.38)$$

the differential equation for the potential is

$$\nabla^2 \psi(\mathbf{r}) = \frac{1}{\lambda^2} \psi(\mathbf{r}) \qquad (6.39)$$

For a spherically symmetric potential, (6.39) is

$$\frac{d^2}{dr^2} [r\psi(r)] = \frac{r\psi(r)}{\lambda^2} \qquad (6.40)$$

The physically significant solution to this equation is

$$\psi(r) = \frac{C}{r} \exp\left(\frac{-r}{\lambda}\right) \qquad (6.41)$$

where the constant of integration,

$$C = \frac{Zq^2}{4\pi\epsilon} \qquad (6.42)$$

for ionized impurity scattering. From (6.41) we see that the accumulation or depletion of charge carriers produces an exponential decay of the scattering potential with a characteristic length λ. This characteristic length is controlled by n^*, the effective electron concentration.

6.2.1 Degenerate Statistics

To determine $n^*(\mathbf{r})$ and, subsequently, n^*, it is necessary to introduce an energy band formalism for nonuniform materials. This formalism is summarized in Fig. 6.4, where we show the terminology for (a) a semiconductor

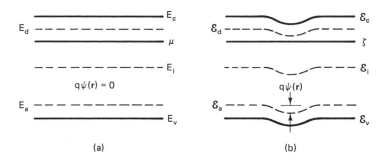

Figure 6.4 Energy band diagrams and terminology for (a) uniform material and (b) the same material with potential perturbations.

with uniform charge carrier and ionized impurity distributions, and (b) the same semiconductor with nonuniform distributions caused by perturbing potentials.

For a material with uniform doping and band structure as in Fig. 6.4(a), the Fermi energy is flat in thermal equilibrium and equal to the chemical potential energy, μ. For the same material with perturbing potentials due to impurities and phonons as in Fig. 6.4(b), the Fermi energy is flat, also in thermal equilibrium, and equal to the electrochemical potential energy,

$$\zeta = \mu(\mathbf{r}) - q\psi(\mathbf{r}) \tag{6.43}$$

All the other energies have the form

$$\mathscr{E}_c(\mathbf{r}) = E_c - q\psi(\mathbf{r}), \quad \text{etc.} \tag{6.44}$$

Since the perturbed material is in thermal equilibrium, we can simply modify the equilibrium distributions for n, p, N_a^-, and N_d^+ derived in Chapter 4 with this formalism to obtain $n(\mathbf{r})$, $p(\mathbf{r})$, $N_a^-(\mathbf{r})$, and $N_d^+(\mathbf{r})$.

From (4.68) and (4.65) the total electron concentration is

$$n(\mathbf{r}) = N_c F_{1/2}[\eta_c(\mathbf{r})] \tag{6.45}$$

where

$$\eta_c(\mathbf{r}) = \frac{\mu - E_c + q\psi(\mathbf{r})}{kT} \tag{6.46}$$

Equations (4.78) and (4.76) give the total hole concentration,

$$p(\mathbf{r}) = N_v F_{1/2}[\eta_v(\mathbf{r})] \tag{6.47}$$

where

$$\eta_v(\mathbf{r}) = \frac{E_v - \mu - q\psi(\mathbf{r})}{kT} \tag{6.48}$$

The total ionized acceptor concentration from (4.98) is

$$N_a^-(\mathbf{r}) = \frac{N_a}{1 + g_a \exp\{[E_a - \mu - q\psi(\mathbf{r})]/kT\}} \tag{6.49}$$

while the ionized donor concentration from (4.97) is

$$N_d^+(r) = \frac{N_d}{1 + g_d \exp\{[\mu - E_d + q\psi(\mathbf{r})]/kT\}} \tag{6.50}$$

Equations (6.45) and (6.47) for the electron and hole concentrations can be simplified in the following manner. From (4.67) the Fermi–Dirac integral is

$$F_{1/2}[\eta(\mathbf{r})] = \frac{2}{\sqrt{\pi}} \int_0^\infty \frac{x^{1/2}\, dx}{1 + \exp[x - \eta(\mathbf{r})]} \tag{6.51}$$

where $\eta(\mathbf{r})$ is given by either (6.46) or (6.48). The exponential in the denominator in (6.51) involving $q\psi(\mathbf{r})/kT$ is then expanded in a Taylor series and the denominator is divided into the numerator. Neglecting terms in $\psi^2(\mathbf{r})$ and higher, which is equivalent to the assumption $q\psi(\mathbf{r}) \ll kT$, (6.45) and (6.47) become

$$n(\mathbf{r}) = n + \frac{q\psi(\mathbf{r})}{kT} \frac{dn}{d\eta_c} \tag{6.52}$$

and

$$p(\mathbf{r}) = p - \frac{q\psi(\mathbf{r})}{kT} \frac{dp}{d\eta_v} \tag{6.53}$$

From the recurrence relationship (5.83),

$$\frac{dn}{d\eta_c} = N_c F_{-1/2}(\eta_c) = n \frac{F_{-1/2}(\eta_c)}{F_{1/2}(\eta_c)} \tag{6.54}$$

and

$$\frac{dp}{d\eta_v} = N_v F_{-1/2}(\eta_v) = p \frac{F_{-1/2}(\eta_v)}{F_{1/2}(\eta_v)} \tag{6.55}$$

Equations (6.49) and (6.50) for the ionized acceptor and donor concentrations can be simplified in a similar manner to obtain

$$N_a^-(\mathbf{r}) = N_a^- + \frac{N_a^0 N_a^-}{N_-} \frac{q\psi(\mathbf{r})}{kT} \tag{6.56}$$

and

$$N_d^+(r) = N_d^+ - \frac{N_d^0 N_d^+}{N_d} \frac{q\psi(\mathbf{r})}{kT} \tag{6.57}$$

The effective electron concentration, n^*, which controls the screening is obtained as follows: (6.52), (6.53), (6.56), and (6.57) are substituted into (6.35). Equation (6.31) for charge neutrality is applied to (6.35) and n^* is then obtained from (6.36) as

$$n^* = \frac{dn}{d\eta_c} + \frac{dp}{d\eta_v} + \frac{N_a^0 N_a^-}{N_a} + \frac{N_d^0 N_d^+}{N_d} \tag{6.58}$$

Equation (6.58) is valid for screening in degenerately doped semiconductors under the assumption that the perturbation energy is substantially less than the thermal energy.

6.2.2 Nondegenerate Statistics

When the doping of the material is such that the Fermi energy is greater than $\mathscr{E}_v + 4kT$ and less than $\mathscr{E}_c - 4kT$, η is negative, and $F_{1/2}(\eta) = F_{-1/2}(\eta)$. From (6.54), (6.55), and (6.58) the effective screening concentration is

$$n^* = n + p + \frac{N_a^0 N_a^-}{N_a} + \frac{N_d^0 N_d^+}{N_d} \tag{6.59}$$

where $N_a^0 = N_a - N_a^-$ and $N_d^0 = N_d - N_d^+$. Eliminating the neutral concentrations,

$$n^* = n + p + N_a^- \left(1 - \frac{N_a^-}{N_a}\right) + N_d^+ \left(1 - \frac{N_d^+}{N_d}\right) \tag{6.60}$$

For nondegenerate n- or p-type material, (6.60) can be further simplified. In n-type material, for example, all the acceptors will usually be ionized, so $N_a^- = N_a$. The acceptors are ionized by electrons from the donors, with the remaining electrons from the donors contributing to conduction. Therefore, $N_d^+ = n + N_a$. Using these arguments and space charge neutrality, in (6.60) the effective electron screening concentration is

$$n^* = n + \frac{(n + N_a)(N_d - N_a - n)}{N_d} \tag{6.61}$$

For p-type material, the effective hole screening concentration is

$$p^* = p + \frac{(p + N_d)(N_a - N_d - p)}{N_a} \tag{6.62}$$

Since N_d and N_a are usually constant in a material, (6.61) and (6.62) are useful in examining screening under conditions where n or p vary.

6.3 COLLISION INTEGRAL

Boltzmann's equation for the time rate of change of the electron distribution function under the influence of internal and applied forces is

$$\frac{\partial f}{\partial t} = -\frac{1}{\hbar} \mathbf{F} \cdot \nabla_k f - \mathbf{v} \cdot \nabla_r f + \frac{\partial f}{\partial t}\bigg|_c \qquad (5.4)$$

In Chapter 5 the first and second terms on the right-hand side of this equation were evaluated under the assumption that the third term could be put in the form

$$\frac{\partial f}{\partial t}\bigg|_c = \frac{-(f - f_0)}{\tau_m} \qquad (5.6)$$

That is, we assumed that the time rate of change of the distribution function due to collisions (the collision term) could be described by a momentum relaxation time, τ_m. In this section we examine the conditions under which this assumption is valid and show how τ_m can be obtained from the scattering potentials derived in Section 6.1. An equation relating the collision term or momentum relaxation time to the basic scattering process is called a *collision integral*. The scattering process itself can be described, quantum mechanically, by a matrix element or, classically, by a differential scattering cross section. We examine these two treatments and develop the relationship between them.

6.3.1 Quantum Treatment

The Hamiltonian for an electron undergoing a scattering process is

$$\mathbf{H} = \mathbf{H}_0 + \Delta \mathbf{U} \qquad (6.63)$$

where $\mathbf{H}_0$ is the unperturbed energy operator and $\Delta \mathbf{U}$ is one (or more) of the scattering potentials, in operator form, derived in Section 6.1. Since the process evolves in time, Schrödinger's equation is

$$(\mathbf{H}_0 + \Delta \mathbf{U})\psi(t) = i\hbar \frac{\partial \psi(t)}{\partial t} \qquad (6.64)$$

Solutions to (6.64) are obtained by constructing time-dependent wavefunctions from a set of time-independent Bloch wavefunctions,

$$\psi(t) = \sum_k A_k(t)\psi_k \exp\left[\frac{(-i\mathscr{E}_k t)}{\hbar}\right] \qquad (7.13)$$

where ψ_k are given by (2.10).

This scattering problem is formally equivalent to the optical transition problem described in Chapter 7. For an electron that is scattered from a state with wavevector **k** to one with wavevector **k'**, the scattering rate is

$$S_{kk'} \equiv \frac{|A_k(t)|^2}{t} \tag{6.65}$$

Using (7.36) this can be written as

$$S_{kk'} = \frac{2\pi}{\hbar} |H_{kk'}|^2 \delta(\mathscr{E}_k - \mathscr{E}_{k'}) \tag{6.66}$$

where from (7.19) the matrix element is

$$H_{kk'} = \frac{1}{N} \int_v \psi_k^* \Delta U \psi_{k'} \, dr \tag{6.67}$$

In (6.67), N is the number of primitive or Wigner–Seitz unit cells and V is the crystal volume.

For an electron to be scattered from an initial state **k** to one of the ($N - 1$) states **k'**, the initial state **k** must be occupied and the final state **k'** must be unoccupied. Conversely, an electron in one of the occupied ($N - 1$) states **k'** can be scattered into the unoccupied state **k**. Considering these two competing processes and summing over all ($N - 1$) values of **k'**, the time rate of *increase* of the distribution function due to collisions is

$$\frac{\partial f}{\partial t}\bigg|_c = N_s \sum_{k'}^{N-1} [S_{k'k} f_{k'}(1 - f_k) - S_{kk'} f_k(1 - f_{k'})] \tag{6.68}$$

where N_s is the number of scattering centers and f_k is the nonequilibrium distribution function at energy $\mathscr{E}(\mathbf{k})$. Since the number of unit cells in the crystal, N, is very large, the summation over the ($N - 1$) values of **k'** can be approximated by an integration over the ($N - 1$) ≃ N values of **k'** in the Brillouin zone. From (2.30), (1.20), and (1.12), each value of **k'** occupies a reciprocal volume,

$$\Omega_{k'} = \frac{(2\pi)^3}{V} \tag{6.69}$$

The integral approximation of (6.68) is, therefore,

$$\frac{\partial f}{\partial t}\bigg|_c = \frac{N_s V}{(2\pi)^3} \int_{\Omega_K} [S_{k'k} f_{k'}(1 - f_k) - S_{kk'} f_k(1 - f_{k'}) \, d\mathbf{k}' \tag{6.70}$$

When (6.70) is used in (5.4), an integrodifferential form of Boltzmann's equation is obtained which is quite general and valid for arbitrary degeneracy.

It is instructive to examine (6.70) in thermal equilibrium. Under this condition there is no change in the distribution function and the left-hand side of (6.70) must equal zero. For this to be true for any value of **k'**,

$$S_{k'k} = S_{kk'} \frac{f_{0k}(1 - f_{0k'})}{f_{0k'}(1 - f_{0k})} \tag{6.71}$$

where the subscript zero denotes the equilibrium Fermi–Dirac distribution function,

$$f_{0k} = \left[1 + \exp \left(\frac{\mathscr{E}_k - \mathscr{E}_f}{kT} \right) \right]^{-1} \tag{6.72}$$

If the material is nondegenerate, $\mathscr{E}_k - \mathscr{E}_f \gg kT$, and (6.71) reduces to

$$S_{k'k} = S_{kk'} \exp \left(\frac{\mathscr{E}_{k'} - \mathscr{E}_k}{kT} \right) \tag{6.73}$$

From (6.73) it can be seen that $S_{k'k} = S_{kk'}$ only when $\mathscr{E}_{k'} = \mathscr{E}_k$ or $|\,k'\,| = |\,k\,|$. That is, the scattering rate from a state k to k' is equal to its inverse only for elastic collisions. It is only under this condition that a universal momentum relaxation time can be defined.

Assuming elastic collisions we will now evaluate the nonequilibrium distribution coefficients for an arbitrary force field. With this we can then obtain the relationship between the momentum relaxation time and the matrix element, or using (6.67) the scattering potential, for a scattering process that conserves energy.

Consider an electron with initial wavevector k scattering into a final state k' under the influence of an arbitrary force, G. This force can include electric, magnetic, and thermal fields. From (5.18) the nonequilibrium distribution function in the relaxation time approximation is

$$f_k = f_{0k} + \frac{\partial f_{0k}}{\partial \mathscr{E}_k} \frac{\hbar}{m^*} k \cdot G \tag{6.74}$$

For a collision at the origin of the reciprocal-space coordinate system in Fig.

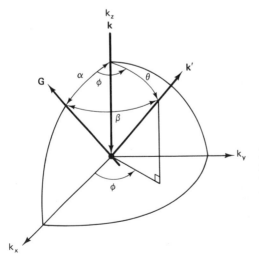

Figure 6.5 Spherical coordinate system in reciprocal space for an electron with wavevector k (along the k_z axis) scattering into a state with wavevector k' in an arbitrary force field G. The scattering center is at the origin. For simplicity the event is rotated so that G has no k_y component.

6.5,

$$\mathbf{k \cdot G} = kG \cos \alpha \tag{6.75}$$

and

$$\mathbf{k' \cdot G} = kG \cos \beta \tag{6.76}$$

The magnitudes of $\mathbf{k}$ and $\mathbf{k'}$ are equal when energy is conserved.

Using (6.74), (6.75), and (6.76), the expression involving distribution functions in the integrand of (6.70) is

$$f_{k'}(1 - f_k) - f_k(1 - f_{k'})$$

$$= \frac{\partial f_0}{\partial \mathscr{E}} \frac{\hbar}{m^*} kG[\cos \beta (1 - \cos \alpha) - \cos \alpha (1 - \cos \beta)] \tag{6.77}$$

Eliminating $\cos \beta$ with the equation for a spherical triangle (see Fig. 6.5) gives

$$\cos \beta = \cos \alpha \cos \theta + \sin \alpha \sin \theta \cos \phi \tag{6.78}$$

Equation (6.77) becomes

$$f_{k'}(1 - f_k) - f_k(1 - f_{k'})$$

$$= \frac{\partial f_0}{\partial \mathscr{E}} \frac{\hbar}{m^*} kG[\sin \alpha \sin \theta \cos \phi - \cos \alpha (1 - \cos \theta)] \tag{6.79}$$

Inserting (6.79) into (6.70), the collision term is

$$\frac{\partial f}{\partial t}\bigg|_c = -\frac{\partial f_0}{\partial \mathscr{E}} \frac{\hbar}{m^*} G \frac{N_s V}{(2\pi)^3} \int_{\Omega_K} S_{kk'}[\cos \alpha (1 - \cos \theta)$$

$$- \sin \alpha \sin \theta \cos \phi]k \, d\mathbf{k'} \tag{6.80}$$

Equation (6.80) can be simplified to give an expression for τ_m in the following manner: First, note that the differential volume in $\mathbf{k}$-space is

$$d\mathbf{k} = k^2 \sin \theta \, d\theta \, d\phi \, dk \tag{6.81}$$

Then, integrate ϕ from 0 to 2π, which eliminates the ϕ-dependent term. Finally, from (6.74) and (6.75),

$$-\frac{\partial f_0}{\partial \mathscr{E}} \frac{\hbar}{m^*} kG \cos \alpha = -(f - f_0) \tag{6.82}$$

Following these steps, (6.80) becomes

$$\frac{1}{\tau_m} = \frac{N_s V}{(2\pi)^2 k} \int_k \int_0^\pi S_{kk'} \sin \theta (1 - \cos \theta) \, d\theta \, k^3 \, dk \tag{6.83}$$

Equation (6.66) for the scattering rate, $S_{kk'}$, shows that the integration over

k is zero except at the point where energy is conserved. Using (6.66) and assuming a parabolic band, the relationship between the momentum relaxation time and the matrix element for the scattering process is

$$\frac{1}{\tau_m} = \frac{N_s V m^{*2} v}{2\pi\hbar^4} \int_0^{\pi} |H_{kk'}|^2 \sin\theta (1 - \cos\theta)\, d\theta \qquad (6.84)$$

6.3.2 Classical Treatment

The classical derivation of the momentum relaxation time proceeds rather simply. For N_s/V scattering centers per unit volume with scattering cross section, σ_m, the mean free time between collisions for an electron with velocity, v, is

$$\frac{1}{\tau_m} = \frac{N_s v}{V} \sigma_m \qquad (6.85)$$

The scattering cross section is determined by setting a scattering center with differential cross section, $\sigma(\theta)$, at the origin in Fig. 6.5. The θ-dependence allows for different scattering mechanisms. An electron scattered by the center into the solid angle (θ, ϕ) *loses* $(1 - \cos\theta)$ of its initial momentum in the incident direction. Taking into account all possible scattering angles yields

$$\sigma_m = \int_{\phi=0}^{2\pi} \int_{\theta=0}^{\pi} \sigma(\theta) \sin\theta (1 - \cos\theta)\, d\theta\, d\phi \qquad (6.86)$$

Using (6.85) and (6.86), the momentum relaxation time is

$$\frac{1}{\tau_m} = \frac{2\pi N_s v}{V} \int_0^{\pi} \sigma(\theta) \sin\theta (1 - \cos\theta)\, d\theta \qquad (6.87)$$

This is the classical collision integral.

Since the quantum and classical integrals have the same angular dependence, a relationship can be obtained between the differential scattering cross section, $\sigma(\theta)$, and the matrix element, $H_{kk'}$, for a given scattering process. Equating (6.84) and (6.87), we obtain

$$\sigma(\theta) = \left(\frac{V m^*}{2\pi\hbar^2} |H_{kk'}| \right)^2 \qquad (6.88)$$

6.4 MATRIX ELEMENTS

In principle the calculation of a matrix element for electron scattering from a given scattering potential using

$$H_{kk'} = \frac{1}{N} \int_V \psi_k^* \,\Delta\mathbf{V}\, \psi_{k'}\, d\mathbf{r} \qquad (6.67)$$

is relatively straightforward. In detail, however, the procedure is often quite laborious, involving a number of approximations and assumptions. Here, we simply indicate the general procedure and refer the reader to the literature for the detailed treatment.

6.4.1 General Procedure

The usual procedure for evaluating a matrix element is first to expand the scattering potential in a Fourier series,

$$\Delta U(\mathbf{r}) = \sum_g A_g \exp(i\mathbf{g}\cdot\mathbf{r}) \qquad (6.89)$$

where the Fourier coefficients are

$$A_g = \frac{1}{V} \int_V \Delta U(\mathbf{r}) \exp(-i\mathbf{g}\cdot\mathbf{r}) \, d\mathbf{r} \qquad (6.90)$$

For Bloch wavefunctions,

$$\psi_k(\mathbf{r}) = \exp(i\mathbf{k}\cdot\mathbf{r}) u_k(\mathbf{r}) \qquad (2.10)$$

$$H_{kk'} = \frac{1}{N} \sum_g \int_V \exp(-i\mathbf{k}\cdot\mathbf{r}) u_k^*(\mathbf{r}) A_g \exp(i\mathbf{g}\cdot\mathbf{r})$$

$$\times \exp(i\mathbf{k}'\cdot\mathbf{r}) u_{k'}(\mathbf{r}) \, d\mathbf{r} \qquad (6.91)$$

$$H_{kk'} = \frac{1}{N} \sum_g A_g \int_V u_k^*(\mathbf{r}) u_{k'}(\mathbf{r}) \exp[i(\mathbf{g} + \mathbf{k}' - \mathbf{k})\cdot\mathbf{r}] \, d\mathbf{r} \qquad (6.92)$$

Since the integral is zero except when

$$\mathbf{g} = \mathbf{k} - \mathbf{k}' \qquad (6.93)$$

(6.92) is

$$H_{kk'} = \frac{A_g}{N} \int_V u_k^*(\mathbf{r}) u_{k'}(\mathbf{r}) \, d\mathbf{r} \qquad (6.94)$$

For parabolic bands $u_k(\mathbf{r}) = u_{k'}(\mathbf{r})$, and the matrix element is simply the Fourier coefficient that satisfies (6.93)

$$H_{kk'} = A_{k-k'} \qquad (6.95)$$

where $A_{k-k'}$ is given by (6.90).

6.4.2 Screening Factor

Since the matrix element for ionized impurity scattering is relatively easy to obtain, we will derive it as an example of the procedure. Also, by comparing the screened and unscreened matrix elements for ionized impurity scattering, the screening factor for a general scattering process can be deduced.

Inserting the unscreened potential for ionized impurity scattering (6.1) into (6.90) yields

$$A_g = \frac{Zq^2}{4\pi\epsilon(0)V} \int_V \exp(-i\mathbf{g}\cdot\mathbf{r}) \frac{d\mathbf{r}}{r} \tag{6.96}$$

For the differential volume element

$$d\mathbf{r} = r^2 \sin\theta \, d\theta \, d\phi \, dr \tag{6.97}$$

$$A_g = \frac{Zq^2}{\epsilon(0)V} \int_0^\infty r \exp(-igr) \, dr \tag{6.98}$$

The integral is evaluated with (3.21) to obtain

$$A_g = \frac{Zq^2}{\epsilon(0)V \mid g \mid^2} \tag{6.99}$$

or from (6.95),

$$H_{kk'} = \frac{Zq^2}{\epsilon(0)V \mid k - k' \mid^2} \tag{6.100}$$

For the screened potential, (6.41),

$$A_g = \frac{Zq^2}{4\pi\epsilon(0)V} \int_V \exp\left(-\frac{r}{\lambda}\right) \exp(-i\mathbf{g}\cdot\mathbf{r}) \frac{d\mathbf{r}}{r} \tag{6.101}$$

Following the procedures above gives us

$$A_g = \frac{Zq^2}{\epsilon(0)V(\mid g \mid^2 + 1/\lambda^2)} \tag{6.102}$$

or

$$H_{kk'} = \frac{Zq^2}{\epsilon(0)V(\mid k - k' \mid^2 + 1/\lambda^2)} \tag{6.103}$$

A comparison of the screened equation (6.103) and the unscreened equation (6.100) indicates that the screening factor for a scattering process in general is

$$\Lambda = \frac{\mid k - k' \mid^2}{\mid k - k' \mid^2 + 1/\lambda^2} \tag{6.104}$$

In Table 6.1 we have summarized the scattering potentials and matrix elements for various scattering mechanisms. Screening can be accounted for by multiplying the matrix elements by (6.104).

TABLE 6.1   Scattering Potentials and Matrix Elements for Various Scattering Mechanisms[a]

Scattering Mechanisms	Scattering Potential	Matrix Element
Impurities		
Ionized	$\dfrac{Zq^2}{4\pi\epsilon(0)r}$	$\dfrac{Zq^2}{\epsilon(0)V\lvert k - k'\rvert^2}$
Neutral	$\dfrac{\hbar^2}{m^*}\left(\dfrac{r_B}{r^5}\right)^{1/2}$	$\dfrac{2\pi\hbar^2}{m^*V}\left(\dfrac{20r_B}{k}\right)^{1/2}$
Acoustic phonons		
Deformation potential	$\mathscr{E}_A\nabla\cdot\mathbf{u}$	$\mathscr{E}_A\left(\dfrac{\hbar}{2V\rho\omega_s}\right)^{1/2}(\mathbf{a}\cdot\mathbf{q}_s)\left(n_q + \dfrac{1}{2} \pm \dfrac{1}{2}\right)^{1/2}$
Piezoelectric	$\dfrac{iqe_{pz}}{\epsilon(0)q_s}\nabla\cdot\mathbf{u}$	$\dfrac{qe_{pz}}{\epsilon(0)}\left(\dfrac{\hbar}{2V\rho\omega_s}\right)^{1/2}\left(n_q + \dfrac{1}{2} \pm \dfrac{1}{2}\right)^{1/2}$
Optical phonons		
Deformation potential	$D\delta u$	$D\left(\dfrac{\hbar}{2V\rho\omega_{LO}}\right)^{1/2}\left(n_q + \dfrac{1}{2} \pm \dfrac{1}{2}\right)^{1/2}$
Polar	$\dfrac{iqe^*}{\omega\epsilon(\infty)q_s}\delta u$	$\dfrac{qe^*}{\Omega\epsilon(\infty)q_s}\left(\dfrac{\hbar}{2V\rho\omega_{LO}}\right)^{1/2}\left(n_q + \dfrac{1}{2} \pm \dfrac{1}{2}\right)^{1/2}$

[a] r_B = Bohr radius; n_q = phonon occupation number; $e^* = \Omega\omega_{LO}\epsilon(\infty)\rho^{1/2}[1/\epsilon(\infty) - 1/\epsilon(0)]^{1/2}$.

6.5 RELAXATION TIMES

With the matrix elements listed in Table 6.1, momentum relaxation times can be calculated from (6.84) for the various scattering mechanisms. Assuming isotropic parabolic energy bands, ionized impurity scattering can be described by the Brooks–Herring equation [H. Brooks, *Adv. Electron. Electron Phys.* 7, 158 (1955)],

$$\frac{1}{\tau_{II}(x)} = \frac{2.41Z^2N_I}{\epsilon_r^2(0)T^{3/2}}\, g(n^*, T, x)\left(\frac{m}{m^*}\right)^{1/2} x^{-3/2} \text{ second}^{-1} \qquad (6.105)$$

where the screening term

$$g(n^*, T, x) = \ln(1 + b) - \frac{b}{1 + b} \qquad (6.106)$$

and

$$b = 4.31 \times 10^{13}\,\frac{\epsilon_r(0)T^2}{n^*}\left(\frac{m^*}{m}\right)x \qquad (6.107)$$

In these equations N_I is the total ionized impurity concentration in cm^{-3} and n^* is the effective screening concentration in cm^{-3} given by (6.58).

For neutral impurity scattering, we use Erginsoy's result [C. Erginsoy, *Phys. Rev. 79*, 1013 (1950)],

$$\frac{1}{\tau_{NI}} = 1.22 \times 10^{-7} \epsilon_r(0) N_N \left(\frac{m}{m^*}\right)^2 \text{ second}^{-1} \qquad (6.108)$$

where N_N is the total neutral impurity concentration in cm^{-3}. Notice that this momentum relaxation time is independent of the carrier energy, x. Usually, neutral impurities have an appreciable effect on carrier scattering only for relatively uncompensated samples at low temperatures.

The momentum relaxation time for deformation potential scattering by acoustic phonons was first calculated by Bardeen and Shockley (J. Bardeen and W. Shockley, *Phys. Rev. 80*, 72 (1950)]. Their result is

$$\frac{1}{\tau_{DA}(x)} = \frac{4.17 \times 10^{19} \mathscr{E}_A^2 T^{3/2}}{C_l} \left(\frac{m^*}{m}\right)^{3/2} x^{1/2} \text{ second}^{-1} \qquad (6.109)$$

for $\mathscr{E}_A$ in eV and C_l in dyn/cm^2. C_l is the spherically averaged longitudinal elastic constant indicated by (6.111) below.

For materials with no inversion symmetry, the acoustic phonons also scatter carriers by means of a piezoelectric interaction. A momentum relaxation time for this process was first formulated by Meijer and Polder [H. J. G. Meijer and D. Polder, *Physica 19*, 255 (1953)]. With spherical averaging of the elastic and piezoelectric constants over a cubic crystal structure [J. D. Zook, *Phys. Rev. 136*, A849 (1964)], this is given by

$$\frac{1}{\tau_{PA}(x)} = 1.05 \times 10^7 h_{14}^2 \left[\left(\frac{3}{C_l} + \frac{4}{C_t}\right)\right] T^{1/2} \left(\frac{m^*}{m}\right)^{1/2} x^{-1/2} \text{ second}^{-1}$$

$$(6.110)$$

In (6.110) $h_{14} = e_{14}/\epsilon(0)$ is the piezoelectric constant in V/cm and the average longitudinal and transverse elastic constants are

$$C_l = \tfrac{1}{5}(3C_{11} + 2C_{12} + 4C_{44}) \qquad (6.111)$$

and

$$C_t = \tfrac{1}{5}(C_{11} - C_{12} + 3C_{44}) \qquad (6.112)$$

in dyn/cm^2. For a hexagonal crystal structure the momentum relaxation time is anisotropic.

For impurities and acoustic phonons the scattering processes are, to a good approximation, elastic. For optical phonons, however, the phonon energy is comparable to the thermal energy of the carriers and the scattering processes are inelastic. Despite this, a momentum relaxation time can still be defined for deformation potential scattering by optical phonons [W. A. Harrison, *Phys. Rev. 104*, 1281 (1956)]. This is given by

$$\frac{1}{\tau_{DO}(x)} = \frac{2.07 \times 10^{19} \mathscr{E}_A^2 T^{1/2} \theta}{C_l[\exp(\theta/T) - 1]} \left(\frac{m^*}{m}\right)^{3/2} \left[\left(x + \frac{\theta}{T}\right)^{1/2}\right.$$

$$\left. + \exp\left(\frac{\theta}{T}\right)\left(x - \frac{\theta}{T}\right)^{1/2}\right] \text{ second}^{-1} \qquad (6.113)$$

for $\mathscr{E}_A$ in eV and C_l in dyn/cm². θ in this equation is the longitudinal optical phonon temperature,

$$\theta \equiv \frac{\hbar\omega_{LO}}{k} \qquad (6.114)$$

$\mathscr{E}_A$ is the acoustic phonon deformation potential constant, which is related to the optical phonon deformation potential constant, D, by

$$\mathscr{E}_A^2 = \frac{C_l D^2}{\rho\omega_{LO}^2} \qquad (6.115)$$

For polar scattering of carriers by optical phonons a universal relaxation time can be defined only for temperatures much less than or much greater than the optical phonon temperature. It is thus necessary to use a variational method to solve the Boltzmann equation and determine the carrier scattering. However, Ehrenreich [H. Ehrenreich, *J. Appl. Phys. 32*, 2155 (1961)] has developed a relaxation time based on a variational calculation for polar scattering which gives the correct solutions to the Boltzmann equation at low and high temperatures. This is given by,

$$\frac{1}{\tau_{PO}(x)} = \frac{1.04 \times 10^{14}[\epsilon_r(0) - \epsilon_r(\infty)]\theta^{1/2}(\theta/T)^r}{\epsilon_r(0)\epsilon_r(\infty)[\exp(\theta/T) - 1]} \left(\frac{m^*}{m}\right)^{1/2} x^{-r} \text{ second}^{-1}$$

$$(6.116)$$

where r varies with (θ/T) as shown in Fig. 6.6.

The results of this section for the momentum relaxation times are summarized in Table 6.2 in the form

$$\tau_i(x) = \tau_i x^{r_i} \qquad (6.117)$$

These expressions are valid only for scattering in isotropic parabolic energy bands. For scattering in more complex bands, see D. L. Rode, *Semiconductors and Semimetals*, Vol. 10, *Transport Phenomena*, ed. R. K. Willardson and A. C. Beer (New York: Academic Press, 1975) or J. D. Wiley (ibid.). It should also be noted that the momentum relaxation times were derived under the assumption that screening can be neglected for *phonon* scattering. This is usually a good assumption for samples with nondegenerate doping.

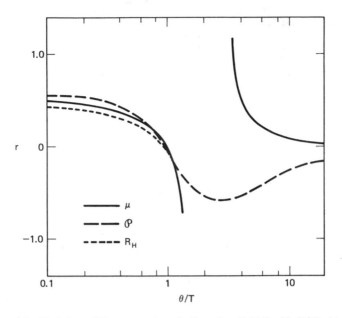

Figure 6.6 Variation of the parameter r in Equation (6.116) with (θ/T) obtained by equating variational solutions for the mobility, μ, thermoelectric power, $\mathcal{P}$, and Hall coefficient, R_H, with the corresponding expressions in the relaxation time approximation. [From H. Ehrenreich, General Electric Research Lab. Rep. No. 61-RL-(27626), June 1961.]

TABLE 6.2 Momentum Relaxation Times and Reduced Energy Dependence for Materials with Isotropic Parabolic Bands[a]

Scattering Mechanisms	τ_i (sec)	r_i
Impurities		
Ionized	$\dfrac{0.414\epsilon_r^2(0)T^{3/2}}{Z^2 N_I(\mathrm{cm}^{-3})g(n^*, T, x)}\left(\dfrac{m^*}{m}\right)^{1/2}$	$\tfrac{3}{2}$
Neutral	$\dfrac{8.16\times10^6}{\epsilon_r(0)N_N(\mathrm{cm}^{-3})}\left(\dfrac{m^*}{m}\right)^2$	0
Acoustic phonons		
Deformation potential	$\dfrac{2.40\times10^{-20}C_l(\mathrm{dyn/cm}^2)}{\mathcal{E}_A^2(\mathrm{eV})T^{3/2}}\left(\dfrac{m}{m^*}\right)^{3/2}$	$-\tfrac{1}{2}$
Piezoelectric	$\dfrac{9.54\times10^{-8}}{h_{14}^2(\mathrm{V/cm})(3/C_l + 4/C_t)T^{1/2}}\left(\dfrac{m}{m^*}\right)^{1/2}$	$\tfrac{1}{2}$
Optical phonons		
Deformation potential	$\dfrac{4.83\times10^{-20}C_l(\mathrm{dyn/cm}^2)[\exp(\theta/T)-1]}{\mathcal{E}_A^2(\mathrm{eV})T^{1/2}\theta}\left(\dfrac{m}{m^*}\right)^{3/2}$	$\cong -\tfrac{1}{2}$
Polar	$\dfrac{9.61\times10^{-15}\epsilon_r(0)\epsilon_r(\infty)[\exp(\theta/T)-1]}{[\epsilon_r(0)-\epsilon_r(\infty)]\theta^{1/2}(\theta/T)^r}\left(\dfrac{m}{m^*}\right)^{1/2}$	$r\left(\dfrac{\theta}{T}\right)$

[a] N_I = concentration of ionized impurities; $g(n^*, T, x) = \ln(1 + b) - b/(1 + b)$; $b = 4.31 \times 10^{13}[\epsilon_r(0)T^2/n^*(\mathrm{cm}^{-3})](m^*/m)x$; N_N = concentration of neutral impurities; $C_l = \tfrac{1}{5}(3C_{11} + 2C_{12} + 4C_{44})$; $C_t = \tfrac{1}{5}(C_{11} - C_{12} + 3C_{44})$; $\theta = \hbar\omega_{LO}/k$.

6.6 COMBINED SCATTERING

In most calculations of transport properties it is necessary to consider several scattering processes at the same time. If these scattering mechanisms are independent of one another, the matrix elements or differential scattering cross sections for each process can be added to obtain the total scattering. In the relaxation time approximation we see from (6.84) or (6.87) that this is equivalent to adding the reciprocal times for each process,

$$\frac{1}{\tau_m(x)} = \sum_i \frac{1}{\tau_i(x)} \tag{6.118}$$

where the $\tau_i(x)$ are given by (6.105), (6.108), (6.109), (6.110), (6.113), and/or (6.116). The desired transport property is then obtained by averaging the appropriate expression involving $\tau_m(x)$ over the electron distribution. From (5.54) this procedure is

$$\langle \tau_m^s(x)x^t \rangle = \frac{2}{3} \frac{\int_0^\infty \tau_m^s(x)(-\partial f_0/\partial x)x^{t+3/2} \, dx}{\int_0^\infty f_0 x^{1/2} \, dx} \tag{6.119}$$

For some combinations of scattering mechanisms it is possible to evaluate (6.119) analytically. The usual procedure is to integrate the numerator by parts and obtain a solution in terms of Fermi–Dirac integrals of order j, which are tabulated in Appendix B. For problems involving ionized impurity scattering this procedure is complicated by the energy or x dependence of the screening term, $g(n^*, T, x)$, given by (6.106) and (6.107). Since it is a slowly varying function of x, however, reasonable approximations can be made. The usual procedure is to evaluate $g(n^*, T, x)$ at a constant energy, $x = x_m$, and remove it from the integral. The value of x_m is determined by the condition that the integrand remaining after the removal of $g(n^*, T, x_m)$ be a maximum [E. M. Conwell and V. F. Weisskopf, *Phys. Rev. 77*, 388 (1950)]. Typically, x_m has a value of about 3 or so.

For most combinations of scattering mechanisms it is necessary to evaluate (6.119) numerically. As an example of this, the temperature dependence of the mobility,

$$\mu = \frac{q\langle \tau_m \rangle}{m^*} \tag{5.50}$$

for high-purity n-type GaAs is shown in Fig. 6.7. Here the mobility for each relevant scattering mechanism was calculated separately from (5.50), (6.105), (6.108), (6.109), (6.110), (6.116), and (6.119) and then combined to compare with experimental data.

As can be seen, the mobility of this GaAs sample is dominated by ionized impurity scattering at low temperatures and by polar optical phonon

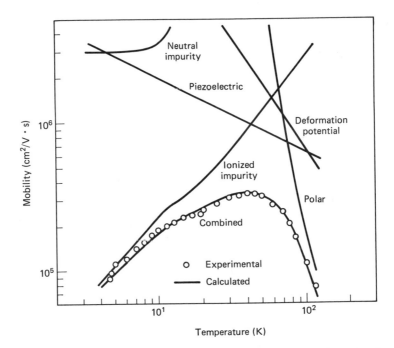

Figure 6.7 Temperature dependence of the mobility for n-type GaAs showing the separate and combined scattering processes. [From C. M. Wolfe, G. E. Stillman, and W. T. Lindley, *J. Appl. Phys. 41*, 3088 (1970).]

scattering at high temperatures. Deformation potential optical phonon scattering is not important for Γ conduction bands [H. Ehrenreich and A. W. Overhauser, *Phys. Rev. 104*, 331 (1956)]. This mobility behavior is typical for polar semiconductors. In these calculations the singly ionized N_I in (6.105) is given by

$$N_I = n + 2N_a \tag{6.120}$$

n^* in (6.107) by (6.61), and N_N in (6.108) by

$$N_N = N_d - N_a - n \tag{6.121}$$

N_d and N_a were obtained by analyzing the experimental temperature dependence of n with (4.90). For n-type material this equation is

$$\frac{n(n + N_a)}{N_d - N_a - n} = \frac{N_c}{g_d} \exp\left(\frac{-\Delta \mathscr{E}_d}{kt}\right) \tag{6.122}$$

The other parameters required in the analysis are typically obtained from other, independent measurements. These are listed for GaAs in Table 6.3 together with the appropriate parameters for other materials for which the analysis above is valid.

TABLE 6.3   **Parameters for Calculating the Transport Properties of n-Type Semiconductors with Isotropic Parabolic Bands**

Material	$\dfrac{m^*}{m}$	$\epsilon_r(0)$	$\epsilon_r(\infty)$	θ (K)	$\mathcal{E}_A$ (eV)	C_l (10^{12} dyn/cm^2)	$h_{14}^2\left(\dfrac{3}{C_l}+\dfrac{4}{C_t}\right)$ (10^3 V^2/dyn)
GaN	0.218	9.87	5.80	1044	8.4	2.65	18.32
GaP	0.13	11.10	9.11	580	13.0	1.66	1.15
GaAs	0.067	12.53	10.90	423	6.3	1.44	2.04
GaSb	0.042	15.69	14.44	346	8.3	1.04	
InP	0.082	12.38	9.55	497	6.8	1.21	0.137
InAs	0.025	14.54	11.74	337	5.8	1.0	0.192
InSb	0.0125	17.64	15.75	274	7.2	0.79	0.409
ZnS	0.312	8.32	5.13	506	4.9	1.28	6.87
ZnSe	0.183	9.20	6.20	360	4.2	1.03	0.620
ZnTe	0.159	9.67	7.28	297	3.5	0.84	0.218
CdS	0.208	8.58	5.26	428	3.3	0.85	32.5
CdSe	0.130	9.40	6.10	303	3.7	0.74	16.7
CdTe	0.096	10.76	7.21	246	4.0	0.70	0.445
HgSe	0.0265	25.6	12.0	268	4	0.80	0.445
HgTe	0.0244	20.0	14.0	199	4	0.61	0.445
PbS		175	17	300	20		
PbSe		250	24	190	24	0.71	
PbTe		400	33	160	25		

PROBLEMS

6.1. In a collision with an acoustic phonon, show that an electron with initial velocity v_i will gain or lose at most only

$$\frac{4u_s}{v_i} - 4\left(\frac{u_s}{V_i}\right)^2$$

of its initial energy, where u_s is the sound velocity.

6.2. An acoustic wave of the form $A\exp[i(\mathbf{q}\cdot\mathbf{r}-\omega t)]$ propagates through an n-type semiconductor with a parabolic band where it produces a variation in the energy of the electrons

$$\mathcal{E} = \mathcal{E}_1 A\exp[i(\mathbf{q}\cdot\mathbf{r}-\omega t)]$$

Since the force exerted on an electron is $\mathbf{F} = -\nabla_r\mathcal{E}$, show that in the relaxation time approximation, a good approximation to the electron distribution is

$$f = f_0 + \frac{\partial f_0}{\partial\mathcal{E}}\frac{i\tau_m\mathbf{v}\cdot\mathbf{q}\mathcal{E}}{1+i\tau_m\mathbf{v}\cdot\mathbf{q}}$$

Does this distribution provide conduction?

6.3. **(a)** Evaluate r_H for ionized impurity scattering using the momentum relaxation time determined in the Brooks–Herring approximation.

   (b) Plot the temperature variation of r_H using parameters appropriate to GaAs.

6.4. Use the Rutherford scattering cross section to derive the mobility for ionized impurity scattering in the Conwell–Weisskopf approximation.

   (a) Discuss the validity of the Born approximation for ionized impurity scattering.

   (b) Discuss the differences between C-W and B-H approximation, particularly in the temperature range where there is carrier freeze-out.

Optical
Properties

When light is incident on a semiconductor, the optical phenomena of absorption, reflection, and transmission are observed. From these optical effects, we obtain much of the information we have concerning the energy band structure and electronic processes in semiconductors. Figure 7.1 shows a hypothetical absorption spectrum as a function of photon energy for a typical semiconductor. As can be seen, a number of processes can contribute to absorption. At high energies photons are absorbed by the transitions of electrons from filled valence band states to empty conduction band states. For energies just below the lowest forbidden energy gap, radiation is absorbed due to the formation of excitons and electron transitions between band and impurity states. The transitions of free carriers within energy bands produce an absorption continuum which increases with decreasing photon energy. Also, the crystalline lattice itself can absorb radiation, with the energy being given off in optical phonons. Finally, at low energies, or long wavelengths, electronic transitions can be observed between impurities and their associated bands.

Many of these processes have important technological applications. For example, intrinsic photodetectors utilize band-to-band absorption, while semiconductor lasers generally operate by means of transitions between impurity and band states. In this chapter we examine these optical processes in detail.

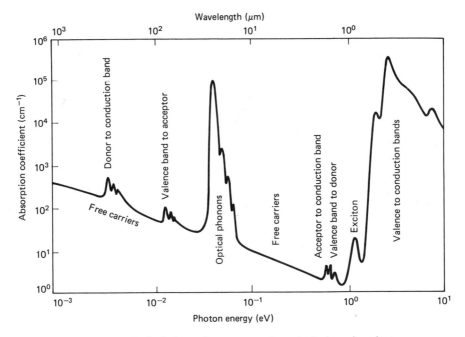

Figure 7.1 Hypothetical absorption spectrum for a typical semiconductor as a function of photon energy.

7.1 ELECTRON–PHOTON INTERACTION

To examine the interaction between an electron and a photon, let us represent the photon by a vector potential, **A**, defined by

$$\mathbf{E} = -\frac{\partial}{\partial t}\mathbf{A} \tag{7.1}$$

$$\mu\mathbf{H} = \boldsymbol{\nabla}_r \times \mathbf{A} \tag{7.2}$$

and

$$\boldsymbol{\nabla}_r \cdot \mathbf{A} = 0 \tag{7.3}$$

We will take the vector potential to have the form of a plane wave,

$$\mathbf{A} = \tfrac{1}{2}A\mathbf{a}\exp\left[i(\mathbf{q}\cdot\mathbf{r} - \omega t)\right] + \tfrac{1}{2}A\mathbf{a}\exp\left[-i(\mathbf{q}\cdot\mathbf{r} - \omega t)\right] \tag{7.4}$$

where **a** is the unit polarization vector in the direction of **E** and **q** is the wavevector. For a photon in a semiconductor, the wavevector is related to the frequency by

$$|\mathbf{q}| = \frac{\omega\eta}{c} \tag{7.5}$$

where c is the velocity of light and η is the refractive index of the material. The energy of the photon is simply

$$\mathscr{E} = \hbar\omega \tag{7.6}$$

Also, since (7.4) is a transverse electromagnetic wave

$$\mathbf{a \cdot q} = 0 \tag{7.7}$$

The classical Hamiltonian of an electron with wavevector $\mathbf{k}$ interacting with a light wave of vector potential $\mathbf{A}$ is

$$\mathbf{H} = \frac{1}{2m}\,(\hbar\mathbf{k} - q\mathbf{A})^2 \tag{7.8}$$

Expanding (7.8), we have

$$\mathbf{H} = \frac{1}{2m}\,(\hbar^2\mathbf{k}^2 - \hbar q\mathbf{k \cdot A} - \hbar q\mathbf{A \cdot k} + q^2\mathbf{A}^2) \tag{7.9}$$

Using the operator form of $\mathbf{k}$ and (7.3), we obtain the quantum mechanical Hamiltonian

$$\mathbf{H} = \frac{1}{2m}\,(-\hbar^2\boldsymbol{\nabla}_r^2 + i2q\hbar\mathbf{A \cdot \boldsymbol{\nabla}}_r + q^2\mathbf{A}^2) \tag{7.10}$$

For low light levels the term with the vector potential squared can be neglected, to obtain

$$\mathbf{H} = \frac{-\hbar^2}{2m}\,\boldsymbol{\nabla}_r^2 + \frac{iq\hbar}{m}\,\mathbf{A \cdot \boldsymbol{\nabla}}_r$$

$$= \mathbf{H}_0 + \mathbf{H}' \tag{7.11}$$

That is, the Hamiltonian consists of a term $\mathbf{H}_0$ which corresponds to the unperturbed electron energy and a term $\mathbf{H}'$ due to the electron–photon interaction.

Since this interaction can result in a change of state for the electron with time, it is necessary to solve the time-dependent Schrödinger equation,

$$(\mathbf{H}_0 + \mathbf{H}')\Psi = i\hbar\,\frac{\partial\Psi}{\partial t} \tag{7.12}$$

Let us construct wavefunction solutions to (7.12) which are linear combinations of the wavefunctions, Ψ_n, for the unperturbed time-independent system

$$\Psi = \sum_n A_n(t)\psi_n \exp\left(\frac{-i\mathscr{E}_n t}{\hbar}\right) \tag{7.13}$$

where the ψ_n satisfy

$$\mathbf{H}_0\psi_n = \mathscr{E}_n\psi_n \tag{7.14}$$

Using (7.13) in (7.12), we obtain

$$\sum_n A_n(\mathbf{H}_0\psi_n + \mathbf{H}'\psi_n) \exp\left(\frac{-i\mathscr{E}_n t}{\hbar}\right)$$

$$= \sum_n \left(A_n\mathscr{E}_n\psi_n + i\hbar\frac{dA_n}{dt}\psi_n\right) \exp\left(\frac{-i\mathscr{E}_n t}{\hbar}\right) \tag{7.15}$$

From (7.14) the first term on the left is equal to the first term on the right in (7.15), and

$$i\hbar \sum_n \frac{dA_n}{dt}\psi_n \exp\left(\frac{-i\mathscr{E}_n t}{\hbar}\right) = \sum_n A_n\mathbf{H}'\psi_n \exp\left(\frac{-i\mathscr{E}_n t}{\hbar}\right) \tag{7.16}$$

If we multiply (7.16) by ψ_m^* exp $(i\mathscr{E}_m t/\hbar)$ and integrate over the volume of the crystal, we have

$$i\hbar \sum_n \frac{dA_n}{dt} \exp\left[\frac{i(\mathscr{E}_m - \mathscr{E}_n)t}{\hbar}\right] \int_V \psi_m^*\psi_n \, d\mathbf{r}$$

$$= \sum_n A_n \exp\left[\frac{i(\mathscr{E}_m - \mathscr{E}_n)t}{\hbar}\right] \int_V \psi_m^*\mathbf{H}'\psi_n \, d\mathbf{r} \tag{7.17}$$

Because of the orthogonality of the unperturbed wavefunctions,

$$\int_V \psi_m^*\psi_n \, d\mathbf{r} = N\delta_{mn} \tag{3.14}$$

(7.17) reduces to

$$i\hbar \frac{dA_m}{dt} = \sum_n A_n H_{mn}(t) \exp\left[\frac{i(\mathscr{E}_m - \mathscr{E}_n)t}{\hbar}\right] \tag{7.18}$$

where

$$H_{mn}(t) \equiv \frac{1}{N} \int_V \psi_m^*\mathbf{H}'\psi_n \, d\mathbf{r} \tag{7.19}$$

is the *matrix element* for an electron transition from state n with energy $\mathscr{E}_n$ to state m with energy $\mathscr{E}_m$ and N is the number of primitive unit cells in the crystal. Equation (7.18) is an exact differential equation for the time-dependent coefficients of the wavefunction of (7.13).

To solve for the coefficients, A_m, we will use first-order perturbation theory. Let us assume that at $t = 0$, the system starts in a time-independent state with energy $\mathscr{E}_0$ and we take

$$A_0(0) = 1, \qquad A_n(0) = 0 \tag{7.20}$$

Equation (7.18) is now

$$i\hbar \frac{dA_m}{dt} = H_{mo}(t) \exp\left[\frac{i(\mathscr{E}_m - \mathscr{E}_0)t}{\hbar}\right] \tag{7.21}$$

Integrating (7.21) with respect to time, the coefficients are given by the expression

$$A_m(t) = \frac{1}{i\hbar} \int_0^t H_{mo}(t) \exp\left[\frac{i(\mathscr{E}_m - \mathscr{E}_0)t}{\hbar}\right] dt \tag{7.22}$$

Since the probability of a transition from state 0 to state m is given by $|A_m(t)|^2$, we must now evaluate the integral in (7.22).

From (7.19) and (7.11) we have

$$H_{mo}(t) = \frac{iq\hbar}{mN} \int_V \psi_m^* (\mathbf{A} \cdot \nabla_r) \psi_0 \, d\mathbf{r} \tag{7.23}$$

Equation (7.4) for the vector potential, $\mathbf{A}$, has two terms: the first corresponds to stimulated absorption and the second corresponds to stimulated emission. For our current purpose, we ignore the stimulated emission term, so that the matrix element is

$$H_{mo}(t) = \frac{iq\hbar A}{2mN} \exp(-i\omega t) \int_V \psi_m^* \exp(i\mathbf{q} \cdot \mathbf{r})(\mathbf{a} \cdot \nabla_r)\psi_0 \, d\mathbf{r} \tag{7.24}$$

$$H_{mo}(t) = H_{m0} \exp(-i\omega t) \tag{7.25}$$

In (7.25) we have separated the matrix element into a time-dependent term and the time-independent term

$$H_{m0} = \frac{iq\hbar A}{2mN} \int_V \psi_m^* \exp(i\mathbf{q} \cdot \mathbf{r})(\mathbf{a} \cdot \nabla_r)\psi_0 \, d\mathbf{r} \tag{7.26}$$

Equation (7.22) now has the form

$$A_m(t) = \frac{H_{m0}}{i\hbar} \int_0^t \exp\left[\frac{i(\mathscr{E}_m - \mathscr{E}_0 - \hbar\omega)t}{\hbar}\right] dt \tag{7.27}$$

which can be integrated to obtain

$$A_m(t) = \frac{H_{m0}}{(\mathscr{E}_m - \mathscr{E}_0 - \hbar\omega)} \left\{ 1 - \exp\left[\frac{i(\mathscr{E}_m - \mathscr{E}_0 - \hbar\omega)t}{\hbar}\right] \right\} \tag{7.28}$$

The transition probability is therefore

$$|A_m(t)|^2 = \frac{4|H_{m0}|^2 \sin^2 \dfrac{(\mathscr{E}_m - \mathscr{E}_0 - \hbar\omega)t}{2\hbar}}{(\mathscr{E}_m - \mathscr{E}_0 - \hbar\omega)^2} \tag{7.29}$$

If we let

$$x = \frac{\mathcal{E}_m - \mathcal{E}_0 - \hbar\omega}{2\hbar} \tag{7.30}$$

the transition probability takes the form

$$|A_m(t)|^2 = \frac{|H_{m0}|^2}{\hbar^2} \frac{\sin^2 xt}{x^2} \tag{7.31}$$

Figure 7.2 shows a plot of the term $(\sin^2 xt)/x^2$ in (7.31). As can be seen, the height of this term increases with t^2, while the width is inversely proportional to t. The area under the curve is

$$\int_{-\infty}^{\infty} \frac{\sin^2 xt}{x^2} dx = \pi t \tag{7.32}$$

Thus, for times sufficiently long that the transition is completed, we can make the approximation

$$\frac{\sin^2 xt}{x^2} \simeq \pi t \delta(x) \tag{7.33}$$

where $\delta(x)$ is the Dirac delta function [L. I. Schiff, *Quantum Mechanics* (New York: McGraw-Hill, 1955), p. 197]. The transition probability is then

$$|A_m(t)|^2 = \frac{\pi |H_{m0}|^2 t}{\hbar^2} \delta \left(\frac{\mathcal{E}_m - \mathcal{E}_0 - \hbar\omega}{2\hbar} \right) \tag{7.34}$$

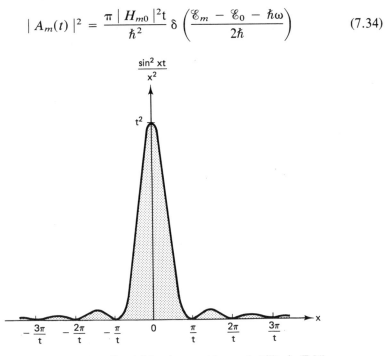

Figure 7.2 Plot of $(\sin^2 xt)/x^2$ for the transition probability in (7.31).

Since

$$\delta(ax) = \frac{1}{a}\,\delta(x) \tag{7.35}$$

(7.34) becomes

$$|A_m(t)|^2 = \frac{2\pi\,|H_{m0}|^2 t}{\hbar}\,\delta(\mathscr{E}_m - \mathscr{E}_0 - \hbar\omega) \tag{7.36}$$

Equation (7.36) tells us that the probability of an electron making a transition from state 0 with energy $\mathscr{E}_0$ to state m with energy $\mathscr{E}_m$ is zero unless the photon energy, $\hbar\omega$, is equal to the difference in energy between the two states. That is, energy must be conserved. We also see that the transition probability increases with time. With (7.36) the transition probability can be evaluated when the *time-dependent* matrix element, H_{m0}, for a given transition is known.

7.2 BAND-TO-BAND ABSORPTION

7.2.1 Direct Transitions

Let us first examine the matrix element for a direct electron transition from a valence band state with wavevector $\mathbf{k}$ to a conduction band state with wavevector $\mathbf{k}'$. As indicated in Fig. 7.3, the initial and final states for a direct transition are uniquely determined by the photon energy, $\hbar\omega$, and the energy band structure. For simplicity we assume parabolic bands. From (7.26) the matrix element is

$$H_{k'k} = \frac{iq\hbar A}{2mN}\int_V \psi_{k'}^* \exp(i\mathbf{q}\cdot\mathbf{r})(\mathbf{a}\cdot\nabla_r)\psi_k\,d\mathbf{r} \tag{7.37}$$

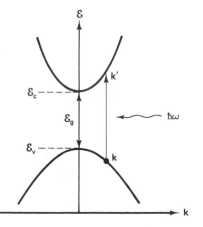

Figure 7.3 Direct optically induced transition of an electron from a valence band state with wavevector $\mathbf{k}$ to a conduction band state with wavevector $\mathbf{k}'$.

where ψ_k and $\psi_{k'}$ are the wavefunctions of the valence and conduction band states, respectively. According to (2.10), these wavefunctions have the form

$$\psi_k = \exp(i\mathbf{k}\cdot\mathbf{r})u_k(\mathbf{r}) \tag{7.38}$$

Using (7.38) in (7.37) and performing the operation, ∇_r, the matrix element is

$$H_{k'k} = \frac{iq\hbar A}{2mN} \int_V \exp[i(\mathbf{k} - \mathbf{k}' + \mathbf{q})\cdot\mathbf{r}]u_{k'}^*[\mathbf{a}\cdot\nabla_r u_k + i(\mathbf{a}\cdot\mathbf{k})u_k]\,d\mathbf{r} \tag{7.39}$$

Since the Bloch functions are periodic in $\mathbf{R}$, the integral over the crystal volume can be taken as a sum of integrals over the primitive unit cell,

$$H_{k'k} = \frac{iq\hbar A}{2mN} \sum_R \exp[i(\mathbf{k} - \mathbf{k}' + \mathbf{q})\cdot\mathbf{R}] \int_\Omega u_{k'}^*[\mathbf{a}\cdot\nabla_r u_k + i(\mathbf{a}\cdot\mathbf{k})u_k]\,d\mathbf{r} \tag{7.40}$$

For conservation of wavevector the sum over $\mathbf{R}$ in (7.40) is simply N and the matrix element is, finally,

$$H_{k'k} = \frac{iq\hbar A}{2m} \int_\Omega u_{k'}^*[\mathbf{a}\cdot\nabla_r u_k + i(\mathbf{a}\cdot\mathbf{k})u_k]\,d\mathbf{r} \tag{7.41}$$

Thus the matrix element consists of two terms: one involving $\nabla_r u_k$ and the other with u_k. Because of the orthogonality of the Bloch function between the bands, the term involving u_k would be zero for $\mathbf{k}$ equal to $\mathbf{k}'$. Therefore, taking into account the small wavevector of the photon, this term is much smaller than the term with $\nabla_r u_k$. For this reason the $\nabla_r u_k$ term in the matrix element of (7.41) is referred to as *allowed*, while the u_k term is referred to as *forbidden*.

Let us first consider the allowed term only. The matrix element for an allowed direct transition is

$$H_{k'k} = \frac{iq\hbar A}{2m} \int_\Omega u_{k'}^*(\mathbf{a}\cdot\nabla_r u_k)\,d\mathbf{r} \tag{7.42}$$

Since

$$\mathbf{p} = \hbar\mathbf{k} = -i\hbar\nabla_r \tag{7.43}$$

we will define a crystal momentum matrix element by

$$\mathbf{p}_{k'k} \equiv -i\hbar \int_\Omega u_{k'}^* \nabla_r u_k\,d\mathbf{r} \tag{7.44}$$

The matrix element in (7.42) then takes the form

$$H_{k'k} = \frac{-qA}{2m}(\mathbf{a}\cdot\mathbf{p}_{k'k}) \tag{7.45}$$

and from (7.36) the probability in the entire crystal that an electron will make a transition from a state with wavevector **k** to a state with wavevector **k'** is

$$| A_{k'}(t) |^2 = \frac{2\pi t}{\hbar} \left(\frac{qA}{2m} \right)^2 (\mathbf{a} \cdot \mathbf{p}_{k'k})^2 \delta(\mathscr{E}_{k'} - \mathscr{E}_k - \hbar\omega) \qquad (7.46)$$

The transition *rate* over the whole crystal is just the transition probability (7.46) divided by t.

To determine the total probability for a band-to-band transition, we assume $\hbar\omega$ to be monochromatic and sum (7.46) over all N allowed values of **k**. Taking into account the volume occupied by each value of **k** from (2.29), the probability that an initial valence band state will be occupied from (4.41), and the probability that a final conduction band state will be unoccupied from (4.42), the total probability for a band-to-band transition is

$$P = \frac{2V}{(2\pi)^3} \int_{\Omega_K} | A_{k'}(t) |^2 f_0 (1 - f_0) \, d\mathbf{k} \qquad (7.47)$$

The factor of 2 in (7.47) accounts for a possible change of spin during absorption. From (7.47) the transition probability per unit volume per unit time or the transition *rate* per unit volume is

$$r = \frac{2}{(2\pi)^3} \int_{\Omega_K} \frac{| A_{k'}(t) |^2}{t} f_0 (1 - f_0) \, d\mathbf{k} \qquad (7.48)$$

Using (7.46) in (7.48) we have

$$r = \frac{2}{4\pi^2 \hbar} \left(\frac{qA}{2m} \right)^2 \int_{\Omega_K} (\mathbf{a} \cdot \mathbf{p}_{k'k})^2 \delta(\mathscr{E}_{k'} - \mathscr{E}_k - \hbar\omega) f_0 (1 - f_0) \, d\mathbf{k} \quad (7.49)$$

Assuming parabolic bands, we can see from Fig. 7.3 that

$$\mathscr{E}_{k'} - \mathscr{E}_k = \mathscr{E}_g + \frac{\hbar^2 k'^2}{2m_e} + \frac{\hbar^2 k^2}{2m_h}$$

$$\simeq \mathscr{E}_g + \frac{\hbar^2 k^2}{2m_r} \qquad (7.50)$$

where m_r is the reduced mass of the electron and hole,

$$m_r = \frac{m_e m_h}{m_e + m_h} \qquad (7.51)$$

Compared with the delta function, the other terms under the integral in (7.49) are slowly varying functions of k. If we take them out of the integral, (7.49) becomes

$$r = \frac{q^2 A^2 \omega f}{16\pi^2 m} f_0 (1 - f_0) \int_{\Omega_K} \delta \left(\mathscr{E}_g + \frac{\hbar^2 k^2}{2m_r} - \hbar\omega \right) 4\pi k^2 \, dk \quad (7.52)$$

where we have defined a dimensionless *oscillator strength*,

$$f \equiv \frac{2(\mathbf{a} \cdot \mathbf{p}_{k'k})^2}{\hbar m \omega} \tag{7.53}$$

The oscillator strength, f, in (7.53) is given approximately by the f-sum rule [A. H. Wilson, *The Theory of Metals* (Cambridge: Cambridge University Press, 1954), p. 47] as

$$f \simeq 1 + \frac{m}{m_h} \tag{7.54}$$

The integral in (7.52) can be evaluated as

$$4\pi \int_{\Omega_K} \delta \left(\mathscr{E}_g + \frac{\hbar k^2}{2m_r} - \hbar \omega \right) k^2 \, dk$$

$$= 4\pi \frac{d}{d(\hbar\omega)} \left| \int_{\Omega_K} k^2 \, dk \right|_{\hbar\omega} = \mathscr{E}_g + \frac{\hbar^2 k^2}{2m_r}$$

$$= \frac{4\pi}{3} \left(\frac{2m_r}{\hbar^2} \right)^{3/2} \frac{d}{d(\hbar\omega)} (\hbar\omega - \mathscr{E}_g)^{3/2}$$

$$= \frac{2\pi}{\hbar^3} (2m_r)^{3/2} (\hbar\omega - \mathscr{E}_g)^{1/2} \tag{7.55}$$

With (7.52) and (7.55) the total allowed transition rate per unit volume between the conduction and valence band is

$$r = \frac{q^2 A^2 \omega f (2m_r)^{3/2}}{8\pi\hbar^3 m} f_0 (1 - f_0)(\hbar\omega - \mathscr{E}_g)^{1/2} \tag{7.56}$$

The absorption coefficient, α, can be obtained from (7.56) by means of

$$\alpha = \frac{r}{\Phi} \tag{7.57}$$

where Φ is the quantum flux or the number of photons crossing the unit area in unit time. From (7.57) we can see that α has units of reciprocal length. The quantum flux, Φ, can be determined from the average value of the Poynting vector $\mathbf{S}$ of the radiation, which is the energy crossing unit area in unit time, by

$$\Phi = \frac{\langle \mathbf{S} \rangle}{\hbar\omega} \tag{7.58}$$

where

$$\mathbf{S} = \mathbf{E} \times \mathbf{H} \tag{7.59}$$

From (7.1), (7.2), and (7.4),

$$\mathbf{E} = A\omega\mathbf{a} \sin (\mathbf{q} \cdot \mathbf{r} - \omega t) \tag{7.60}$$
$$\mu\mathbf{H} = -A(\mathbf{q} \times \mathbf{a}) \sin (\mathbf{q} \cdot \mathbf{r} - \omega t)$$

The average value of $\mathbf{S}$ over a period is, therefore,

$$\langle \mathbf{S} \rangle = \frac{1}{2} |\mathbf{q}| \frac{\omega A^2}{\mu} \tag{7.61}$$

or, using (7.5),

$$\langle \mathbf{S} \rangle = \tfrac{1}{2} \eta \epsilon_0 c \omega^2 A^2 \tag{7.62}$$

Using (7.56), (7.58), and (7.62) in (7.57), the absorption coefficient for allowed direct band-to-band transitions is

$$\alpha = \frac{q^2 (2m_r)^{3/2} f}{4\pi\epsilon_0 \eta c m \hbar^2} f_0 (1 - f_0)(\hbar\omega - \mathscr{E}_g)^{1/2} \tag{7.63}$$

If we assume that all valence band states are full and all conduction band states are empty, α is given by

$$\alpha = 2.7 \times 10^5 \left(\frac{2m_r}{m}\right)^{3/2} \frac{f}{\eta} (\hbar\omega - \mathscr{E}_g)^{1/2} \text{ cm}^{-1} \tag{7.64}$$

for $(\hbar\omega - \mathscr{E}_g)$ in units of eV.

The analysis of absorption for *forbidden* direct transitions is somewhat similar to that given above for allowed transitions. From (7.41) the matrix element is

$$H_{k'k} = \frac{-q\hbar A}{2m} (\mathbf{a} \cdot \mathbf{k}) \int_\Omega u_{k'}^* u_k \, d\mathbf{r} \tag{7.65}$$

Using (7.65) in (7.36) the probability for a forbidden transition between states with wavevectors $\mathbf{k}$ and $\mathbf{k}'$ is

$$| A_{k'}(t) |^2 = \frac{2\pi t}{\hbar} \left(\frac{q\hbar A}{2m}\right)^2 | \mathbf{a} \cdot \mathbf{k} |^2 f' \delta(\mathscr{E}_{k'} - \mathscr{E}_k - \hbar\omega) \tag{7.66}$$

where

$$f' = \left| \int_\Omega u_{k'}^* u_k \, d\mathbf{r} \right|^2 \tag{7.67}$$

For $\mathbf{k}$ not equal to $\mathbf{k}'$, $0 < f' \ll 1$. To determine the transition rate per unit volume, we must integrate (7.66) over the first Brillouin zone as in (7.48). The term $| \mathbf{a} \cdot \mathbf{k} |^2$ in (7.66), however, introduces an additional $\mathbf{k}$ dependence in the integral. We will use the average value of $| \mathbf{a} \cdot \mathbf{k} |$, which is $\frac{1}{3}k^2$, to obtain

$$r = \frac{q^2 A^2 \hbar f'}{24\pi^2 m^2} f_0(1 - f_0) \int_{\Omega_K} \delta \left(\mathscr{E}_g + \frac{\hbar^2 k^2}{2m_r} - \hbar\omega \right) 4\pi k^2 \, dk \quad (7.68)$$

which can be compared to (7.52) for an allowed transition.

When the integral in (7.68) is evaluated in a manner similar to (7.55), the total forbidden transition rate per unit volume is

$$r = \frac{q^2 A^2 f'(2m_r)^{5/2}}{12\pi\hbar^4 m^2} f_0(1 - f_0)(\hbar\omega - \mathscr{E}_g)^{3/2} \quad (7.69)$$

Using (7.69), (7.58), and (7.62) in (7.57), the forbidden absorption coefficient is

$$\alpha = \frac{q^2(2m_r)^{5/2} f'}{6\pi\epsilon_0 \eta c m^2 \hbar^2} f_0(1 - f_0) \frac{(\hbar\omega - \mathscr{E}_g)^{3/2}}{\hbar\omega} \quad (7.70)$$

or

$$\alpha = 1.8 \times 10^5 \left(\frac{2m_r}{m} \right)^{5/2} \frac{f'}{\eta} \frac{(\hbar\omega - \mathscr{E}_g)^{3/2}}{\hbar\omega} \, \text{cm}^{-1} \quad (7.71)$$

for a completely full valence band, a completely empty conduction band, and $\hbar\omega$ and $\mathscr{E}_g$ in eV. By comparing (7.71) with (7.64), we can see that the absorption coefficient for a forbidden transition is much smaller than for an allowed transition. Also, notice that the two expressions have a different dependence on $(\hbar\omega - \mathscr{E}_g)$.

7.2.2 Indirect Transitions

For direct optically induced transitions, we found that the initial and final electron states were uniquely determined by the photon energy, $\hbar\omega$. For indirect optically induced transitions, however, this is not the case. That is, an indirect transition requires the absorption or emission of a phonon to conserve the wavevector. Since the simultaneous absorption of a photon and a phonon is a higher-order process, one would expect indirect transition probabilities to be much less than those for direct transitions. However, because of the additional degree of freedom introduced by the phonon energy, $\hbar\omega_s$, transitions to many more states are possible.

This is illustrated in Fig. 7.4(a), where a valence band electron in initial state 0 can make a transition to final state 1 in the conduction band with the absorption of a photon of energy, $\hbar\omega$, and a phonon of energy, $\hbar\omega_{s1}$, and wavevector, $\mathbf{q}_{s1}$; or to final state 2 with the absorption of a photon of the same energy, $\hbar\omega$, and a phonon of energy, $\hbar\omega_{s2}$, and wavevector, $\mathbf{q}_{s2}$. Figure 7.4(b) shows similar possible transitions from initial states 1 and 2 in the valence band to a final state 0 in the conduction band. We can also envision

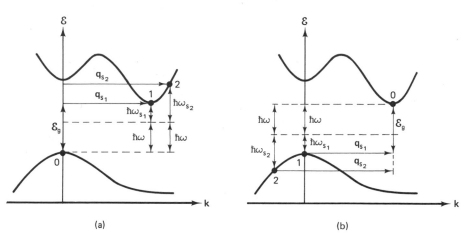

(a)   (b)

Figure 7.4 Indirect optically induced transitions of electrons (a) from initial state 0 in the valence band to final states 1 and 2 in the conduction band and (b) from initial states 1 and 2 in the valence band to final state 0 in the conduction band.

transitions involving the absorption of a photon and the emission of phonons. Thus, for a given photon energy a range of possible energies is available for indirect transitions. Notice also that indirect band-to-band absorption begins at photon energies below the bandgap such that

$$\hbar\omega \geq \mathcal{E}_g - \hbar\omega_s \tag{7.72}$$

where $\hbar\omega_s$ is the phonon energy.

As shown in Fig. 7.5, an indirect transition can be described by a direct transition from state 0 in the valence band to a short-lived virtual state I_c in the conduction band with simultaneous absorption or emission of a phonon to scatter the electron from I_c to conduction band state 1. In a similar manner, the same indirect transition can be described as a transition from 0 to a

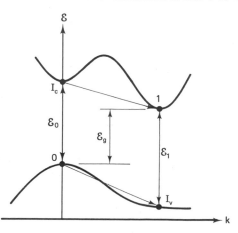

Figure 7.5 Analysis of indirect transitions by means of a direct transition from 0 to a virtual state I_c with simultaneous absorption or emission of a phonon to scatter from I_c to 1, or emission of a phonon to scatter from 0 to virtual state I_v with a simultaneous direct transition from I_v to 1.

virtual state I_v in the valence band with a simultaneous direct transition from I_v to 1. In such a treatment conservation of energy can be relaxed in the transitions to the virtual states because of the short times the electron remains in these states. Of course, energy must be conserved in the complete indirect process.

Indirect transitions described in terms of these virtual states can be analyzed by second-order perturbation theory [R. A. Smith, *Wave Mechanics of Crystalline Solids* (New York: Wiley, 1961)]. In such a treatment, the absorption coefficient for allowed indirect transitions involving conduction band virtual states is given by

$$\alpha_c(\pm\omega_s) = \frac{g_c q^2 m_h^{3/2} f_c \omega_s \mathscr{E}_0 (\hbar\omega \pm \hbar\omega_s - \mathscr{E}_g)^2}{32\pi\epsilon_0 \eta cm m_e^{1/2} \hbar\omega l_c kT (\mathscr{E}_0 - \hbar\omega)^2} \frac{\pm 1}{\exp(\pm\hbar\omega_s/kT) - 1}$$

(7.73)

where the + and − signs are for the absorption and emission of phonons, respectively. The absorption coefficient for allowed indirect transitions involving valence band virtual states is

$$\alpha_v(\pm\omega_s) = \frac{g_c q^2 m_e^{3/2} f_v \omega_s \mathscr{E}_1 (\hbar\omega \pm \hbar\omega_s - \mathscr{E}_g)^2}{32\pi\epsilon_0 \eta cm m_h^{1/2} \hbar\omega l_v kT (\mathscr{E}_1 - \hbar\omega)^2} \frac{\pm 1}{\exp(\pm\hbar\omega_s/kT) - 1}$$

(7.74)

In (7.73) and (7.74), g_c is the number of conduction band minima, f_c and f_v are oscillator strengths for the appropriate transitions, and l_c and l_v are the mean free paths for electron scattering in each band. The total allowed indirect absorption coefficient is

$$\alpha = \alpha_c(+\omega_s) + \alpha_c(-\omega_s) + \alpha_v(+\omega_s) + \alpha_v(-\omega_s) \qquad (7.75)$$

Forbidden indirect transitions can also be analyzed by means of virtual states. In this case the absorption coefficient for transitions through conduction band virtual states is

$$\alpha_c(\pm\omega_s) = \frac{g_c q^2 m_h^{5/2} f_c' \omega_s (\hbar\omega \pm \hbar\omega_s - \mathscr{E}_g)^3}{48\pi\epsilon_0 \eta cm^2 m_e^{1/2} \hbar\omega l_c kT (\mathscr{E}_0 - \hbar\omega)^2} \frac{\pm 1}{\exp(\pm\hbar\omega_s/kT) - 1} \qquad (7.76)$$

with a similar expression for transitions through valence band virtual states.

From (7.63), (7.70), (7.73), and (7.76) it can be seen that the dependence of the absorption coefficient on $(\hbar\omega - \mathscr{E}_g)$ is different for each of the band-to-band absorption processes. That is, direct allowed absorption varies as the $\frac{1}{2}$ power of $(\hbar\omega - \mathscr{E}_g)$, direct forbidden as the $\frac{3}{2}$ power, indirect allowed as the 2 power, and indirect forbidden as the 3 power. It is by this difference that the absorption processes can be distinguished experimentally.

7.2.3 Fermi Energy Dependence

Although band-to-band absorption is an intrinsic process, it can be strongly affected by the doping of the semiconductor. For lightly doped materials the fundamental absorption edge (energy at which appreciable band-to-band absorption begins) is determined by the energy gap, $\mathscr{E}_g$. For heavily doped semiconductors, however, when the Fermi energy lies in the conduction or valence bands, the position of the Fermi level must be taken into account. Figure 7.6 shows such a situation in an n-type, direct energy gap semiconductor. Since the Fermi level, ζ, is in the conduction band, essentially all states in the valence band and from the bottom of the conduction band to $\zeta - 4kT$ in the conduction band are occupied with electrons. If a photon of energy, $\hbar\omega = \mathscr{E}_g$, is incident on the material, there are no unoccupied states in the conduction band to which a transition can be made. Thus no band-to-band absorption occurs. As can be seen, no absorption is obtained until

$$\hbar\omega = \mathscr{E}_g + \frac{\hbar^2 k^2}{2}\left(\frac{1}{m_e} + \frac{1}{m_h}\right) \tag{7.77}$$

and the fundamental absorption edge has been shifted from $\hbar\omega = \mathscr{E}_g$ to this value by the doping. This is referred to as the Burstein–Moss shift [E. Burstein, *Phys. Rev. 93*, 632 (1954); T. S. Moss, *Proc. Phys. Soc. London B76*, 775 (1954)].

An expression for the fundamental absorption edge in terms of the

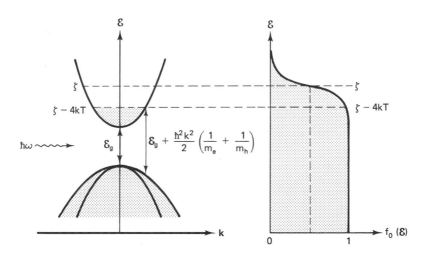

Figure 7.6 Diagram showing how the fundamental absorption edge of an n-type semiconductor is shifted to higher energy by doping.

Fermi energy can be obtained from (7.77) in the form

$$\hbar\omega = \mathcal{E}_g + \frac{\hbar^2 k^2}{2m_e}\left(1 + \frac{m_e}{m_h}\right) \tag{7.78}$$

Using (2.118), equation (7.78) becomes

$$\hbar\omega = \mathcal{E}_g + (\mathcal{E} - \mathcal{E}_c)\left(1 + \frac{m_e}{m_h}\right) \tag{7.79}$$

or

$$\hbar\omega = \mathcal{E}_g + (\zeta - \mathcal{E}_c - 4kT)\left(1 + \frac{m_e}{m_h}\right) \tag{7.80}$$

Equations (7.63) and (7.70) for direct band-to-band absorption can be modified to account for the Burstein–Moss shift by determining the probability of an unoccupied conduction band state. From (4.42) and (7.79),

$$1 - f_0 = \frac{1}{1 + \exp\left[(\zeta - \mathcal{E})/kT\right]} \tag{7.81}$$

$$1 - f_0 = \frac{1}{1 + \exp\left[\dfrac{\zeta - \mathcal{E}_c}{kT} - \dfrac{(\hbar\omega - \mathcal{E}_g)m_h}{(m_e + m_h)kT}\right]} \tag{7.82}$$

$$1 - f_0 = \frac{1}{1 + \exp\left[\dfrac{(\zeta - \mathcal{E}_c)m_e - (\hbar\omega - \mathcal{E}_g)m_r}{m_e kT}\right]} \tag{7.83}$$

where the reduced mass, m_r, is given by

$$m_r = \frac{m_e m_h}{m_e + m_h} \tag{7.84}$$

Thus the shift of absorption with doping for n-type material can be accounted for with (7.83).

7.2.4 Temperature Dependence

Another factor that controls the fundamental absorption edge is the temperature of the sample. This is reflected primarily in the expansion and contraction of the lattice with temperature and its effect on the energy gap. This dependence can be seen in Fig. 2.14. Since the temperature dependence of the energy gap varies considerably among semiconductors, it is best determined from experimental results. For Si an empirical fit to the experimental data is given by

$$\mathcal{E}_g(T) = 1.165 - 2.84 \times 10^{-4}T \text{ in eV} \tag{7.85}$$

where T is in K. For Ge the temperature dependence of the energy gap is

$$\mathscr{E}_g(T) = 0.742 - 3.90 \times 10^{-4}T \tag{7.86}$$

The form of the temperature dependence for the III–V compound semiconductors is somewhat different. For GaAs the experimental data are best fit by

$$\mathscr{E}_g(T) = 1.522 - \frac{5.8 \times 10^{-4}T^2}{T + 300} \tag{7.87}$$

while GaP is

$$\mathscr{E}_g(T) = 2.338 - \frac{6.2 \times 10^{-4}T^2}{T + 460} \tag{7.88}$$

Notice that in (7.85) to (7.88) the energy gap decreases with increasing temperature. Although this behavior holds for most semiconductors, for some materials such as the IV–VI compounds the energy gap increases with increasing temperature.

7.2.5 Electric Field Dependence

The electric field dependence of the fundamental absorption edge is referred to as the Franz–Keldysh effect [W. Franz, *Z. Naturforsch. 13a*, 484 (1958); L. V. Keldysh, *Sov. Phys.-JETP 7*, 788 (1958)]. As indicated in Fig. 7.7, this electroabsorption process can be thought of as photon-assisted tunneling through the energy gap. That is, the electron wavefunctions in the valence and conduction bands have an exponentially decaying amplitude in the energy gap. In the presence of an electric field a valence band electron must tunnel through a triangular barrier to reach the conduction band. When there is no photon absorption as in Fig. 7.7(a), the height of this barrier is $\mathscr{E}_g$ and its thickness, t, can be determined from

$$q\mathbf{E} = \nabla_r\mathscr{E} = \frac{\mathscr{E}_g}{t} \tag{7.89}$$

In one dimension the barrier thickness is

$$t = \frac{\mathscr{E}_g}{qE} \tag{7.90}$$

However, with photon absorption as in Fig. 7.7(b), the barrier height is reduced to $\mathscr{E}_g - \hbar\omega$ and the barrier thickness becomes

$$t(\hbar\omega) = \frac{\mathscr{E}_g - \hbar\omega}{qE} \tag{7.91}$$

Obviously, the tunneling probability is considerably enhanced with photon absorption and depends on the electric field as well as the photon energy.

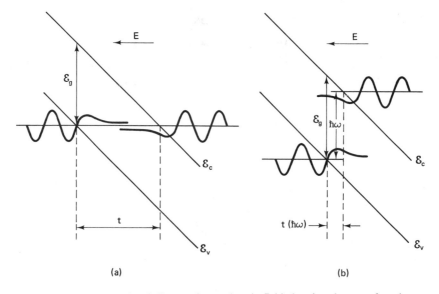

Figure 7.7 Energy band diagram in an electric field showing the wavefunction overlap (a) without and (b) with absorption of a photon of energy $\hbar\omega$.

An analysis [K. Tharmalingham, *Phys. Rev. 130,* 2204 (1963)] of this photon-assisted tunneling process indicates that the electroabsorption coefficient for a direct energy gap is given by

$$\alpha(\hbar\omega, E) = 1.0 \times 10^4 \frac{f}{n} \left(\frac{2m_r}{m}\right)^{4/3} E^{1/3} \int_\beta^\infty |\, Ai(z)\,|^2 \, dz \text{ in cm}^{-1} \quad (7.92)$$

where

$$\beta = 1.1 \times 10^5 \left(\frac{2m_r}{m}\right)^{1/3} \frac{\mathscr{E}_g - \hbar\omega}{E^{2/3}} \quad (7.93)$$

for E in volts/cm and $\mathscr{E}_g - \hbar\omega$ in eV. In (7.92) and (7.93) $Ai(z)$ is the Airy function, f is the oscillator strength given by (7.54), and m_r is the reduced electron–hole mass given by (7.84). Figure 7.8 shows calculated values of the electroabsorption coefficient for GaAs, assuming a uniform electric field.

7.2.6 Magnetic Field Dependence

When a magnetic field is applied to a semiconductor, the components of electron motion perpendicular to the field describe circular orbits at the cyclotron frequency, ω_c. From (5.147) the cyclotron frequency for parabolic bands is

$$\omega_c = \frac{qB}{m^*} \quad (7.94)$$

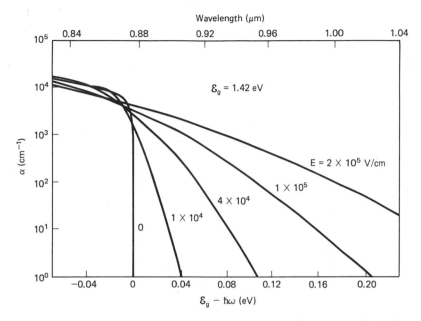

Figure 7.8 Electric field and photon energy dependence of the band-to-band absorption for GaAs. [After G. E. Stillman and C. M. Wolfe, "Avalanche Photodiodes" in *Semiconductors and Semimetals*, Vol. 12, *Infrared Detectors II*, ed. R. K. Willardson and A. C. Beer (New York: Academic Press, 1977), p. 291.]

An analysis of this effect in the effective mass approximation shows that the electron wavefunctions are described by Schrödinger's equation for a simple harmonic oscillator. As shown in Fig. 7.9, the allowed energy levels are quantized into Landau levels given by

$$\mathscr{E} = \mathscr{E}_c + \frac{\hbar^2 k^2}{2m_e} + \hbar\omega_{ce}\left(n + \frac{1}{2}\right) \tag{7.95}$$

for the conduction band, and by

$$\mathscr{E} = \mathscr{E}_v - \frac{\hbar^2 k^2}{2m_h} - \hbar\omega_{ch}\left(n + \frac{1}{2}\right) \tag{7.96}$$

for the valence bands. In (7.95) and (7.96) **k** is in a direction perpendicular to the field and n takes on all integer values, including zero. From (7.94), (7.95), and (7.96) the magnetic field dependence of the absorption edge or energy gap is simply

$$\mathscr{E}_g(B) = \mathscr{E}_g(0) + \frac{q\hbar B}{2m_r} \tag{7.97}$$

where m_r is the reduced electron–hole mass.

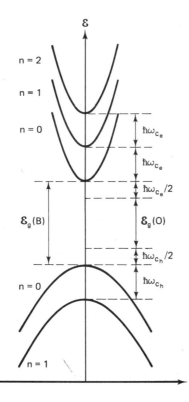

Figure 7.9 Splitting of the conduction and valence bands into Landau levels by a magnetic field.

7.3 EXCITON ABSORPTION

As indicated in Fig. 7.1, some structure in the absorption spectra of semi-conductors is often observed just below the fundamental absorption edge. This structure is due to exciton absorption. This phenomenon can be understood in the following way. The energy band calculations in Chapter 2 were for one electron only, where the rest of the crystal was treated as a periodic potential. In this case there are no allowed energy states between the minima of the conduction bands and the maxima of the valence bands. In the creation of an electron–hole pair, as in band-to-band absorption, we looked at a perturbation of these one-electron energy bands. We ignored, however, the Coulombic interaction between the electron and the hole. For sufficiently low thermal energy this Coulombic attraction binds the electron and hole together to produce the quasi-particle known as an exciton. This is the first excited state of the one-electron energy bands.

In Chapter 3 we used the effective mass approximation to examine a similar problem: the interaction between an electron and a donor atom. The only difference between that problem and this one is that here the positively charged hole is free to move. Thus this problem is exactly analogous to the hydrogen atom with a reduced mass corresponding to the relative motion of the electron and hole. From (3.33) the n binding energies of the exciton are given by

$$\mathscr{E}_{xn} = 13.6 \left(\frac{1}{n\epsilon_r}\right)^2 \left(\frac{m_r}{m}\right) \qquad \text{in eV} \qquad (7.98)$$

where the reduced mass is

$$\frac{1}{m_r} = \frac{1}{m_e} + \frac{1}{m_h} \qquad (7.99)$$

and ϵ_r is the dielectric constant of the semiconductor. In a similar manner the orbital radii of the exciton are

$$r_{xn} = 0.53n^2\epsilon_r \left(\frac{m}{m_r}\right) \qquad \text{in Å} \qquad (7.100)$$

As indicated in Fig. 7.10, an exciton is free to move throughout the crystal with a dispersion relationship given by

$$\mathscr{E}_{xn}(\mathbf{k}) = \mathscr{E}_{xn} + \frac{\hbar^2 k^2}{2M} \qquad (7.101)$$

Since this represents the collective motion of the electron and the hole, the appropriate effective mass for (7.101) is

$$M = m_e + m_h \qquad (7.102)$$

To determine quantitatively the amount of absorption due to exciton for-

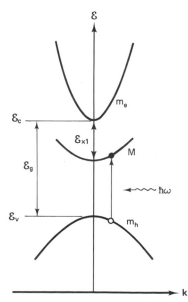

Figure 7.10 Creation of an exciton by photon absorption below the fundamental absorption edge.

mation, we can use the same procedure as for band-to-band absorption. The appropriate modification of (7.64) for exciton absorption is

$$\alpha_n = 2.7 \times 10^5 \left(\frac{2m_r}{m}\right)^{3/2} \frac{f}{n} (\hbar\omega - \mathscr{E}_g - \mathscr{E}_{xn})^{1/2} \text{ cm}^{-1} \qquad (7.103)$$

In this case the reduced mass is

$$\frac{1}{m_r} = \frac{1}{M} + \frac{1}{m_h} \qquad (7.104)$$

7.4 FREE CARRIER ABSORPTION

When the energy of the incident radiation is too small to create electron–hole pairs or excitons, other absorption processes can occur. The absorption of radiation by electrons in the conduction bands or by holes in the valence bands produces an absorption background below the fundamental absorption edge which increases with wavelength as indicated in Fig. 7.1. In this process the electric field of the incident radiation accelerates the free carriers, which, in turn, are decelerated by collisions with the lattice. Thus the energy of the radiation field is converted to heat.

This problem can be examined semiclassically by using Maxwell's equations to characterize the radiation in the semiconductor:

$$\nabla \times \mathbf{H} = \sigma^*\mathbf{E} + \frac{\partial \mathbf{D}}{\partial t}, \qquad \nabla \cdot \mathbf{H} = 0$$

$$\nabla \times \mathbf{E} = -\mu_0 \frac{\partial \mathbf{H}}{\partial t}, \qquad \nabla \cdot \mathbf{D} = 0 \qquad (7.105)$$

$$\mathbf{D} = \epsilon\mathbf{E}$$

In these equations, we have assumed there are no space-charge regions in the semiconductor, so the divergence of the electric field is zero. Also, a complex conduction current, $\sigma^*\mathbf{E}$, is included in the equation for the curl of the magnetic field. From (5.52) and (5.145) the high-frequency conductivity is

$$\sigma^* = \frac{q^2 n \langle \tau_m^* \rangle}{m^*} \qquad (7.106)$$

where

$$\langle \tau_m^* \rangle = \left\langle \frac{\tau_m}{1 + \omega^2\tau_m^2} \right\rangle + i \left\langle \frac{\omega\tau_m^2}{1 + \omega^2\tau_m^2} \right\rangle \qquad (7.107)$$

For $\omega\tau_m \gg 1$,

$$\langle \tau_m^* \rangle = \left\langle \frac{1}{\omega^2\tau_m} \right\rangle + \frac{i}{\omega} \tag{7.108}$$

and

$$\sigma^* = \frac{q^2 n}{m^*\omega^2} \left\langle \frac{1}{\tau_m} \right\rangle + i \frac{q^2 n}{m^*\omega}$$

$$= \sigma_r + i\sigma_i \tag{7.109}$$

The magnetic field can be eliminated from (7.105) to obtain an equation for a propagating radiation wave with losses,

$$\nabla^2 \mathbf{E} = \epsilon\mu_0 \frac{\partial^2 \mathbf{E}}{\partial t^2} + \sigma^*\mu_0 \frac{\partial \mathbf{E}}{\partial t} \tag{7.110}$$

Choosing a traveling-wave solution of the form

$$\mathbf{E} = \mathbf{a}E \exp\left[i(\mathbf{q}\cdot\mathbf{r} - \omega t)\right] \tag{7.111}$$

the wavevector $\mathbf{q}$ must satisfy

$$q = \frac{\omega}{c}\left(\epsilon_r + i\frac{\sigma^*}{\epsilon_0\omega}\right)^{1/2} \tag{7.112}$$

where $c = (\epsilon_0\mu_0)^{-1/2}$ is the free-space velocity of light. Since in free space,

$$q = \frac{\omega}{c} = \frac{2\pi}{\lambda} \tag{7.113}$$

(7.112) indicates that in the semiconductor the phase velocity and wavelength of the radiation are reduced by a complex refractive index

$$n^* \equiv \left(\epsilon_r + i\frac{\sigma^*}{\epsilon_0\omega}\right)^{1/2} \tag{7.114}$$

Setting

$$n^* = n + ik \tag{7.115}$$

and using (7.109) in (7.114), we obtain

$$n^2 - k^2 = \epsilon_r - \frac{\sigma_i}{\epsilon_0\omega} \tag{7.116}$$

$$2nk = \frac{\sigma_r}{\epsilon_0\omega} \tag{7.117}$$

For wave propagation in the z direction, (7.111) is

$$\mathbf{E} = \hat{x}E \exp\left[i(qz - \omega t)\right] \tag{7.118}$$

where

$$q = \frac{\omega}{c} (\eta + ik) \tag{7.119}$$

Therefore, (7.118) has the form

$$\mathbf{E} = \hat{x}E \exp \left(\frac{-\omega k z}{c} \right) \exp \left[i \left(\frac{\omega \eta z}{c} - \omega t \right) \right] \tag{7.120}$$

In (7.120) it can be seen that the velocity and wavelength of the radiation in the semiconductor are decreased by η, while the electric field is attenuated by

$$\frac{\omega k}{c} = \frac{\sigma_r}{2\eta c \epsilon_0} \tag{7.121}$$

Since the intensity or power of the radiation is proportional to $| \mathbf{E}^2 |$, the absorption coefficient for free carriers is

$$\alpha = \frac{\sigma_r}{\eta c \epsilon_0} \tag{7.122}$$

or using (7.109), we obtain

$$\alpha = \frac{q^2 n}{\eta c \epsilon_0 m^* \omega^2} \left\langle \frac{1}{\tau_m} \right\rangle \tag{7.123}$$

In terms of the free-space wavelength, (7.123) becomes

$$\alpha = \frac{q^2 n \lambda^2}{4\pi^2 \eta c^3 \epsilon_0 m^*} \left\langle \frac{1}{\tau_m} \right\rangle \tag{7.124}$$

Thus, free carrier absorption depends on the concentration of free carriers, n, the process by which they convert the light into heat through τ_m, and it increases with the square of the free-space wavelength, λ.

From (7.109), (7.116), and (7.117) we see that the free carriers also lower the refractive index,

$$\eta^2 = \epsilon_r - \frac{q^2 n}{\epsilon_0 m^* \omega^2} + k^2 \tag{7.125}$$

If we define a plasma frequency,

$$\omega_p^2 \equiv \frac{q^2 n}{\epsilon m^*} \tag{7.126}$$

(7.125) becomes

$$\eta^2 = \epsilon_r \left[1 - \left(\frac{\omega_p}{\omega} \right)^2 \right] + \frac{\epsilon_r^2}{4\eta^2} \left(\frac{\omega_p}{\omega} \right)^4 \left\langle \frac{1}{\omega \tau_m} \right\rangle^2 \tag{7.127}$$

For $\omega\tau_m \gg 1$, the last term on the right-hand side is small and

$$\eta^2 = \epsilon_r \left[1 - \left(\frac{\omega_p}{\omega} \right)^2 \right] \tag{7.128}$$

In terms of the free-space wavelength,

$$\eta^2 = \epsilon_r - \frac{q^2 n\lambda^2}{4\pi^2\epsilon_0 m^* c^2} \tag{7.129}$$

Although this effect is relatively small in semiconductors, heavily doped regions can be used to guide light into higher index, more lightly doped regions for some optoelectronic applications.

7.5 REFLECTIVITY

Equations (7.105) can also be used to determine the reflectivity of an ideal semiconductor surface. Let us consider a plane wave normally incident in the positive z direction on a semiconductor surface at $z = 0$. A semiconductor with complex refractive index given by (7.114) occupies the space with positive z values. At the surface, part of the incident radiation is reflected in the minus z direction and the rest penetrates the semiconductor, where its velocity and wavelength are reduced by η and its intensity attenuated by absorption.

For the region outside the semiconductor ($z < 0$), the electric field of the radiation is

$$E_x(z) = E_i \exp\left[i\omega \left(\frac{z}{c} - t \right) \right] + E_r \exp\left[-i\omega \left(\frac{z}{c} + t \right) \right] \tag{7.130}$$

where E_i is the magnitude of the incident electric field and E_r is the magnitude of the reflected electric field. From (7.105) the corresponding expression for the magnetic field for $z < 0$ is

$$\mu_0 c H_y(z) = E_i \exp\left[i\omega \left(\frac{z}{c} - t \right) \right] - E_r \exp\left[-i\omega \left(\frac{z}{c} + t \right) \right] \tag{7.131}$$

In the semiconductor ($z > 0$) the expression for the transmitted electric field is

$$E_x(z) = E_t \exp\left[i\omega \left(\frac{\eta^* z}{c} - t \right) \right] \tag{7.132}$$

and the magnetic field from (7.105) is

$$\mu_0 c H_y(z) = \eta^* E_t \exp\left[i\omega \left(\frac{\eta^* z}{c} - t \right) \right] \tag{7.133}$$

At the surface ($z = 0$) the electric and magnetic fields are continuous. Setting (7.130) equal to (7.132) and (7.131) equal to (7.133) at $z = 0$, we obtain

$$E_t = E_i + E_r \tag{7.134}$$
$$n^* E_t = E_i - E_r$$

Eliminating E_t from these equations yields

$$\frac{E_r}{E_i} = \frac{1 - n^*}{1 + n^*} \tag{7.135}$$

Since the intensity of the light depends on the square of the electric field, the reflectivity is defined as

$$R \equiv \left(\frac{1 - n^*}{1 + n^*} \right)^2 \tag{7.136}$$

Using (7.115), we have

$$R = \frac{(n - 1)^2 + k^2}{(n + 1)^2 + k^2} \tag{7.137}$$

7.6 LATTICE ABSORPTION

Figure 7.1 shows an absorption band due to optical phonons which is almost as strong as the absorption due to electron transitions from the valence to conduction bands. This absorption is due to electric dipole coupling between photons and phonons which arises from the motion of charged atoms in the crystal. (The structure on the high-energy side is due to multiple phonon absorption.) Thus this absorption is strong for semiconductors with an ionic component of bonding and weak for those with purely covalent bonding. Since the photon is a transverse electromagnetic wave, it couples most strongly to the transverse optical phonons. For photon frequencies between the transverse and longitudinal optical phonon frequencies, the reflection is very high. If the radiation from a hot body is reflected several times from an ionic crystal, the dominant remaining frequencies will be those in this band of energy. For this reason this absorption is called the "restrahlen" or "residual ray" band.

Let us consider the interaction between a traveling light wave with electric field given by (7.111) and an ionic crystal with a basis of two atoms per primitive unit cell. Due to an electronegativity difference, the cations have an effective charge $+e^*$ and the anions $-e^*$. If we consider only interactions between ions in the unit cell, the equations of motion for the displacement of the ions from their Bravais lattice sites are, from (3.72),

$$M_1 \frac{\partial^2 \mathbf{u}_1}{\partial t^2} = -\frac{1}{2} \alpha(\mathbf{u}_1 - \mathbf{u}_2) + e^*\mathbf{E}_1$$

$$M_2 \frac{\partial^2 \mathbf{u}_2}{\partial t^2} = -\frac{1}{2} \alpha(\mathbf{u}_2 - \mathbf{u}_1) - e^*\mathbf{E}_1$$

(7.138)

In these equations α is the coupling coefficient between ions. The effect of the photon electric field $\mathbf{E}$ on the displacement is included by way of the local field $\mathbf{E}_1$ at the primitive or Wigner–Seitz unit cell.

Defining the relative displacement between ions as $\delta\mathbf{u} \equiv \mathbf{u}_1 - \mathbf{u}_2$, the coupled equations of motion (7.138) become

$$\frac{\partial^2 \delta\mathbf{u}}{\partial t^2} = -\frac{\alpha}{2M_r} \delta\mathbf{u} + \frac{e^*}{M_r} \mathbf{E}_1$$

(7.139)

where the reduced ion mass is

$$\frac{1}{M_r} = \frac{1}{M_1} + \frac{1}{M_2}$$

(7.140)

Considering a solution for (7.139) of the form

$$\delta\mathbf{u} = \mathbf{b}\, \delta u \exp\left[i(\mathbf{q}_s \cdot \mathbf{r} - \omega_s t)\right]$$

(7.141)

conservation of energy and momentum requires that the frequencies and wavevectors of the photon (7.111) and phonons (7.141) be equal. That is, $\omega_s = \omega$ and $q_s = q \simeq 0$. Thus the photon interaction is with the optical branch of the phonon dispersion curves shown in Fig. 3.23. Under these conditions (7.139) becomes

$$-\omega^2 \delta\mathbf{u} = -\frac{\alpha}{2M_r} \delta\mathbf{u} + \frac{e^*}{M_r} \mathbf{E}_1$$

(7.142)

or

$$\delta\mathbf{u} = \frac{e^*/M_r}{\omega_0^2 - \omega^2} \mathbf{E}_1$$

(7.143)

where

$$\omega_0^2 \equiv \frac{\alpha}{2M_r}$$

(7.144)

From (7.143) we see that the relative displacement between ions, $\delta\mathbf{u}$, resonates at the photon frequency $\omega = \omega_0$, and the lattice absorbs energy from the light wave. The interaction between a photon and the lattice in this frequency range is sometimes treated as a quasi-particle known as a "polariton."

The ability of the ions to respond to electromagnetic radiation constitutes an ionic contribution to the polarization of the crystal in addition to the contribution from the atomic cores. For N/V Wigner–Seitz cells per unit volume this ionic polarization is

$$\mathbf{P}_i = \frac{Ne^*}{V}\,\delta\mathbf{u} \tag{7.145}$$

Substituting (7.143) in this equation gives

$$\mathbf{P}_i(\omega) = \frac{Ne^{*2}/M_r V}{\omega_0^2 - \omega^2}\,\mathbf{E}_1 \tag{7.146}$$

Notice that this frequency-dependent ionic polarization produces a frequency-dependent permittivity or dielectric constant given by

$$\mathbf{D}(\omega) = \epsilon(\omega)\mathbf{E} = \epsilon_0\mathbf{E} + \mathbf{P}(\omega) \tag{7.147}$$

where $\mathbf{P}(\omega)$ is the total polarization of the crystal. At low frequencies the permittivity

$$\epsilon(0) = \epsilon_r(0)\epsilon_0 \tag{7.148}$$

is due to both ionic and atomic polarization. At high frequencies the ions can no longer respond to the field and the permittivity,

$$\epsilon(\infty) = \epsilon_r(\infty)\epsilon_0 \tag{7.149}$$

is due to atomic polarization only. Thus the total polarization is

$$\mathbf{P}(\omega) = \mathbf{P}_i(\omega) + \mathbf{P}(\infty) \tag{7.150}$$

where from (7.147),

$$\mathbf{P}(\infty) = [\epsilon(\infty) - \epsilon_0]\mathbf{E} \tag{7.151}$$

The polarization of the lattice also screens the applied electric field from the unit cell, so that the local field and the applied field are different. In cubic crystals the relationship is

$$\mathbf{E}_1 = \mathbf{E} + \frac{1}{3\epsilon_0}\,\mathbf{P} \tag{7.152}$$

For simplicity in the analysis to follow, we will assume that $\mathbf{E}_1 = \mathbf{E}$. With this assumption we use (7.146) and (7.151) in (7.150) and insert (7.150) in (7.147), to obtain

$$\epsilon(\omega)\mathbf{E} = \frac{Ne^{*2}/M_r V}{\omega_0^2 - \omega^2}\,\mathbf{E} + \epsilon(\infty)\mathbf{E} \tag{7.153}$$

Defining a frequency,

$$\omega_1^2 \equiv \frac{Ne^{*2}}{M_r V \epsilon(\infty)} \tag{7.154}$$

(7.153) gives

$$\epsilon_r(\omega) = \epsilon_r(\infty) \left(1 + \frac{\omega_1^2}{\omega_0^2 - \omega^2} \right) \tag{7.155}$$

and

$$\epsilon_r(0) = \epsilon_r(\infty) \left(\frac{\omega_0^2 + \omega_1^2}{\omega_0^2} \right) \tag{7.156}$$

as the relationship between the low- and high-frequency dielectric constants. Although the frequencies ω_0 and ω_1 in (7.155) and (7.156), as defined by (7.144) and (7.154), appear to be somewhat arbitrary, they are intimately related to the transverse and longitudinal optical phonon frequencies.

To examine these relations, let us first consider a transverse lattice wave in the form of (7.141). Assuming a transverse displacement $\delta \mathbf{u}_{TO}$ with frequency ω_{TO} in the equation of motion (7.139), we obtain

$$\omega_{TO}^2 \, \delta \mathbf{u}_{TO} = \omega_0^2 \, \delta \mathbf{u}_{TO} - \frac{e^*}{M_r} \mathbf{E} \tag{7.157}$$

Taking the curl of this equation yields

$$\nabla \times \delta \mathbf{u}_{TO} = i\mathbf{q} \times \delta \mathbf{u}_{TO} \neq 0 \tag{7.158}$$

since the displacement is perpendicular to the direction of propagation. For the electric field associated with the displacement, however,

$$\nabla \times \mathbf{E} = 0 \tag{7.159}$$

Thus

$$\omega_{TO}^2 = \omega_0^2 \tag{7.160}$$

and the transverse phonon frequency is not affected by the electric field.

Next consider a longitudinal wave with displacement $\delta \mathbf{u}_{LO}$ and frequency ω_{LO}. From the equation of motion, (7.139),

$$\omega_{LO}^2 \, \delta \mathbf{u}_{LO} = \omega_0^2 \, \delta \mathbf{u}_{LO} - \frac{e^*}{M_r} \mathbf{E} \tag{7.161}$$

Taking the divergence of this equation gives

$$\nabla \cdot \delta \mathbf{u}_{LO} = i\mathbf{q} \cdot \delta \mathbf{u}_{LO} \neq 0 \tag{7.162}$$

since the displacement is in the direction of propagation. For the electric flux density associated with the displacement,

$$\nabla \cdot \mathbf{D} = 0 \tag{7.163}$$

Using (7.143) in (7.153), we have

$$\mathbf{D} = \frac{Ne^*}{V} \delta \mathbf{u}_{LO} + \epsilon(\infty)\mathbf{E} \tag{7.164}$$

or

$$\nabla \cdot \mathbf{E} = \nabla \cdot \left[\frac{\mathbf{D}}{\epsilon(\infty)} - \frac{Ne^*}{V\epsilon(\infty)} \delta \mathbf{u}_{LO} \right]$$

$$= - \frac{Ne^*}{V\epsilon(\infty)} \nabla \cdot \delta \mathbf{u}_{LO} \tag{7.165}$$

and (7.161) is

$$\omega_{LO}^2 = \omega_0^2 + \omega_1^2 \tag{7.166}$$

Thus the longitudinal optical phonon frequency is higher than the transverse frequency because of the field interaction.

Using (7.166) and (7.160) in (7.156), we find that the dielectric constants are related to the optical phonon frequencies by

$$\frac{\epsilon_r(0)}{\epsilon_r(\infty)} = \left(\frac{\omega_{LO}}{\omega_{TO}} \right)^2 \tag{7.167}$$

This equation is known as the Lyddane–Sachs–Teller relation [R. H. Lyddane, R. G. Sachs, and E. Teller, *Phys. Rev. 59*, 613 (1941)]. From (7.155) the frequency dependence of the dielectric constant is

$$\epsilon_r(\omega) = \epsilon_r(\infty) \frac{\omega_{LO}^2 - \omega^2}{\omega_{TO}^2 - \omega^2} \tag{7.168}$$

Notice from this equation that the dielectric constant is infinite for $\omega = \omega_{TO}$, zero for $\omega = \omega_{LO}$, and negative for ω between ω_{TO} and ω_{LO}. This behavior is shown in Fig. 7.11(a).

This behavior of the dielectric constant strongly affects the reflectivity of the crystal. Neglecting the carrier concentration in (7.125) gives us

$$\epsilon_r = \eta^2 - k^2 \tag{7.169}$$

and the reflectivity is

$$R = \frac{(\eta - 1)^2 + k^2}{(\eta + 1)^2 + k^2} \tag{7.137}$$

In the region between ω_{TO} and ω_{LO} where ϵ_r is negative, k^2 must be much greater than η^2 according to (7.168). From (7.137) this produces a reflectivity

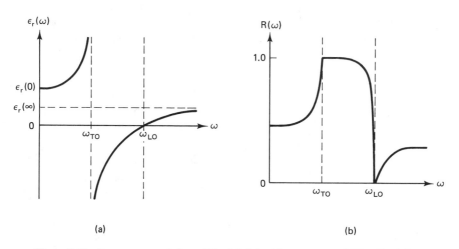

Figure 7.11   Frequency variation of the (a) dielectric constant and (b) reflectivity in the vicinity of the optical phonon frequencies.

near unity which accounts for the "restrahlen" band. The behavior of the reflectivity is illustrated schematically in Fig. 7.11(b).

Using the Lyddane–Sachs–Teller relation, an equation relating the effective charge of the ions to more easily measurable quantities can be obtained. Equation (7.154) gives

$$e^{*2} = \frac{M_r V \epsilon(\infty)}{N} \omega_1^2 \tag{7.170}$$

where from (7.160) and (7.166),

$$\omega_1^2 = \omega_{LO}^2 - \omega_{TO}^2 \tag{7.171}$$

Using (7.167) in (7.171), we have

$$e^{*2} = \frac{M_r V}{N} \omega_{LO}^2 \epsilon^2(\infty) \left[\frac{1}{\epsilon(\infty)} - \frac{1}{\epsilon(0)} \right] \tag{7.172}$$

Since the mass density (g/cm³)

$$\rho = \frac{M_r}{\Omega} = \frac{M_r N}{V} \tag{7.173}$$

(7.172) can be expressed as

$$e^* = \Omega \omega_{LO} \epsilon(\infty) \rho^{1/2} \left[\frac{1}{\epsilon(\infty)} - \frac{1}{\epsilon(0)} \right]^{1/2} \tag{7.174}$$

In terms of the transverse optical frequency, (7.174) has the somewhat simpler form

$$e^* = \Omega \omega_{TO} \rho^{1/2} [\epsilon(0) - \epsilon(\infty)]^{1/2} \tag{7.175}$$

Since as indicated in Chapter 6 the definition of this quantity is not unique, (7.175) is called the Born effective ionic charge.

7.7 IMPURITY ABSORPTION

The last absorption process we discuss is that due to direct band-to-impurity transitions. From the discussion in Section 7.2 the transition probability has the same form as for direct band-to-band transitions. From (7.63) and (7.53) the proportionality of the absorption coefficient for an electron transition from an initial to a final state is

$$\alpha_{if} \propto N_I f_i (1 - f_f) \frac{(\hbar\omega - \Delta\mathscr{E}_{if})^{1/2}}{\hbar\omega} \tag{7.176}$$

where N_I is the total concentration of impurities, f_i the probability that the initial state is full, $1 - f_f$ the probability that the final state is empty, and $\Delta\mathscr{E}_{if}$ the energy difference between the initial and final state.

As an example, consider the transition of an electron from the valence band to an empty donor in n-type material. In n-type material the valence band is full, so $f_i = 1$. The concentration of empty donors is

$$N_I(1 - f_f) = N_d^+ \tag{7.177}$$

where N_d^+ is given by (4.97). Substituting (4.97) into (7.170) and using the appropriate notation,

$$\alpha_{vd} \propto \frac{N_d}{1 + g_d \exp[(\mathscr{E}_f - \mathscr{E}_d)/kT]} \frac{[\hbar\omega - (\mathscr{E}_d - \mathscr{E}_v)]^{1/2}}{\hbar\omega} \tag{7.178}$$

Similar expressions can be obtained for transitions involving full donors from (4.88) and full acceptors from (4.93).

PROBLEMS

7.1. Determine the energy at which appreciable band-to-band absorption begins at 300 K for a p-type GaAs sample with a hole concentration of 1×10^{20} cm^{-3}. Take into account both the light- and heavy-hole bands, which are degenerate at Γ.

7.2. Determine the free-carrier absorption at 300 K and 0.905 μm (free-space wavelength) for a GaAs sample with $n = N_d - N_a = 9.8 \times 10^{17}$ cm^{-3} and $N_a = 9.8 \times 10^{17}$ cm^{-3}. For this sample ionized impurity scattering is the dominant scattering mechanism.

7.3. In the density-of-states representation we have a hypothetical intrinsic semiconductor (direct) as shown in Fig. P7.3 and have no problems with k-selection rules relative to generation or recombination transitions.
(a) At low temperature we pump a thin sample of this material and measure

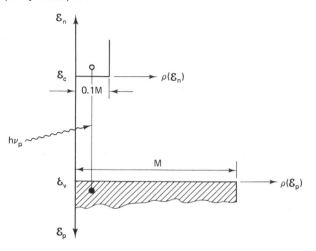

Figure P7.3

its transmission in order to determine the absorption coefficient $\alpha = \alpha(\mathscr{E}_p)$ $= \alpha(h\nu_p)$. The magnitude of the absorption depends on the joint density of upper and lower states separated by $h\nu_p$, appropriate filling factors (probabilities), and an assumed constant coupling between valence and conduction band states. How does α vary as a function of $\mathscr{E}_p$?

   (b) Suppose that the pump intensity I_0 can be increased to very high levels. At larger and larger I_0 and fixed $\mathscr{E}_p = h\nu_p$, how does the absorption of the thin sample change (vary)? Justify your answer.

7.4. In a semiconductor laser Fabry–Perot resonator of length L the number M of half-wavelengths in the structure is $M(\lambda'_M/2) = L$, where $\lambda' \equiv \lambda/\eta$, λ is the free-space wavelength and η is the index of refraction of the semiconductor medium. (Index of refraction is the ratio of the speed of light in free space to that in the medium.) Find the spacing $d\lambda$ between adjacent wavelength modes in the resonator. Remember that the index of refraction, $\eta = \eta(\lambda)$ is a function of wavelength.

7.5. Given a semiconductor laser of length L and index of refraction η, where η is a function of λ, $\eta(\lambda)$.

   (a) Derive the formula for the spectral separation of the laser modes, $d\lambda$.

   (b) Suppose that one end of the original cavity is coated with an antireflective coating, and an external mirror is placed a distance l from this end. What is the formula for $d\lambda$ in this case?

   (c) Define the quantity $\eta - \lambda(d\eta/d\lambda) = \Delta$ as the index dispersion expression. What is the expression for this quantity from part (a)? Part (b)? (In terms of L, l, λ, and $d\lambda$)

   (d) If the typical mode spacing for no external mirror is 3 Å and a typical value of Δ is 6, what is a practical limit on l/L such that a monochrometer with a resolution of 0.3 Å can still "see" separate modes in the case of the arrangement in part (b)?

The quantity Δ, called the index dispersion, is useful in interpreting laser mode data and inferring behavior of density of states and absorption.

Excess
Carriers

Previous chapters of this book have dealt primarily with distributions of electrons and holes determined by thermal excitation in intrinsic material or by doping in extrinsic material. In Chapter 4, (4.73) and (4.79) gave the relationship between these free electron and hole concentrations and the Fermi energy. Also, a mass-action law relating these concentrations was derived [equation (4.83)]. From thermodynamic considerations we found that the Fermi energy was equal to the equilibrium electrochemical potential (energy) in the presence of any internal electrostatic potential. In Chapter 5 we extended these concepts in the presence of externally applied forces. A nonequilibrium electrochemical potential and its relationship to charge current density, equation (5.81), was obtained by superimposing an external electrostatic potential on any internal one. It was also assumed that neither internal nor externally applied forces changed the carrier distributions produced by net doping or intrinsic excitation.

Many of the most important effects in semiconductor materials and devices, however, are caused by the changes in electron and hole concentrations produced by internal or externally applied forces or electromagnetic radiation. These forces or excitations can produce carrier concentrations above or below those produced by doping or thermal excitations, and we refer to the changes in either case as excess carrier concentrations. For electrons this concept can be represented by

$$n(\mathbf{r}, t) = n_0(\mathbf{r}) + \delta n(\mathbf{r}, t) \tag{8.1}$$

with a similar expression for holes. In this equation, $n(\mathbf{r}, t)$ is the total electron concentration which can be determined from Hall or differential capacitance measurements, $n_0(\mathbf{r})$ the electron concentration produced by the net doping $(N_d - N_a)$ or thermal excitation (n_i), and $\delta n(\mathbf{r}, t)$ the excess electron concentration produced by the internal or externally applied forces or radiation. In general, the net doping or intrinsic concentration can vary spatially, so $n_0(\mathbf{r})$ is usually a function of $\mathbf{r}$ only, while external forces or radiation can vary with distance and time to produce corresponding changes in $n(\mathbf{r}, t)$ through $\delta n(\mathbf{r}, t)$.

It should be emphasized that excess electron and hole distributions can exist in either equilibrium (internal forces but no external forces or radiation) or nonequilibrium (external forces or radiation). For this purpose we will use the following nomenclature. In equilibrium with no excess carrier distribution (uniform material), there is no external or internal electrostatic potential (except for the periodic crystal potential), and from (4.40) the Fermi energy ($\mathscr{E}_f$) is equal to the chemical potential (μ). When an external potential is applied to a sample with no excess carriers, the Fermi energy is equal to the electrochemical potential (ζ) and a charge current flows that is proportional to the gradient of the electrochemical potential ($\nabla \zeta$).

In equilibrium with an excess carrier distribution (nonuniform material), there is an internal electrostatic potential and the Fermi energy is equal to the electrochemical potential (ζ). When an external potential is applied to a sample with excess carriers, the excess carrier distributions are changed so that the electrochemical potential can no longer be used to describe the distribution of both electrons and holes. Under these conditions we introduce the concept of quasi-Fermi energies for electrons (ζ_n) and holes (ζ_p). Also, when radiation is applied to a sample with no excess carriers, the radiation can produce excess carriers and again quasi-Fermi energies are used to describe the carrier distributions. These quasi-Fermi energies have no apparent thermodynamic interpretation.

This discussion is summarized schematically in Table 8.1. By the term "uniform material" we mean material with no spatial variations in doping or energy band structure. Notice that when it is necessary to use quasi-Fermi energies, the mass-action law between electrons and holes is no longer

TABLE 8.1   **Equilibrium and Nonequilibrium Nomenclature**

Material	No Applied Potential	Applied Potential	Applied Radiation
Uniform	$\mathscr{E}_f = \mu$	$\mathscr{E}_f = \zeta$	ζ_n, ζ_p
	$n_0 p_0 = n_i^2$	$n_0 p_0 = n_i^2$	$np \neq n_i^2$
Nonuniform	$\mathscr{E}_f = \zeta$	ζ_n, ζ_p	ζ_n, ζ_p
	$np = n_i^2$	$np \neq n_i^2$	$np \neq n_i^2$

valid. The only nonequilibrium situation indicated where the mass-action law can be used is when small electric fields are applied to uniform samples. This is a near-equilibrium situation. For nonuniform materials in equilibrium the mass-action law is valid, but n is not necessarily equal to n_0 or p equal to p_0. These concepts are discussed in more detail in the following sections.

8.1 RECOMBINATION

Excess electron and hole distributions in semiconductors can be produced by incident radiation with energy greater than the minimum energy gap, $\mathscr{E}_g$, or by the application of electric fields to nonuniform samples. In this section we examine the kinetics of several mechanisms by which these excess distributions return to their equilibrium values. For simplicity these electron–hole recombination processes will first be considered in the absence of current: that is, with no externally applied forces.

8.1.1 Band-to-Band Recombination

The dominant recombination mechanism in semiconductors with direct energy gaps (conduction band minima and valence band maxima at the same **k** value) is typically band-to-band with photon emission corresponding to the gap energy. In equilibrium the electron and hole concentrations are average concentrations produced by two opposing processes: thermal generation and recombination. The thermal generation rate, G_t, in electron–hole pairs per cm³·s is mostly independent of n and p because of the large number of bound electrons in the valence band and empty states in the conduction band. G_t depends only on temperature and the intrinsic material properties. Thus

$$G_t = G_t(T)$$

The recombination rate, R, however, should depend on the concentrations of electrons in the conduction band and holes in the valence band since both an electron and a hole are required for recombination. Thus

$$R = R(n, p, T)$$

The functional dependence of R on n and p can be deduced from the following argument: If there were no holes, R would have to be zero. Also, if there were no electrons, R would have to be zero. The simplest functional form that has this property is the product np, or

$$R(n, p, T) = r(T)np \qquad (8.2)$$

where $r(T)$ is a rate constant with units of cm³/s. In equilibrium the thermal generation and recombination rates must be equal, and

$$G_t(T) = R(n_0, p_0, T) = r(T)n_0p_0 = r(T)n_i^2 \tag{8.3}$$

Now, let us examine a nonequilibrium situation, where, for example, above bandgap radiation produces excess electron–hole pairs. In the steady state, an excess generation rate G results where

$$G + G_t(T) = R(n, p, T) \tag{8.4}$$

When the external stimulus is turned off, $G = 0$ and the nonequilibrium electron and hole distributions decay to their equilibrium values at a rate determined by the excess recombination rate,

$$R(n, p, T) - G_t(T) = -\frac{\partial}{\partial t}[n(\mathbf{r}, t)]$$

$$= -\frac{\partial}{\partial t}[p(\mathbf{r}, t)] \tag{8.5}$$

From (8.1) the time dependence of the total concentrations is due to the time dependence of the excess concentrations, so

$$R(n, p, T) - G_t(T) = -\frac{\partial}{\partial t}[\delta n(\mathbf{r}, t)] = -\frac{\partial}{\partial t}[\delta p(\mathbf{r}, t)] \tag{8.6}$$

Using (8.1), (8.2), and (8.3) in (8.6), we obtain

$$r(n_0\,\delta p + p_0\,\delta n + \delta n\,\delta p) = -\frac{\partial}{\partial t}(\delta n) = -\frac{\partial}{\partial t}(\delta p) \tag{8.7}$$

This is the differential equation that describes the decay of the excess carrier concentrations. For simplicity the explicit dependence of the various concentrations on $\mathbf{r}$ and t have been omitted.

Since (8.7) is difficult to solve in general, we make the assumption that $\delta p(\mathbf{r}, t) = \delta n(\mathbf{r}, t)$ everywhere. This assumption will be examined in some detail later. Equation (8.7) then becomes

$$-\frac{\partial}{\partial t}(\delta p) = r(n_0 + p_0 + \delta p)\,\delta p \tag{8.8}$$

By separation of variables, we obtain

$$\int_0^t \frac{\frac{\partial}{\partial t}(\delta p)\,dt}{(n_0 + p_0 + \delta p)\,\delta p} = -r\int_0^t dt = = -rt \tag{8.9}$$

The left-hand side of this equation is an exact differential, which when integrated gives

$$\frac{1}{n_0 + p_0} \left[\ln \frac{\delta p}{n_0 + p_0 + \delta p} \right]_0^t = -rt \qquad (8.10)$$

Solving (8.10) explicitly for $\delta p(t)$, we have

$$\delta p(t) = \frac{(n_0 + p_0)\, \delta p(0)}{[n_0 + p_0 + \delta p(0)]\, \exp\,[r(n_0 + p_0)t] - \delta p(0)} \qquad (8.11)$$

A similar expression can be obtained for $\delta n(t)$. It should be emphasized that (8.11) is valid for arbitrary values of n_0, p_0, and $\delta p(0)$.

A much simpler expression can be obtained for the decay of non-equilibrium carrier distributions in band-to-band recombination when the excess concentrations are much smaller than the equilibrium concentrations: that is, when $\delta p(0) = \delta n(0) \ll n_0 + p_0$. Under these conditions (8.11) reduces to

$$\delta p(t) = \delta p(0) \exp\,[-r(n_0 + p_0)t]$$

$$= \delta p(0) \exp\left(-\frac{t}{\tau_p}\right) \qquad (8.12)$$

where

$$\tau_p \equiv \frac{1}{r(n_0 + p_0)} \qquad (8.13)$$

Thus, for low excess concentrations, the excess electron and hole concentrations decay exponentially to zero with an *excess carrier lifetime* defined by (8.13). Notice also that the excess recombination rate from (8.6) is

$$R - G_t = \frac{\delta p(0)}{\tau_p} \exp\left(-\frac{t}{\tau_p}\right)$$

$$= \frac{\delta p(t)}{\tau_p} \qquad (8.14)$$

When the excess concentrations are much larger than the equilibrium concentrations, or $\delta p(0) = \delta n(0) \gg n_0 + p_0$, we have, from (8.9),

$$\delta p(t) = \frac{\delta p(0)}{1 + rt\, \delta p(0)} \qquad (8.15)$$

This is referred to as *quadratic recombination*. Equation (8.15) represents a hyperbolic decay of excess carriers. If we assume that the concept of an excess carrier lifetime is valid for quadratic band-to-band recombination, then from (8.6), (8.8), and (8.14), we have

$$R \pm G_t = -\frac{\partial}{\partial t}\,(\delta p) = r(\delta p)^2 = \frac{\delta p}{\tau_p} \qquad (8.16)$$

or

$$\tau_p(t) = \frac{1}{r\,\delta p(t)} \tag{8.17}$$

The explicit dependence of $\tau_p(t)$ on t depends on $\delta p(t)$, where $\delta p(t)$ is given by (8.15). Thus, in quadratic recombination the carrier lifetime varies with t until the linear recombination of (8.13) becomes dominant.

Combining (8.13) and (8.17), the excess carrier lifetimes, in general, for band-to-band recombination are

$$\frac{1}{\tau_n(t)} = \frac{1}{\tau_p(t)} = r[n_0 + p_0 + \delta p(t)] \tag{8.18}$$

since from (8.6), (8.7), and (8.14) the reciprocal lifetimes add.

8.1.2 Auger Recombination

For heavily doped, direct bandgap semiconductors one of the more important competing processes to band-to-band recombination (where photons are emitted) is Auger recombination. In this process the energy and momentum produced by electron–hole recombination are given to a second electron or hole, which then loses this energy and momentum by the emission of phonons. This process is illustrated in Fig. 8.1.

The analysis of Auger recombination is very similar to that for band-to-band recombination. In equilibrium the thermal Auger generation and

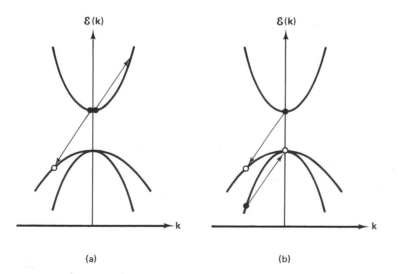

(a)   (b)

Figure 8.1   Auger recombination in which (a) an electron recombines with a hole giving energy and momentum to an electron and (b) an electron recombines with a hole giving energy and momentum to a hole.

recombination rates must be equal. For the process indicated in Fig. 8.1(a) these rates must be proportional to $n_0^2 p_0$ since two electrons and one hole are involved, while for the process shown in Fig. 8.1(b) these rates must depend on $n_0 p_0^2$. Thus, analogous to (8.3),

$$G_t = R = r_1 n_0^2 p_0 + r_2 n_0 p_0^2 \qquad (8.19)$$

With a steady-state generation of excess carriers, we have

$$G + G_t = R = r_1 n^2 p + r_2 n p^2 \qquad (8.20)$$

When the excess generation G is turned off, the excess carriers then recombine in a manner described by

$$R - G_t = r_1(n^2 p - n_0^2 p_0) + r_2(n p^2 - n_0 p_0^2)$$

$$= -\frac{\partial}{\partial t}(\delta p) \qquad (8.21)$$

With (8.21) and (8.14) and the assumption $\delta n = \delta p$ everywhere, we can define lifetimes for Auger recombination as

$$\frac{1}{\tau_n(t)} = \frac{1}{\tau_p(t)} = r_1[(2p_0 + n_0)n_0 + (2n_0 + p_0)\,\delta p(t) + \delta p^2(t)]$$

$$+ r_2[(2n_0 + p_0)p_0 + (2p_0 + n_0)\,\delta p(t) + \delta p^2(t)] \qquad (8.22)$$

8.1.3 Band-to-Impurity Recombination

Band-to-impurity recombination and trapping phenomena are important effects in semiconductors with indirect energy gaps and those with high concentrations of deep or tightly bound nonhydrogenic centers. The analysis of excess carrier trapping and recombination involving such centers is referred to as the Hall–Shockley–Read theory [R. N. Hall, *Phys. Rev. 87*, 387 (1952); W. Shockley and W. T. Read, *Phys. Rev. 87*, 835 (1952)].

The possible electron and hole transitions involving these deep centers are shown in Fig. 8.2. The process marked r_c is the capture of a conduction band electron by a center at $\mathscr{E}_t$ in the forbidden gap, while g_c is the emission of an electron from the center into the conduction band. The process labeled r_v is the emission of an electron from a center into an empty state in the valence band (or the capture of a valence band hole by a center), while g_v is the capture of a valence band electron by a center (or the emission of a hole from the center into the valence band). If the processes r_c and g_c have a higher probability of occurring than r_v and g_v, the center acts as an electron trap. Conversely, if r_v and g_v have a higher probability than r_c and g_c, the center acts as a hole trap. The center enhances recombination (acts as a recombination center) when r_c and r_v have a higher probability than g_c and g_v.

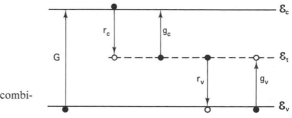

Figure 8.2 Band-to-impurity recombination and trapping processes.

Consider N_t centers at energy $\mathcal{E}_t$ with a probability of occupancy at equilibrium,

$$f_0 = \left[1 + \exp\left(\frac{\mathcal{E}_t - \zeta}{kT} \right) \right]^{-1} \tag{8.23}$$

For a steady-state generation rate of electron–hole pairs, G, the rate of electron capture by the center, r_c, must be proportional to the number of electrons in the conduction band and the number of empty centers,

$$r_c \propto nN_t(1 - f)$$

where f is the nonequilibrium distribution function. Physically, we expect the proportionality factor to have the form $v_{th}\sigma_n$, where

$$v_{th} = \sqrt{\frac{3kT}{m}} \simeq 10^7 \text{ cm/s} \tag{8.24}$$

The v_{th} in this equation is the thermal velocity of an electron and σ_n is the capture cross section of the center for an electron. Actually, the mass in (8.24) should be the effective mass. It is more convenient in the analysis, however, to absorb the ratio of the masses in σ_n and determine σ_n empirically. σ_n is of the order of atomic dimensions $\simeq 10^{-15}$ cm². The rate of electron capture is then

$$r_c = v_{th}\sigma_n nN_t(1 - f) \tag{8.25}$$

electrons/cm³·s. The rate of electron emission from the center into the conduction band, g_c, is proportional to the number of centers occupied, or

$$g_c = e_n N_t f \tag{8.26}$$

where e_n is the emission probability. In a similar manner, the rate of valence band hole capture, r_v, is proportional to p and the number of centers occupied, or

$$r_v = v_{th}\sigma_p pN_t f \tag{8.27}$$

The rate of hole emission from the center into the valence band, g_v, is proportional to the number of centers unoccupied,

$$g_v = e_p N_t(1 - f) \tag{8.28}$$

In the steady state, (8.25) to (8.28) are related by

$$r_c = g_c + G \tag{8.29}$$

and

$$r_v = g_v + G \tag{8.30}$$

The emission probabilities, e_n and e_p, can be evaluated in equilibrium. For $G = 0$, $r_c = g_c$ and from (8.25) and (8.26),

$$e_n = v_{th}\sigma_n n_0 \frac{1 - f_0}{f_0} \tag{8.31}$$

Using (8.23) and

$$n_0 = N_c \exp\left(\frac{\zeta - \mathscr{E}_c}{kT}\right) \tag{8.32}$$

we find that

$$e_n = v_{th}\sigma_n n_t \tag{8.33}$$

In this equation it is convenient to define a quantity

$$n_t \equiv N_c \exp\left(\frac{\mathscr{E}_t - \mathscr{E}_c}{kT}\right) \tag{8.34}$$

which is the concentration of electrons that would be in conduction band if $\zeta = \mathscr{E}_t$. From (8.30) in equilibrium $r_v = g_v$. With this and (8.27) and (8.28),

$$e_p = v_{th}\sigma_p p_0 \frac{f_0}{1 - f_0} \tag{8.35}$$

or

$$e_p = v_{th}\sigma_p p_t \tag{8.36}$$

Again, it is convenient to define a quantity

$$p_t \equiv N_v \exp\left(\frac{\mathscr{E}_v - \mathscr{E}_t}{kT}\right) \tag{8.37}$$

The nonequilibrium distribution function f can be obtained by eliminating G from (8.29) and (8.30), with the result

$$r_c - r_v = g_c - g_v \tag{8.38}$$

Using (8.25) to (8.28) in (8.38) and solving for f, we find that

$$f = \frac{\sigma_n n + \sigma_p p_t}{\sigma_n(n + n_t) + \sigma_p(p + p_t)} \tag{8.39}$$

With this expression the excess recombination rate, $R - G_t$, for band-to-impurity recombination can be obtained.

The excess recombination rate can be deduced from (8.29) and (8.30) to be

$$R - G_t = G = r_c - g_c = r_v - g_v \qquad (8.40)$$

Inserting (8.25) and (8.26) or (8.27) and (8.28), and using (8.39) for f, the excess recombination rate is

$$R - G_t = \frac{\sigma_n \sigma_p v_{th} N_t (np - n_i^2)}{\sigma_n (n + n_t) + \sigma_p (p + p_t)} \qquad (8.41)$$

With the definitions,

$$\frac{1}{\tau_{p0}} \equiv \sigma_p v_{th} N_t \quad \text{and} \quad \frac{1}{\tau_{n0}} \equiv \sigma_n v_{th} N_t \qquad (8.42)$$

(8.41) becomes

$$R - G_t = \frac{np - n_i^2}{\tau_{p0}(n + n_t) + \tau_{n0}(p + p_t)} = -\frac{\partial}{\partial_t}(\delta p) \qquad (8.43)$$

If the assumption is made that $\delta n = \delta p$ everywhere, we can also obtain excess carrier lifetimes for band-to-impurity recombination. With this assumption,

$$\frac{1}{\tau_n(t)} = \frac{1}{\tau_p(t)} = \frac{n_0 + p_0 + \delta p(t)}{\tau_{p0}[n_0 + n_t + \delta p(t)] + \tau_{n0}[p_0 + p_t + \delta p(t)]} \qquad (8.44)$$

where an exact solution for $\delta p(t)$ can be found from (8.43).

Figure 8.3 shows schematically the variation of excess carrier lifetime with doping from (8.44) assuming that $\delta p \ll n_0 + p_0$. The carrier lifetime has a maximum of $\tau_{p0} + \tau_{n0}$ when ζ is near $\mathscr{E}_t$. Under these conditions the

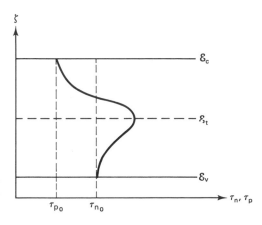

Figure 8.3 Variation of the excess carrier lifetimes with doping for band-to-impurity recombination.

centers are about half full, and the capture of either electrons or holes does not dominate the recombination process. When the material is doped n-type, however, ζ moves toward $\mathscr{E}_c$ and the lifetime approaches τ_{p0}. Under these conditions the centers are mostly full, and the hole capture process becomes more efficient, dominating the process. When the sample is p-type, the lifetime approaches τ_{n0} for similar reasons.

8.1.4 Surface Recombination

In addition to the bulk recombination mechanisms, recombination at the surface of a semiconductor is usually enhanced by surface states. This has important consequences for the formation of ohmic or rectifying contacts. For ohmic contacts it is desirable to have a high surface recombination rate so that it is not possible to sustain excess carrier distributions in their vicinity. This can usually be achieved with a rough, abraded surface. For rectifying contacts the opposite is true and surface recombination can be minimized with a highly polished, damage-free surface.

Consider the surface at $x = 0$ in Fig. 8.4 and assume that there are N_{ts} surface states per cm^2 at energy $\mathscr{E}_t$ in the forbidden gap. From (8.41) the excess recombination rate per unit area is

$$(R - G_t)_s = \frac{\sigma_n \sigma_p v_{\text{th}} N_{ts}(n_0 + p_0 + \delta p)\, \delta p(0)}{\sigma_n(n + n_t) + \sigma_p(p + p_t)} \tag{8.45}$$

where all the concentrations are evaluated at $x = 0$. For hole conservation, the number of holes recombining per unit surface area per second, which is $(R - G_t)_s$, must be equal to the number per unit area per second arriving at the surface from the bulk. If we assume no electric fields in the surface

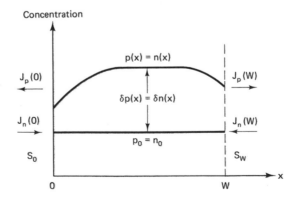

Figure 8.4   Diagram of the carrier concentrations and surface currents produced by surface recombination. For simplicity of illustration the sample is assumed to be intrinsic.

region, the hole current at the surface is

$$J_p(0) = -qD_p \frac{\partial p(0)}{\partial x}$$

$$= -qD_p \frac{\partial}{\partial x} [\delta p(0)] \tag{8.46}$$

The number of holes arriving per unit area per second is, therefore,

$$D_p \frac{\partial}{\partial x} [\delta p(0)] = \left[\frac{\sigma_n \sigma_p v_{th} N_{ts}(n_0 + p_0 + \delta p)}{\sigma_n(n + n_t) + \sigma_p(p + p_t)} \right] \delta p(0) \tag{8.47}$$

Notice that the quantity in the square brackets on the right-hand side of (8.47) has units of cm/s. This is referred to as the *surface recombination velocity*, which we have labeled as S_0 in Fig. 8.4. From (8.47),

$$D_p \frac{\partial}{\partial x} [\delta p(0)] = S_0 \, \delta p(0) \tag{8.48}$$

and with (8.46),

$$J_p(0) = -qS_0 \, \delta p(0) \tag{8.49}$$

Although the surface recombination velocity has units of velocity, it is regarded as a positive scalar quantity. The minus sign in (8.49) reflects the fact that S_0 and $\delta p(0)$ are positive while $J_p(0)$, as shown in Fig. 8.4, is in the negative x-direction. A similar analysis for $J_n(0)$ gives

$$J_n(0) = qS_0 \, \delta n(0) \tag{8.50}$$

On the other hand, for the surface at $x = W$,

$$J_p(W) = qS_W \, \delta p(W) \tag{8.51}$$

while

$$J_n(W) = -qS_W \, \delta n(W) \tag{8.52}$$

If the excess concentrations on either surface happen to be negative, the concentration gradients and thus the current directions would be reversed. Notice that in all cases the electron and hole currents at the surface are equal and opposite. Thus there is no net current flow in the sample.

8.2 CARRIER CONTINUITY

In the preceding section the effect of current flow on the distribution of charge carriers was neglected. We will now examine recombination with current flow. Consider a closed surface A around an arbitrary volume V in a semiconductor. The number of holes in this volume can be increased by the generation of holes in the volume and by the flow of holes into the volume through the enclosing area. On the other hand, the number of holes can be decreased by recombination in the volume and by a flow of holes out of the volume through the enclosing area. The net rate of increase in holes in this volume is therefore given by

$$\int_V \frac{\partial p(\mathbf{r}, t)}{\partial t} \, dV = \int_V [G_t + G(\mathbf{r}, t) - R(\mathbf{r}, t)] \, dV - \frac{1}{q} \oint_A \mathbf{J}_p(\mathbf{r}, t) \cdot d\mathbf{A}$$

(8.53)

The minus sign in the last term on the right-hand side of this equation is required because $d\mathbf{A}$ is a vector normal to the surface in the outward direction. $G(\mathbf{r}, t)$ in the first term on the right-hand side includes all generation processes except thermal. From the divergence theorem, however,

$$\oint_A \mathbf{J}_p(\mathbf{r}, t) \cdot d\mathbf{A} = \int_V \nabla \cdot \mathbf{J}_p(\mathbf{r}, t) \, dV$$

(8.54)

so that

$$\int_V \frac{\partial p(\mathbf{r}, t)}{\partial t} = \int_V \left[G_t + G(\mathbf{r}, t) - R(\mathbf{r}, t) - \frac{1}{q} \nabla \cdot \mathbf{J}_p(\mathbf{r}, t) \right] dV \quad (8.55)$$

Since this result must hold for any volume, we have

$$\frac{\partial p(\mathbf{r}, t)}{\partial t} = G_t + G(\mathbf{r}, t) - R(\mathbf{r}, t) - \frac{1}{q} \nabla \cdot \mathbf{J}_p(\mathbf{r}, t)$$

(8.56)

Using (8.1) and a suitably defined excess carrier lifetime, we obtain the *continuity equation for holes*,

$$\frac{\partial}{\partial t} [\delta p(\mathbf{r}, t)] = -\frac{1}{q} \nabla \cdot \mathbf{J}_p(\mathbf{r}, t) - \frac{\delta p(\mathbf{r}, t)}{\tau_p(t)} + G(\mathbf{r}, t)$$

(8.57)

Taking into account the fact that a net electric current flow out of the volume increases the number of electrons in the volume, we obtain a similar *continuity equation for electrons*,

$$\frac{\partial}{\partial t} [\delta n(\mathbf{r}, t)] = +\frac{1}{q} \nabla \cdot \mathbf{J}_n(\mathbf{r}, t) - \frac{\delta n(\mathbf{r}, t)}{\tau_n(t)} + G(\mathbf{r}, t)$$

(8.58)

With these continuity equations, we can now examine the assumption used in the preceding section that $\delta p(\mathbf{r}, t) = \delta n(\mathbf{r}, t)$ everywhere. For this purpose consider a uniformly doped n-type sample and assume that it is possible to inject a uniform excess majority carrier density $\delta n(t)$ so that $\delta n(t) \ll n_0$. Under these conditions, Poisson's equation, the current equation, and the continuity equation are

$$\nabla \cdot \mathbf{E}(\mathbf{r}, t) = \frac{-q}{\epsilon} \delta n(t)$$

$$\mathbf{J}_n(\mathbf{r}, t) = q\mu_n n(t)\mathbf{E}(\mathbf{r}, t) \tag{8.59}$$

$$\frac{d}{dt}[\delta n(t)] = + \frac{1}{q}\nabla \cdot \mathbf{J}_n(\mathbf{r}, t) - \frac{\delta n(t)}{\tau_n}$$

There is no diffusion term in the current equation because the concentration of electrons is uniform. Substituting the current equation and then Poisson's equation into the continuity equation, the time dependence of the excess majority carrier concentration is governed by

$$\frac{d}{dt}[\delta n(t)] = - \left[\frac{q\mu_n n_0}{\epsilon} + \frac{1}{\tau_n}\right] \delta n(t) \tag{8.60}$$

Notice that $q\mu_n n_0$ is the conductivity, σ_n, of the sample. If the injection process is turned off at $t = 0$, the excess majority carriers decay according to

$$\delta n(t) = \delta n(0) \exp\left[-\left(\frac{\sigma_n}{\epsilon} + \frac{1}{\tau_n}\right)t\right] \tag{8.61}$$

Thus the excess majority carriers decrease with two characteristic times: τ_n due to recombination and ϵ/σ_n, which is referred to as the *dielectric relaxation time*,

$$\tau_d \equiv \frac{\epsilon}{\sigma_n} = \frac{\epsilon}{q\mu_n n_0} \tag{8.62}$$

The decay process is dominated by the smaller of the two characteristic times. Since n_0 is the majority carrier concentration, in most cases τ_d is much smaller than τ_n and the excess majority carriers are removed from the sample at the dielectric relaxation time before they have time to recombine. Physically, what happens is that the excess majority carriers produce an electric field which terminates on the contacts or surfaces. This induced electric field then sweeps the excess majority carriers out of the sample at the dielectric relaxation time, reducing the field to zero.

Let us now continue our argument by assuming that it is possible to inject a uniform concentration of excess *minority* carriers, $\delta p(t)$, into the same uniform n-type sample. Set $p_0 \ll \delta p(t) \ll n_0$. The relevant equations are

$$\nabla \cdot \mathbf{E}(\mathbf{r}, t) = \frac{q}{\epsilon} \delta p(t)$$

$$\mathbf{J}_p(\mathbf{r}, t) = q\mu_p p(t)\mathbf{E}(\mathbf{r}, t) \tag{8.63}$$

$$\frac{d}{dt}[\delta p(t)] = -\frac{1}{q}\nabla \cdot \mathbf{J}_p(\mathbf{r}, t) - \frac{\delta p(t)}{\tau_p}$$

Again, there is no diffusion term because the concentrations are uniform. Performing substitutions as before gives us

$$\frac{d}{dt}[\delta p(t)] = -\left[\frac{q\mu_p \delta p(t)}{\epsilon} + \frac{1}{\tau_p}\right]\delta p(t) \tag{8.64}$$

This equation has the same form as (8.8), with the solution being

$$\delta p(t) = \frac{\delta p(0)}{(1 + \tau_p/\tau_d)\exp(t/\tau_p) - \tau_p/\tau_d} \tag{8.65}$$

In this equation we have defined a dielectric relaxation time for the excess minority carriers as

$$\tau_d \equiv \frac{\epsilon}{q\mu_p \,\delta p(0)} \tag{8.66}$$

Since $\delta p(0)$ is relatively small, τ_d for the excess minority carriers is usually much larger than τ_p and (8.65) becomes

$$\delta p(t) = \delta p(0)\exp\left(-\frac{t}{\tau_p}\right) \tag{8.67}$$

Thus the excess minority carriers decay by recombination.

With what do they recombine? When we consider the first part of this discussion, the answer is apparent. The injection of excess minority holes produces an electric field that terminates on the contacts or surfaces. This electric field causes excess majority electrons to move into the sample at their very short dielectric relaxation time. After this short time, $\delta n = \delta p$ internally everywhere and the excess electrons and holes decay at the longer excess carrier lifetime. This is the concept of internal *space-charge neutrality*. Uniform materials are usually space-charge neutral for times greater than the majority carrier dielectric relaxation time. It should be emphasized, however, that materials which have nonuniform doping or energy band structure can have strong space-charge regions.

As an indication of the relative sizes of these characteristic dielectric relaxation and recombination times, consider injecting 10^{12} cm^{-3} excess holes into an n-type GaAs sample with $n_0 = 10^{14}$ cm^{-3}. We will assume the following material parameters: $\tau_n \approx \tau_p \approx 10^{-9}$ s; $n_i \approx 10^7$ cm^{-3}; $\epsilon_r = 12.5$; $\mu_n = 8 \times 10^3$ cm^2/V·s; and $\mu_p = 5 \times 10^2$ cm^2/V·s.

$$p_0 = \frac{n_i^2}{n_0} = \frac{(10^7)^2}{10^{14}} = 1 \text{ cm}^{-3}$$

Since $1 \ll 10^{12} \ll 10^{14}$ cm^{-3}, $p_0 \ll \delta p(0) \ll n_0$. For the majority carriers,

$$\tau_{dn} = \frac{\epsilon_r \epsilon_0}{q\mu_n n_0} = \frac{(12.5)(8.85 \times 10^{-14})}{(1.6 \times 10^{-19})(8 \times 10^3)(10^{14})} \approx 10^{-11} \text{ s}$$

For the excess minority carriers,

$$\tau_{d\delta p} = \frac{\epsilon_r \epsilon_0}{q\mu_p \, \delta p(0)} = \frac{(12.5)(8.85 \times 10^{-14})}{(1.6 \times 10^{-19})(500)(10^{12})} \simeq 10^{-8} \text{ s}$$

We then have the characteristic times in the order,

$$10^{-11} \ll 10^{-9} \ll 10^{-8} \quad \text{or} \quad \tau_{dn} \ll \tau_n = \tau_p \ll \tau_{d\delta p}$$

Thus, in this example the material is space-charge neutral for times greater than 10^{-11} s. Notice, however, that the characteristic times in this example fall within only three orders of magnitude. Thus it would not be very difficult to find another reasonable example where the ordering of the dielectric relaxation and recombination times was different.

8.3 DIFFUSION AND DRIFT

8.3.1 Ambipolar Diffusion

In the preceding section we considered the uniform injection of excess carriers in a semiconductor to illustrate in a simple manner the important concepts of *dielectric relaxation* and *space-charge neutrality*. Usually, excess carriers are injected into a sample with some spatial dependence. When excess electrons and holes are injected together, as they are in optical absorption, they diffuse in their concentration gradients. Because of differences in diffusion coefficients, some spatial separation of charge is obtained, which induces an internal electric field. This induced field is in a direction that retards the diffusion of the faster carrier and enhances the diffusion of the slower carrier, until the electrons and holes diffuse together. This process is called *ambipolar diffusion*.

Under these conditions, it is necessary to take into account both the drift and diffusion of carriers. The current equations are

$$\mathbf{J}_p(\mathbf{r}, t) = q\mu_p p(\mathbf{r}, t)\mathbf{E}(\mathbf{r}, t) - qD_p \, \nabla p(\mathbf{r}, t)$$
$$\mathbf{J}_n(\mathbf{r}, t) = q\mu_n n(\mathbf{r}, t)\mathbf{E}(\mathbf{r}, t) + qD_n \, \nabla n(\mathbf{r}, t) \tag{8.68}$$

and Poisson's equation is

$$\nabla \cdot \mathbf{E}(\mathbf{r}, t) = \frac{q}{\epsilon}[\delta p(\mathbf{r}, t) - \delta n(\mathbf{r}, t)] \tag{8.69}$$

We assume uniformly doped material, so

$$n(\mathbf{r}, t) = n_0 + \delta n(\mathbf{r}, t)$$
$$p(\mathbf{r}, t) = p_0 + \delta p(\mathbf{r}, t) \tag{8.70}$$

Using (8.68), (8.69) and (8.70) in the continuity equations (8.57) and (8.58), we obtain

$$\frac{\partial}{\partial t} (\delta p) = -\frac{1}{\tau_{dp}} (\delta p - \delta n) - \mu_p \mathbf{E} \cdot \boldsymbol{\nabla}(\delta p) + D_p \boldsymbol{\nabla}^2(\delta p) - \frac{\delta p}{\tau_p} + G \quad (8.71)$$

$$\frac{\partial}{\partial t} (\delta n) = +\frac{1}{\tau_{dn}} (\delta p - \delta n) + \mu_n \mathbf{E} \cdot \boldsymbol{\nabla}(\delta n) + D_n \boldsymbol{\nabla}^2(\delta n) - \frac{\delta n}{\tau_n} + G \quad (8.72)$$

where

$$\tau_{dp} \equiv \frac{\epsilon}{q\mu_p p} \quad \text{and} \quad \tau_{dn} \equiv \frac{\epsilon}{q\mu_n n} \quad (8.73)$$

In these equations the dependencies of $\delta p(\mathbf{r}, t)$ and $\mathbf{E}(\mathbf{r}, t)$ on $\mathbf{r}$ and t are implicit.

To solve these equations it is necessary to make some reasonable physical assumptions. Although the assumptions $\delta p \simeq \delta n \gg \delta p - \delta n$ are usually good, in (8.71) and (8.72) the terms $(\delta p - \delta n)$ are multiplied by the reciprocals of the dielectric relaxation times. Thus the first term on the right-hand side of both equations can be very large. Fortunately, this term can be eliminated by multiplying (8.71) by $n\mu_n$ and (8.72) by $p\mu_p$ and adding, to obtain

$$\frac{\partial}{\partial t} (\delta p) = D^* \boldsymbol{\nabla}^2(\delta p) + \mu^* \mathbf{E} \cdot \boldsymbol{\nabla}(\delta p) + G - \frac{\delta p}{\tau_p} \quad (8.74)$$

In this equation, which is called the *ambipolar diffusion equation,* we have assumed that $\delta p \simeq \delta n$ and used the Einstein relationship to define an effective diffusion coefficient and mobility,

$$D^* \equiv \frac{(p + n)D_p D_n}{pD_p + nD_n} \quad (8.75)$$

$$\mu^* \equiv \frac{(p_0 - n_0)\mu_p \mu_n}{p\mu_p + n\mu_n} \quad (8.76)$$

Equations (8.74), (8.75), and (8.76) describe the time evolution and propagation of excess electron and hole distributions which interact by way of drift and diffusion. In certain important practical cases these equations are greatly simplified. For example, under low injection conditions in an n-type sample, $n_0 \gg p_0$, $\delta p \simeq \delta n$; (8.75) reduces to $D^* = D_p$; and (8.76) becomes $\mu^* = -\mu_p$. In p-type material $p_0 \gg n_0$, $\delta n \simeq \delta p$, and $D^* = D_n$, $\mu^* = \mu_n$. Thus, for extrinsic material under low-level injection, the problem of the propagation of excess majority and minority carrier distributions is reduced to a consideration of excess minority carriers only. In intrinsic material under low-level injection, $p_0 = n_0 = n_i \gg \delta n \simeq \delta p$, and $D^* = 2D_p D_n/(D_p + D_n)$, $\mu^* = 0$. That is, the propagation of the excess carrier distributions are controlled by diffusion. It can also be seen that under high-level injection, propagation in extrinsic materials is controlled by diffusion.

8.3.2 Applied Electric Fields

In the discussion above the effect of the electric field induced by excess carriers was examined. We now examine the effects of an externally applied electric field. Consider a constant electric field applied to a uniform n-type sample normal to a surface at which electron–hole pairs are generated optically. The direction of the field is into the sample, so that the minority holes drift and diffuse into the bulk while the electrons are swept into the surface contact. Application of the hole current and continuity equations gives

$$\frac{\partial}{\partial t}(\delta p) = D_p \nabla^2 (\delta p) - \mu_p p \nabla \cdot \mathbf{E} - \mu_p \mathbf{E} \cdot \nabla(\delta p) + G - \frac{\delta p}{\tau_p} \qquad (8.77)$$

We will assume that a steady-state photon flux is strongly absorbed at the surface generating $\delta p(0)$ electron–hole pairs at $x = 0$. This allows us to set $G(x) = 0$ and $\partial(\delta p)/\partial t = 0$ in (8.77). We further assume that the field induced by the excess carriers is small in comparison to the constant applied field, so $\nabla \cdot \mathbf{E} = 0$. Because the field is normal to the surface, the problem is one-dimensional and (8.77) reduces to

$$\frac{d^2}{dx^2}(\delta p) - \frac{\mu_p E}{D_p}\frac{d}{dx}(\delta p) - \frac{\delta p}{D_p \tau_p} = 0 \qquad (8.78)$$

This equation has a solution of the form

$$\delta p(x) = \delta p(0) \exp(Ax) \qquad (8.79)$$

Substituting (8.79) into (8.78), we obtain

$$\left(A^2 - \frac{\mu_p E}{D_p} A - \frac{1}{D_p \tau_p}\right) \delta p(x) = 0 \qquad (8.80)$$

Since this must be valid for any value of $\delta p(x)$, the terms multiplying $\delta p(x)$ are equal to zero and

$$A = \frac{\mu_p E}{2D_p} \pm \left[\left(\frac{\mu_p E}{2D_p}\right)^2 + \frac{1}{D_p \tau_p}\right]^{1/2} \qquad (8.81)$$

Physically, we know that $p(x)$ must decrease as x increases, so the solution to this problem is

$$\delta p(x) = \delta p(0) \exp\left\{x \frac{\mu_p E}{2D_p} - x\left[\left(\frac{\mu_p E}{2D_p}\right)^2 + \frac{1}{D_p \tau_p}\right]^{1/2}\right\} \qquad (8.82)$$

First consider the distribution of excess holes when the field is zero. Equation (8.82) is then

$$\delta p(x) = \delta p(0) \exp\left[-\frac{x}{(D_p \tau_p)^{1/2}}\right] \qquad (8.83)$$

Thus the excess holes diffuse and decay by recombination with the majority electrons in a characteristic distance,

$$L_p \equiv (D_p\tau_p)^{1/2} \tag{8.84}$$

This characteristic distance is referred to as the *hole diffusion length*. The behavior of the excess holes with zero field is shown in Fig. 8.5(a). We next consider the distribution of excess holes for large electric fields. For this purpose (8.82) is manipulated to have the form

$$\delta p(x) = \delta p(0) \exp\left\{\frac{x}{L_p}[\alpha - (\alpha^2 + 1)^{1/2}]\right\} \tag{8.85}$$

where

$$\alpha \equiv \frac{\mu_p\tau_p E}{2L_p} \tag{8.86}$$

For E large or $\alpha \gg 1$,

$$(\alpha^2 + 1)^{1/2} = \alpha + \frac{1}{2\alpha} - \frac{1}{8\alpha^3} + \cdots$$

$$\simeq \alpha + \frac{1}{2\alpha} \tag{8.87}$$

Using the approximation of (8.87) in (8.85),

$$\delta p(x) = \delta p(0) \exp\left(\frac{-x}{\mu_p\tau_p E}\right) \tag{8.88}$$

Here we find drift and recombination of excess holes with a characteristic length $\mu_p\tau_p E$, as illustrated in Fig. 8.5(b).

Although in general it is necessary to take into account the effects of both diffusion and drift, some problems can be simplified by comparing the

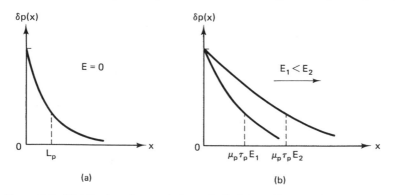

Figure 8.5 Distribution of excess holes optically injected by a steady-state source (a) when the electric field is zero and (b) when the electric field is large.

characteristic lengths for diffusion, L_p, and drift, $\mu_p\tau_pE$. Drift can be ignored and a solution similar to (8.83) can be obtained if $\mu_p\tau_pE \ll L_p$, while diffusion can be neglected and a solution like (8.88) can be used for $L_p \ll \mu_p\tau_pE$.

The problem above can also be analyzed for a pulsed radiation source generating $\delta p(0, 0)$ electron–hole pairs at $x = 0$ and $t = 0$. The partial differential equation that determines the distribution of excess holes is

$$D_p \frac{\partial^2}{\partial x^2}(\delta p) - \mu_p E \frac{\partial}{\partial x}(\delta p) - \frac{\delta p}{\tau_p} = \frac{\partial}{\partial t}(\delta p) \tag{8.89}$$

Taking the Laplace transform with the initial condition $\delta p(0, 0)$ the solution is found to be a Gaussian distribution,

$$\delta p(x, t) = \frac{\delta p(0, 0)}{(4\pi D_p t)^{1/2}} \exp\left[\frac{-(x - \mu_p Et)^2}{4D_p t} - \frac{t}{\tau_p}\right] \tag{8.90}$$

The total excess hole concentration in the sample at time t is

$$\delta P(t) = \int_0^\infty \delta p(x, t)\, dx = \delta p(0, 0) \exp\left(-\frac{t}{\tau_p}\right) \tag{8.91}$$

The distributions of excess holes at several times without and with an applied field are shown in Fig. 8.6.

For a general pulsed source $\delta p(\xi, \tau)$ distributed in space and time, the solution would be

$$\delta p(x, t) = \int_0^t \int_0^\infty \frac{\delta p(\xi, \tau)}{[4\pi D_p(t - \tau)]^{1/2}}$$
$$\cdot \exp\left\{\frac{-[(x - \xi) - \mu E(t - \tau)]^2}{4D_p(t - \tau)} - \frac{t - \tau}{\tau_p}\right\} d\xi\, d\tau \tag{8.92}$$

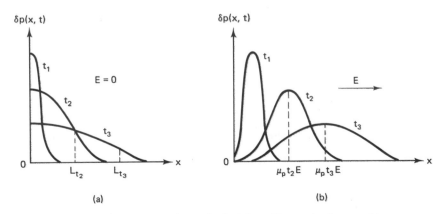

Figure 8.6   Distribution of excess holes optically injected by a pulsed source (a) when the electric field is zero and (b) when the electric field is large.

8.3.3 Surface Recombination

In most practical situations when excess carriers are created in a sample by light absorbed at or near a surface, the distribution of excess carriers is strongly affected by surface recombination. Consider the same problem as in the preceding section, except that now we will relax the assumption of strongly absorbed light and provide for a nonuniform generation of electron–hole pairs in the bulk of the semiconductor.

Let a steady-state light source with power density P and energy $\hbar\omega$ be incident on the surface of the n-type sample with a flux density $\phi_q = P/\hbar\omega$ photons per $cm^2 \cdot s$. The surface has a reflectivity $R(\hbar\omega)$ and a surface recombination velocity S. The optical absorption coefficient of the semiconductor is $\alpha(\hbar\omega)$. With these parameters the photon flux at any point x in the sample is

$$\phi(x) = (1 - R)\phi_q \exp(-\alpha x) \equiv \phi \exp(-\alpha x) \qquad (8.93)$$

The equation governing the distribution of excess carriers is

$$\frac{d^2}{dx^2}(\delta p) - \frac{\mu_p E}{D_p}\frac{d}{dx}(\delta p) - \frac{\delta p}{L_p^2} + \frac{G}{D_p} = 0 \qquad (8.94)$$

For simplicity, we will assume that $\mu_p \tau_p E \ll L_p$, and the drift term can be neglected. If each proton in the sample creates one electron–hole pair, the generation rate is

$$G(x) = \alpha\phi(x) = \alpha\phi \exp(-\alpha x) \qquad (8.95)$$

The equation to be solved is therefore

$$\frac{d^2}{dx^2}(\delta p) - \frac{\delta p}{L_p^2} = -\frac{\alpha\phi}{D_p}\exp(-\alpha x) \qquad (8.96)$$

with the surface recombination boundary condition given by (8.49), $J_p(0) = -qS\,\delta p(0)$.

The solution to (8.96) has the form

$$\delta p(x) = A \exp\left(-\frac{x}{L_p}\right) + B \exp(-\alpha x) \qquad (8.97)$$

where A is evaluated from the boundary condition and B is chosen to satisfy the differential equation. Substituting (8.97) into (8.96), we find that

$$B = \frac{\alpha\phi\tau_p}{1 - (\alpha L_p)^2} \qquad (8.98)$$

The effect of the electric field on the excess minority carriers has been assumed to be small, so the current equation has only a diffusion component,

and

$$-\frac{J_p(0)}{q} = D_p \frac{d}{dx} [\delta p(0)] = S \, \delta p(0) \qquad (8.99)$$

is the boundary condition. Using (8.97) and its derivative evaluated at $x = 0$ in (8.99), we arrive at the solution,

$$\delta p(x) = \frac{\alpha \phi \tau_p}{1 - (\alpha L_p)^2} \left[\exp(-\alpha x) - \frac{s + \alpha D_p}{s + D_p/L_p} \exp\left(-\frac{x}{L_p}\right) \right] \qquad (8.100)$$

Figure 8.7 illustrates the results of (8.100) under several limiting assumptions. In Fig. 8.7(a) and (b) the surface recombination velocity is assumed to be zero, so there are surface effects. In part (a) the light is assumed to be weakly absorbed and the excess hole distribution simply reflects the photon absorption profile in the sample. Part (b) is for strongly absorbed light and shows the diffusion process. In Fig. 8.7(c) and (d) the surface recombination is assumed to be infinite. As can be seen, this reduces the excess hole concentration at the surface to zero and removes most of the

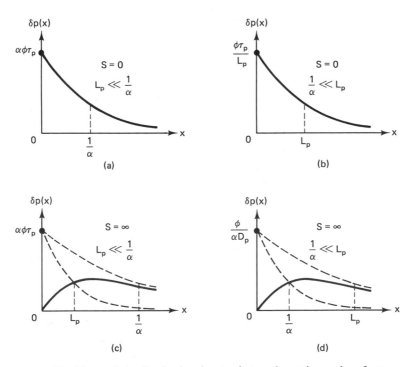

Figure 8.7 Excess hole distribution due to photon absorption and surface recombination according to (8.100).

excess holes near the surface for both weakly (c) and strongly (d) absorbed radiation.

8.4 NONUNIFORM MATERIAL

In the preceding discussions we have examined in some detail excess carrier distributions in materials with uniform doping and energy band structure. We assumed that the excess carriers were created by incident radiation or by some other nonequilibrium process. In materials with nonuniform doping or energy bands, however, excess carriers and space-charge regions can be obtained in equilibrium with no externally applied forces or excitations. These excess carriers and space-charge regions are produced by the transfer of charge from one region of a nonuniform semiconductor to another. The transfer of charge is required, thermodynamically, to minimize the free energy of the system and produce a constant electrochemical potential.

8.4.1 Spatially Varying Energy Bands

Consider a semiconductor, with a band structure determined by the crystal potential, which is subjected to internal or external force fields. These force fields can act in two ways: (1) they can affect the crystal potential of the unit cell, thereby changing the band structure; or (2) they can produce a potential energy which is superimposed on the energy band structure.

Let $E_{n0}(\mathbf{k})$ be the band structure of a uniform semiconductor in the absence of force fields other than the crystal potential, where n is the band index, and let

$$E_n(\mathbf{k}, \mathbf{r}) = E_{n0}(\mathbf{k}) + \delta E_n(\mathbf{r}) \qquad (8.101)$$

be the band structure in the presence of force fields. $\delta E_n(\mathbf{r})$ is thus the change in band structure produced by the forces. If we let $\Psi(r)$ be the potential energy of the forces, the total energy presented in the usual energy band diagram is

$$\mathscr{E}_n(\mathbf{k}, \mathbf{r}) = E_n(\mathbf{k}, \mathbf{r}) + \Psi(\mathbf{r}) \qquad (8.102)$$

Equation (8.102) provides for both possible effects of a force field on an otherwise uniform semiconductor.

If we consider a semiconductor that is nonuniform in the absence of force fields, however, the unperturbed band structure must be $E_{n0}(\mathbf{k}, \mathbf{r})$. This could be achieved by inserting different basis atoms in the primitive unit cells. The position dependence of the band structure is then due to the crystal potential which varies from one unit cell to another. Under these conditions, however, the periodicity of the crystal is obviously lost and the concept of an energy band structure in the usual sense is not apparent. On the other hand, if we define the energy band structure $E_{n0}(\mathbf{k}, \mathbf{r})$ as that which would be obtained if the unit cell at $\mathbf{r}$ were periodically repeated, we can include the effects of force fields in the same manner as above for uniform materials.

This treatment of nonuniform energy bands has been justified more rigorously [A. H. Marshak and K. M. van Vliet, *Solid-State Electron. 21*, 417 (1978)] in the following manner. The exact Schrödinger equation for the crystal in a force field is

$$\mathbf{H}\phi(\mathbf{r}) = \mathcal{E}\phi(\mathbf{r}) \qquad (8.103)$$

where as in (3.10) the Hamiltonian is given by

$$\mathbf{H} = -\frac{\hbar^2}{2m}\nabla^2 + U(\mathbf{r}) + \Psi(\mathbf{r}) \qquad (8.104)$$

In (8.104) $U(\mathbf{r})$ is the crystal potential energy and $\Psi(\mathbf{r})$ is the potential energy of the force fields, which is assumed to vary slowly over a unit cell.

Equation (8.103) can be transformed into an equivalent Schrödinger equation,

$$\mathbf{H}_n A(\mathbf{r}) = \mathcal{E}_n A(\mathbf{r}) \qquad (8.105)$$

in a manner similar to the way (3.9) was transformed into (3.31). $A(\mathbf{r})$ are Wannier functions. In (8.105), however, the effective Hamiltonian is

$$\mathbf{H}_n = -\frac{\hbar^2}{2m_n^*(\mathbf{r})}\nabla^2 + \Psi(\mathbf{r}) \qquad (8.106)$$

where n is the band index. Equation (8.106) has the form

$$\mathbf{H}_n = E_n(-i\nabla, \mathbf{r}) + \Psi(\mathbf{r}) \qquad (8.107)$$

where the $\mathbf{r}$ dependence is represented by the spatially varying effective mass $m_n^*(\mathbf{r})$ of band n. Using the correspondence principle, we replace the operator $-i\nabla$ with $\mathbf{k}$ in (8.107) to obtain

$$\mathbf{H}_n(\mathbf{k}, \mathbf{r}) = E_n(\mathbf{k}, \mathbf{r}) + \Psi(\mathbf{r}) = \mathcal{E}_n(\mathbf{k}, \mathbf{r}) \qquad (8.108)$$

where $\mathcal{E}_n(\mathbf{k}, \mathbf{r})$ is the total energy of band n.

8.4.2 Applied Forces

Let us consider what effects some commonly encountered force fields have on energy band structure. If a nonuniform strain field is applied to a uniform sample with band structure $E_{n0}(\mathbf{k})$, it produces a nonuniform lattice spacing and modifies the energy bands by $\delta E_n(\mathbf{r})$. In addition, the potential energy of the strain field $\Psi(\mathbf{r})$ is added to the band structure to produce a total energy,

$$\mathscr{E}_n(\mathbf{k}, \mathbf{r}) = E_{n0}(\mathbf{k}) + \delta E_n(\mathbf{r}) + \Psi(\mathbf{r}) \qquad (8.109)$$

Electric fields can also produce changes in lattice spacing and modify the energy bands. This can occur in semiconductors which are piezoelectric. Usually, these changes are small and $E_n(\mathbf{k}) = E_{n0}(\mathbf{k})$ for uniform samples. The potential energy of the electrostatic field $\Psi(\mathbf{r}) = -q\psi(\mathbf{r})$, however, is added to the force-free energy bands to produce a total energy,

$$\mathscr{E}_n(\mathbf{k}, \mathbf{r}) = E_n(\mathbf{k}) - q\psi(\mathbf{r}) \qquad (8.110)$$

for materials that are uniform in the absence of force fields. For materials that are nonuniform in the absence of force fields, the corresponding relationship is given by

$$\mathscr{E}_n(\mathbf{k}, \mathbf{r}) = E_n(\mathbf{k}, \mathbf{r}) - q\psi(\mathbf{r}) \qquad (8.111)$$

Equations (8.108) or (8.110) and (8.111) provide a complete state-space description of the energy bands. In reciprocal or k-space, (8.111) becomes

$$\mathscr{E}_n(\mathbf{k}) = E_n(\mathbf{k}, \mathbf{r}_0) - q\psi(\mathbf{r}_0) \qquad (8.112)$$

where $\mathbf{r}_0$ is some fixed point in real space. Thus the electrostatic potential produces no structure in $\mathscr{E}(\mathbf{k})$ diagrams. It simply shifts the entire diagram relative to the reference energy. In real or r-space, (8.111) is

$$\mathscr{E}_n(\mathbf{r}) = E_n(\mathbf{k}_e\mathbf{r}) - q\psi(\mathbf{r}) \qquad (8.113)$$

where $\mathbf{k}_e$ is some fixed point in reciprocal space. Usually, $\mathbf{k}_e$ is located at band extrema, so for the lowest-energy conduction band and highest valence band, (8.113) becomes

$$\begin{aligned} \mathscr{E}_c(\mathbf{r}) &= E_c(\mathbf{r}) - q\psi(\mathbf{r}) \\ \mathscr{E}_v(\mathbf{r}) &= E_v(\mathbf{r}) - q\psi(\mathbf{r}) \end{aligned} \qquad (8.114)$$

respectively. We see, then, that an electrostatic potential in an $\mathscr{E}(\mathbf{r})$ diagram simply shifts all the force-free energy levels, including the band extrema, by $-q\psi(\mathbf{r})$.

As an example, we consider the effect of a constant electric field on the one-electron energy bands of a uniform material. In the absence of the electric field the energy band diagram is as shown in Fig. 8.8(a). The material is uniform, so there is no internal electrostatic potential, and the energy levels are horizontal. In this diagram several quantities have been defined. First, we established a reference state, E_0, for the energy band diagram. This reference, E_0, is the *infinite or force-free vacuum* level. It is defined as the energy an electron would have when it is completely free from the influence of the material or forces applied to the material. $\mathscr{E}_l$ is the *local vacuum level* or vacuum level in the presence of forces. It is the energy an electron would have at rest outside the material. In the absence of forces

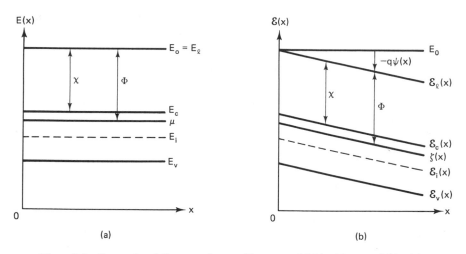

Figure 8.8   Energy band diagrams for a uniform material (a) without and (b) with a constant applied electric field. The diagrams are referenced to E_0, the field-free or infinite vacuum level at $x = 0$. χ is the electron affinity and Φ is the work function of the material.

$E_0 = \mathscr{E}_l = E_l$. χ is referred to as the *electron affinity*. It is the energy required to excite an electron from the bottom of the conduction band to the local vacuum level. The electron affinity takes into account work against the image force (many-body effect) and any surface dipole moments, since these effects are not accounted for in the one-electron energy bands. Φ is the *work function* of the semiconductor. It is the change in free energy of the material when an electron is added or taken away from the system. Since the work function changes with doping in semiconductors, it is more convenient to use the electron affinity to characterize the material.

Figure 8.8(b) shows what happens to the energy bands when a constant electric field is applied. We have *arbitrarily* set the zero for electrostatic potential at $x = 0$. Since the electric field is constant, the potential increases linearly with x. We have shown $-q\psi(x)$, which has a negative gradient, so $\psi(x)$ has a positive gradient. Since

$$\mathbf{E} = -\nabla\psi \tag{8.115}$$

the electric field is in the negative x direction. Notice that the change in energy levels between Fig. 8.8(a) and (b) is just due to $-q\psi(x)$. To construct Fig. 8.8 we used (8.114):

$$\zeta(x) = \mu - q\psi(x) \tag{8.116}$$

which is the definition of the electrochemical potential; and

$$\mathscr{E}_l(x) = E_0 - q\psi(x) \tag{8.117}$$

which is the definition of the local vacuum level.

The motion of electrons and holes in an energy band diagram with a constant electric field is shown in Fig. 8.9. The holes are accelerated in the direction of the electric field while the electrons move in the opposite direction. Both maintain constant total energy $\mathscr{E}$, exchanging potential energy $-q\psi$ for kinetic energy $\frac{1}{2}m^*v^2$, until they suffer a collision. During a collision they lose part or all of their total energy to the scattering center, and then begin again. This nonequilibrium process continues repeatedly while the external field is applied. Notice, however, that the change carriers have their equilibrium concentrations during this process.

8.4.3 Nonuniform Doping

The simplest situation in which excess carriers are produced in equilibrium is for nonuniformly doped semiconductors. The procedure by which this takes place is illustrated in Fig. 8.10. Figure 8.10(a) shows two samples with the same energy band structure but different doping, which are not in electrical contact. The diagram indicates a p-type sample on the left and an n-type sample on the right. Both materials are in equilibrium with equilibrium concentrations of majority carrier holes and electrons produced by the fixed negatively charged acceptors and positively charged donors, respectively. Since each sample is in equilibrium with no internal electrostatic potential, the chemical potentials are uniform. The hole and electron concentrations in the p-type material are

$$p_{p0} = n_i \exp\left(\frac{E_i - \mu_1}{kT}\right)$$

$$n_{p0} = n_i \exp\left(\frac{\mu_1 - E_i}{kT}\right) \tag{8.118}$$

and the electron and hole concentrations in the n-type material are

$$n_{n0} = n_i \exp\left(\frac{\mu_2 - E_i}{kT}\right)$$

$$p_{n0} = n_i \exp\left(\frac{E_i - \mu_2}{kT}\right) \tag{8.119}$$

The zero subscript indicates equilibrium concentrations in the absence of fields.

Figure 8.10(b) illustrates the energy band diagram after electrical contact is made. To mimimize the free energy, electrons are transferred from the n-type material to the p-type, producing a deficit of electrons on the n-type side of the junction and an excess of electrons (a deficit of holes due to recombination) on the p-type side. This transfer continues until the free energy is minimized. This corresponds to a uniform electrochemical potential, ζ, and equilibrium throughout the combined material. As indicated sche-

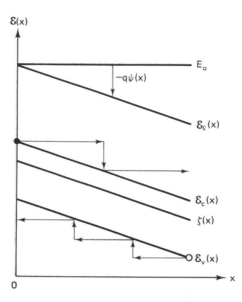

Figure 8.9 Diagram showing the motion
of electrons and holes with a constant
electric field in the negative x direction.

matically in Fig. 8.10(b), this transfer of electrons "uncovers" donor atoms
on the n-type side and acceptor atoms on the p-type side. This produces a
space-charge region and an internal potential in the vicinity of the junction
region of this *p-n homojunction*.

The equilibrium hole and electron concentrations now vary with dis-

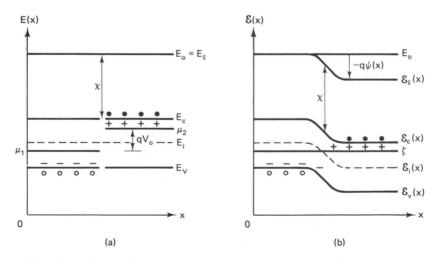

Figure 8.10 Energy band diagrams for nonuniform doping (a) before contact between two uniform regions and (b) after contact and electron transfer from right
to left to establish equilibrium.

tance as

$$p(x) = n_i \exp\left[\frac{\mathscr{E}_i(x) - \zeta}{kT}\right] \tag{8.120}$$

and

$$n(x) = n_i \exp\left[\frac{\zeta - \mathscr{E}_i(x)}{kT}\right) \tag{8.121}$$

If these two equations are multiplied together, we can see that the mass-action law is valid over the entire sample, including near the junction where there are equilibrium excess electron and hole distributions. Thus

$$p(x)n(x) = [p_{p0} + \delta p(x)][n_{p0} + \delta n(x)]$$

$$= [p_{n0} + \delta p(x)][n_{n0} + \delta n(x)]$$

$$= n_i^2 \tag{8.122}$$

for the whole semiconductor.

The spatial dependence of $p(x)$ and $n(x)$ can be related to the spatial dependence of the potential by using (8.113) in (8.120) and (8.121). Making this substitution yields

$$p(x) = n_i \exp\left[\frac{E_i - \zeta - q\psi(x)}{kT}\right] \tag{8.123}$$

$$n(x) = n_i \exp\left[\frac{\zeta - E_i + q\psi(x)}{kT}\right] \tag{8.124}$$

These equations can be related to (8.118) and (8.119) by referring to Fig. 8.10(a) and (b). In the figure we have arbitrarily set $\psi(0) = 0$. Thus, in the p-region,

$$E_i - \zeta = E_i - \mu_1 \tag{8.125}$$

In the n-region, we have, from Fig. 8.10,

$$E_i - \zeta = (\mu_2 - \mu_1) - (\mu_2 - E_i)$$

$$= qV_0 + E_i - \mu_2 \tag{8.126}$$

With these identities (8.123) and (8.124) become

$$p(x) = p_{p0} \exp\left[\frac{-q\psi(x)}{kT}\right] \tag{8.127}$$

$$= p_{n0} \exp\left[\frac{qV_0 - q\psi(x)}{kT}\right] \tag{8.128}$$

and

$$n(x) = n_{n0} \exp \left[\frac{q\psi(x) - qV_0}{kT} \right] \qquad (8.129)$$

$$= n_{p0} \exp \left[\frac{q\psi(x)}{kT} \right] \qquad (8.130)$$

8.4.4 Nonuniform Band Structure

A nonuniform energy band structure in and by itself does not produce excess carrier distributions. Figure 8.11(a) shows the energy band diagram for two materials with different electron affinity, χ, and energy gaps, $\mathscr{E}_g = E_c - E_v$, which are not in electrical contact. If the surface properties and doping in these two materials were such that the chemical potentials, μ_1 and μ_2, were equal, during and after contact there would be no transfer of charge. Under these conditions the energy band diagram would be the same before and after contact, with discontinuities in χ, E_c, E_i, and E_v. This would be a fortuitous occurrence, however, and in general there will be a difference in chemical potentials before contact, as indicated in Fig. 8.11(a).

After contact, with the difference in chemical potentials shown, electrons are transferred from right to left to minimize the free energy and produce a uniform electrochemical potential, ζ, in equilibrium. As indicated schematically for this *n-n heterojunction,* this electron transfer "uncovers"

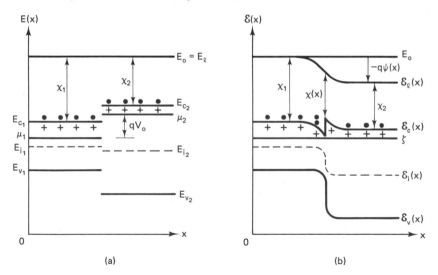

Figure 8.11 Energy band diagrams for nonuniform band structure and doping (a) before contact between two uniform regions and (b) after contact and electron transfer from right to left to establish equilibrium.

donors on the right-hand side and creates excess electrons on the left-hand side. Thus, an internal potential is induced in the same manner as for a p-n homojunction. For the heterojunction the discontinuities in χ, $\mathscr{E}_c$, $\mathscr{E}_i$, and $\mathscr{E}_v$ are the same as the discontinuities before electron transfer: that is, they are equal to $\chi_2 - \chi_1$, $E_{c2} - E_{c1}$, $E_{i2} - E_{i1}$, and $E_{v2} - E_{v1}$, respectively.

The spatial dependence of the electron and hole concentrations in regions 1 and 2 can be related to the spatial dependence of the electrostatic potential in the same manner as for a homojunction. From Fig. 8.11 and similar reasoning, we find that

$$n_1(x) = n_{10} \exp\left[\frac{q\psi(x)}{kT}\right] \tag{8.131}$$

$$p_1(x) = p_{10} \exp\left[\frac{-q\psi(x)}{kT}\right] \tag{8.132}$$

and

$$n_2(x) = n_{20} \exp\left[\frac{q\psi(x) - qV_0}{kT}\right] \tag{8.133}$$

$$p_2(x) = p_{20} \exp\left[\frac{qV_0 - q\psi(x)}{kT}\right] \tag{8.134}$$

Again, the mass-action concept is valid for the entire sample, except in this case the mass-action laws are different for the two regions because of different band structure. Notice, also, that there are discontinuities in the electron and hole concentrations,

$$\frac{n_1(x')}{n_2(x')} = \frac{n_{10}}{n_{20}} \exp\left(\frac{qV_0}{kT}\right) \tag{8.135}$$

$$\frac{p_1(x')}{p_2(x')} = \frac{p_{10}}{p_{20}} \exp\left(\frac{qV_0}{kT}\right) \tag{8.136}$$

at the heterojunction interface that do not exist in the homojunction.

8.5 THERMODYNAMIC FORCES

In Chapter 5 we found that the electron current density produced by an applied electric field in a uniform, isothermal, nondegenerate sample is given by

$$\mathbf{J}_n(\mathbf{r}) = qn\mu_n\mathbf{E}(\mathbf{r}) \tag{8.137}$$

This current is referred to as a *drift* current. When the sample is nonuniformly doped, an additional force term arises due to the electron concentration gradient. From (5.87) the electron current density in this case is

$$\mathbf{J}_n(\mathbf{r}) = qn(\mathbf{r})\mu_n\mathbf{E}(\mathbf{r}) + kT\mu_n\,\boldsymbol{\nabla}n(\mathbf{r}) \qquad (8.138)$$

The additional term in (8.138) is called a *diffusion current*. In general, however, the relationship between the current density and the forces in a nonuniform, isothermal semiconductor with an electrostatic potential is given
by (5.81),

$$\mathbf{J}_n(\mathbf{r}) = qn(\mathbf{r})\mu_n\mathbf{E}(\mathbf{r}) + n(\mathbf{r})\mu_n\,\boldsymbol{\nabla}\mu(r) \qquad (8.139)$$

    Equation (8.139) can be put in a more concise form using (8.115) and
(8.116) to obtain

$$\mathbf{J}_n(\mathbf{r}) = n(\mathbf{r})\mu_n\,\boldsymbol{\nabla}\zeta(\mathbf{r}) \qquad (8.140)$$

For hole current in the absence of excess carriers, we also have

$$\mathbf{J}_p(\mathbf{r}) = p(\mathbf{r})\mu_p\,\boldsymbol{\nabla}\zeta(\mathbf{r}) \qquad (8.141)$$

In these equations $n(\mathbf{r})$ and $p(\mathbf{r})$ are the total electron and hole concentrations. Although when n and p vary with distance, μ_n and μ_p also vary, the
mobility changes are small in comparison to the concentration changes. For
this reason we ignore changes in μ_n and μ_p. The point of this discussion,
however, is that when (8.140) and (8.141) are applied to materials with nonuniform energy band structure, additional current terms arise.

    Consider the relationship between the carrier concentrations of a non-
degenerate heterostructure and the electrochemical potential. From (4.73)
and (4.79) we find that

$$\zeta(\mathbf{r}) = \mathscr{E}_c(\mathbf{r}) + kT\ln\frac{n(\mathbf{r})}{N_c(\mathbf{r})} \qquad (8.142)$$

for electrons and

$$\zeta(r) = \mathscr{E}_v(\mathbf{r}) - kT\ln\frac{p(\mathbf{r})}{N_v(\mathbf{r})} \qquad (8.143)$$

for holes. For nonuniform band structures, (8.106) indicates that the effective
masses vary spatially, so both $N_c(\mathbf{r})$ and $N_v(\mathbf{r})$ can vary with $\mathbf{r}$. Substituting
(8.114) into (8.142) and (8.143), we have

$$\zeta(\mathbf{r}) = E_c(\mathbf{r}) - q\psi(\mathbf{r}) + kT\ln n(\mathbf{r}) - kT\ln N_c(\mathbf{r}) \qquad (8.144)$$

$$\zeta(\mathbf{r}) = E_v(\mathbf{r}) - q\psi(\mathbf{r}) - kT\ln p(\mathbf{r}) + kT\ln N_v(\mathbf{r}) \qquad (8.145)$$

    Notice that all four terms in (8.144) and (8.145) can depend on $\mathbf{r}$. Thus
the general electrochemical force $\boldsymbol{\nabla}\zeta(\mathbf{r})$ used to determine the current density
can also have four components. The first terms, $\boldsymbol{\nabla}E_c(\mathbf{r})$ or $\boldsymbol{\nabla}E_v(\mathbf{r})$, are due
to spatial variations in the crystal potential. These are essentially electric
field forces where the field is produced by the crystal potential. We will

refer to them as *crystal potential forces*. The second terms, $-q\,\nabla\psi(\mathbf{r})$, are the usual electric field forces produced by excess charge in the material or by an applied electrostatic potential. The third terms, $kT\,\nabla\ln n(\mathbf{r})$ and $-kT\,\nabla\ln p(\mathbf{r})$, are the usual diffusion forces caused by carrier concentration gradients. The fourth terms, $-kT\,\nabla\ln N_c(\mathbf{r})$ and $kT\,\nabla\ln N_v(\mathbf{r})$, however, are new and depend on the effective density of states. Notice that for both electrons and holes, these terms tend to force the carriers into regions with higher density of states. From the derivation of the carrier distributions in Chapter 4, we see that this process tends to maximize the entropy and, from (4.35), minimize the free energy of the system. Therefore, the fourth terms in (8.144) and (8.145) can be considered *entropy forces*.

As an example of the effects of these forces, consider the heterostructure shown in Fig. 8.12. The diagram of part (a) shows the nonequilibrium energy levels before charge transfer with no electric field, and part (b) indicates the equilibrium levels after charge transfer which produces an internal electric field. First look at the nonequilibrium charge transfer process in part (a). Consider the hole at x_1. There are no electric field or crystal field forces acting on the hole. Does it move? Yes! Both the diffusion force and the entropy force cause it to move to the right. Consider the electron at x_1. The crystal field and the diffusion forces move it to the left, although the entropy force is in the opposite direction. Thus both electron and hole transfer occurs in the vicinity of x_1. Now consider the electron and the hole at x_2. Again, there is no electric field, but the crystal field forces tend to move both carriers to the right. However, both the diffusion and entropy forces tend to move both the electron and the hole to the left. These forces balance each other and there is no charge transfer in the vicinity of x_2.

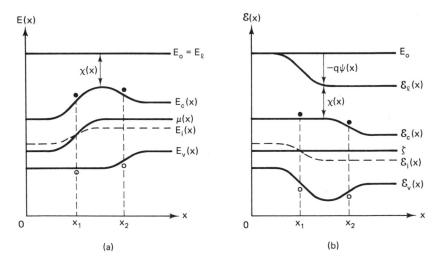

Figure 8.12 Energy band diagrams for a general heterostructure (a) before charge transfer and (b) after charge transfer in equilibrium.

Now look at the equilibrium diagram in part (b). Consider the hole at x_1. Before charge transfer a hole at this point would move to the right due to diffusion and entropy forces. Now there is an electric field at x_1 which counterbalances the other forces. The same is true for the electron at x_1. Before charge transfer, the crystal field and the diffusion forces tended to move it to the left, while the entropy force tended to move it to the right. Now, the electric field, which tends to move it to the right, counterbalances the other three forces. Since there was no charge transfer in the vicinity of x_2, the forces there remain unchanged. The only difference between now and the nonequilibrium condition is that now the electron and hole at x_2 lie in a constant internal potential, V_0. This is because we arbitrarily set the potential to zero at $x = 0$.

8.6 QUASI-FERMI ENERGIES

In equilibrium with excess carriers or in nonequilibrium with no excess carriers, the electron and hole concentrations can be characterized with a single parameter: the Fermi energy. The Fermi energy is identical to the chemical potential, μ, in the absence of an electrostatic potential and equal to the electrochemical potential, ζ, in the presence of a potential. Under either of these conditions the mass-action law for electrons and holes is valid. The action of external stimuli, however, can change the electron and hole concentrations so that $np \neq n_i^2$. Under these conditions no single parameter can be used to characterize both n and p. Thus, to maintain the same formalisms for samples with nonequilibrium excess carrier distributions, it is necessary to develop some new concepts.

Two methods for keeping the same description of carrier distributions are illustrated in Fig. 8.13 for a uniform sample with nonequilibrium excess electron and hole distributions in the center. Figure 8.13(a) shows the concept of energy gap changes in a region of excess carrier distributions. The idea here is that excess carriers can be roughly described by a distribution function such as (4.41) in which the lattice temperature T is replaced by an effective temperature T_e. For tetrahedrally coordinated semiconductors, the energy gaps, $\mathscr{E}_g(T)$, decrease with increasing lattice temperature [C. D. Thurmond, *J. Electrochem. Soc.* **122**, 1133 (1975)]. If $\mathscr{E}_g(T)$ is replaced by $\mathscr{E}_g(T_e)$, the energy gap will decrease for T_e greater than T.

The carrier concentrations can then be described by

$$n(T_e) = N_c(T_e) \exp \left[\frac{\mu - \mathscr{E}_c(T_e)}{kT_e} \right]$$

$$p(T_e) = N_v(T_e) \exp \left[\frac{\mathscr{E}_v(T_e) - \mu}{kT_e} \right]$$

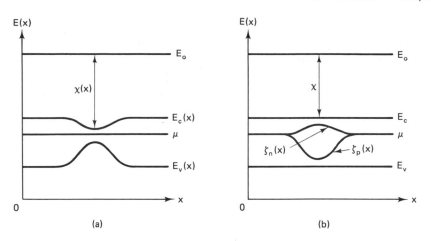

Figure 8.13 Methods of maintaining the same formalism for samples with non-equilibrium excess carrier distributions. Part (a) illustrates the concept of energy gap changes with effective carrier temperature. Part (b) indicates the quasi-Fermi level concept.

and

$$n(T_e)p(T_e) = N_c(T_e)N_v(T_e) \exp\left[\frac{-\mathscr{E}_g(T_e)}{kT_e}\right]$$

$$= n_i^2(T_e)$$

where T_e is adjusted to satisfy the equations. In these equations T_e varies with x, and we assume that the electron and hole distributions have the same effective temperature. Thus, with this concept, the equations have the form as for equilibrium distributions. In addition, a change in energy band structure with electron temperature has some physical appeal. A rigorous thermodynamic description of this concept has been developed [J. A. van Vechten and M. Wautelet, *Phys. Rev. B23,* 521 (1981)].

The conventional method of describing excess carrier distributions, however, is illustrated in Fig. 8.13(b). This is the idea of quasi-Fermi levels. In this treatment quasi-Fermi energies for electrons, ζ_n, and for holes, ζ_p, are defined as those quantities that satisfy

$$n(\mathbf{r}) \equiv N_c(\mathbf{r}) \exp\left[\frac{\zeta_n(\mathbf{r}) - \mathscr{E}_c(\mathbf{r})}{kT}\right] \tag{8.146}$$

$$p(\mathbf{r}) \equiv N_v(\mathbf{r}) \exp\left[\frac{\mathscr{E}_v(\mathbf{r}) - \zeta_p(\mathbf{r})}{kT}\right] \tag{8.147}$$

Notice that the mass-action law is no longer valid and

$$n(\mathbf{r})p(\mathbf{r}) = n_i^2(\mathbf{r}) \exp\left[\frac{\zeta_n(\mathbf{r}) - \zeta_p(\mathbf{r})}{kT}\right] \tag{8.148}$$

In a similar manner, instead of (8.140) and (8.141), the general forces driving the electron and hole currents for nonequilibrium excess carrier distributions are the gradients of the quasi-Fermi energies:

$$\mathbf{J}_n(\mathbf{r}) = n(\mathbf{r})\mu_n \, \nabla \zeta_n(\mathbf{r}) \qquad (8.149)$$

$$\mathbf{J}_p(\mathbf{r}) = p(\mathbf{r})\mu_p \nabla \zeta_p(\mathbf{r}) \qquad (8.150)$$

This is the formalism we will use. It should be pointed out, however, that although the quasi-Fermi levels indicate the concentration of electrons in the conduction bands or holes in the valence bands, they do not reveal anything about the distribution of carriers. Thus the quasi-Fermi levels have no apparent thermodynamic significance.

PROBLEMS

8.1. Suppose that a narrow pulse of minority carriers is generated at a *point* within a semiconductor bar and drifts in an electric field, E, applied to the bar. What is the ratio of the sidewise spread of the carriers to the drift distance, where the drift distance is expressed in volts?

8.2. Obtain an explicit expression for the decay of excess holes $\delta p(t)$ through N_t (cm^{-3}) generation–recombination centers of energy $\mathscr{E}_t$ assuming only that $\delta n = \delta p$ everywhere.

8.3. Derive an expression for the excess recombination rate $R - G_t$ for a sample with N_0 (cm^{-2} eV^{-1}) generation–recombination centers distributed uniformly in energy throughout the forbidden gap.

8.4. For homogeneous materials in general show that the assumption of space-charge neutrality, $\delta n \simeq \delta p$, is valid when the Debye length squared is much less than the diffusion length squared. The Debye length is given by $\lambda^2 = \epsilon kT / q^2 n$, and so on.

8.5. A steady-state excess carrier distribution is generated in an n-type sample by a plane source in the yz plane at $x = 0$. A uniform electric field E is applied in the $+x$ direction. The excess minority carrier concentration is obtained from the reverse saturation current in two probes located at $x = a$ and $x = b$, where $b > a$. Show that the diffusion length is

$$L_p^2 = \frac{(b-a)^2}{\ln \dfrac{\delta p(a)}{\delta p(b)} \left[\ln \dfrac{\delta p(a)}{\delta p(b)} + \dfrac{qE(b-a)}{kT} \right]}$$

8.6. Show that if $f(x, t)$ satisfies the equation

$$D \frac{\partial^2 f}{\partial x^2} - \frac{f}{\tau} = \frac{\partial f}{\partial t}$$

then $f(\xi, t)$ where $\xi = x - \mu E t$ must be a solution to

$$D \frac{\partial^2 f}{\partial x^2} - \mu E \frac{\partial f}{\partial x} - \frac{f}{\tau} = \frac{\partial f}{\partial t}$$

8.7. A homogeneous n-type sample with absorption coefficient α is irradiated with a uniform incident photon flux ϕ_q through a surface with reflectivity R and surface recombination velocity S. Obtain an expression for the change in conductance $\delta G/G_0$ produced by the light. Assume that $\alpha L_p \gg 1$ and plot $\delta G/G_o$ versus d/L_p for $0 < d/L_p \le 5$, $S = 0$, and $S = \infty$.

8.8. Obtain expressions for the potential $\psi(x)$ and excess electron concentration $\delta n(x)$ in otherwise intrinsic material produced by a Gaussian ion implantation

$$n_0(x) = \frac{\phi}{\sqrt{2\pi}\,\Delta R} \exp\left[-\frac{1}{2}\left(\frac{x-R}{\Delta R}\right)^2\right]$$

where ϕ is ion dose, R is the range, and ΔR is the straggle of ions. Assume that $n_0(R) \gg n_i$.

8.9. An n-type heterostructure of thickness L is fabricated so that $\chi(x) = 4.0 - 0.4 \sin(\pi x/L)$ eV and $\mathscr{E}_g(x) = 1.4 + 0.4 \sin(\pi x/L)$ eV at 300 K. The electron concentration at the surface $x = 0$ is 1×10^{16} cm^{-3}. From k·p theory the conduction band effective mass is given by

$$\frac{m}{m^*(x)} \simeq 1 + \frac{20 \text{ eV}}{\mathscr{E}_g(x)}$$

(a) If no space charge is formed in the sample, what is the electron concentration at the center $x = L/2$?

(b) If the electron concentration is uniform, what is the electrostatic potential at the center $x = L/2$?

8.10. Steady-state radiation with photon energy 2 eV and power density 3.2×10^{-3} W/cm^2 is incident on a semiconductor surface with zero reflectivity and infinite surface recombination velocity. The material is uniform with an absorption coefficient of 10^4 cm^{-1}, a hole mobility of 500 cm^2/V·s, and an excess hole recombination lifetime of 10^{-9} s at 300 K. A uniform electric field of 10^4 V/cm is applied normal to the surface to sweep the excess holes into the sample.

(a) Using suitable assumptions, obtain an expression for the excess hole concentration as a function of distance in the sample.

(b) What is the excess hole concentration at 1×10^{-4} cm from the surface?

8.11. Differential capacitance measurements indicate that a diffused n-type sample has a total electron concentration at 300 K given by

$$n(x) = n \exp(-\alpha x^2) + n_0$$

where $n = 10^{18}$ cm^{-3}, $n_0 = 10^{17}$ cm^{-3}, and $\alpha = 10^{10}$ cm^{-2}.

(a) By trial and error determine the point at which the electric field induced by the excess electrons is a maximum.

(b) Obtain values for the maximum field and the built-in potential.

8.12. Obtain a general expression for the continuity of holes which does not assume linear recombination. Express your answer in terms of the linear band-to-band direct recombination lifetime. Assume only that $\delta n = \delta p$.

8.13. In a semiconductor with a uniform equilibrium concentration n_0, show that a small fluctuation of majority carrier concentration $\delta n(x_0, 0)$ changes in space and time according to

$$\delta n(\xi, t) = \frac{\delta n(0, 0)}{(4\pi Dt)^{1/2}} \exp\left(\frac{-\xi^2}{4Dt} - \frac{t}{\eta}\right)$$

where $\xi = (x - x_0) + \mu E t$ and $\eta = \epsilon\tau/(\sigma\tau + E)$. What happens to the fluctuation when η is negative?

8.14. Consider an n-type bar as shown in Fig. P8.14 and biased to give nominally a drift field E_0 and a longitudinal current I. We introduce excess electrons and holes (low-level) into a region as shown and assume that carrier recombination and diffusion are negligible. The region of excess carriers drifts to the right down the bar. How much is the field reduced in the region of excess carriers? Is the flux of electrons to the left in the excess carrier region as large as the electron flux to the left of the region?

8.15. Consider an n-type semiconductor bar that is rather long in both the plus and minus directions and in which at $t = 0$ we generate p_0 excess holes at $n = 0$. Assume that the excess carrier lifetime is infinite so that the excess carriers $p'(x, t)$ simply diffuse. That is, they obey the equation

$$\frac{\partial^2 p'(x, t)}{\partial x^2} = \frac{1}{D_p} \frac{\partial p'(x, t)}{\partial t}$$

Find, as directly as possible, the mean-square diffusion distance $\langle x^2 \rangle$ of the excess carriers.

8.16. Materials such as GaAs have a very short carrier lifetime ($\approx 10^{-9}$ s) which is difficult to measure. The interferometer scheme shown in Fig. P8.16 proves to be a useful arrangement for measurement of short lifetimes in *direct* materials. The mirror and the sample (near one another) receive two beams split from a common, modulated ($\cos \omega_0 t$), coherent laser beam. When the sample is replaced with a reflecting surface (dummy sample) and the movable mirror is set at the zero (0) position, the photodetector registers a null (two signals: same ω_0, same amplitude, and 180° phase difference). When the dummy reflector is replaced with a thin sample that is excited by one of the incident beams, the sample radiates recombination radiation at retarded phase (time) that can be canceled at the null detector if the attenuator is adjusted and the movable mirror is displaced a distance x as shown so as also to delay the reference beam (i.e., increase its phase delay and null the interferometer). Find

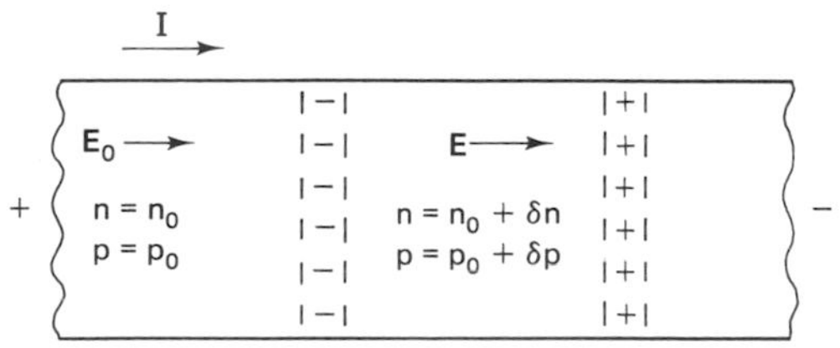

Figure P8.14

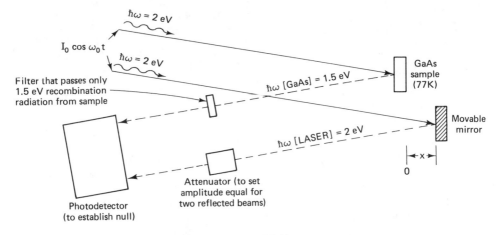

Figure P8.16

the relationship between the carrier lifetime τ and x. (*Hint:* Assume that the sample is thin, or that only a thin layer of it is excited, and consider what the excess carriers do. Do not get confused by the fact that two different wavelength photons strike the detector; their modulation frequency, ω_0, is the same.)

Heterostructures

The abrupt *p-n* junction is probably the most important concept in semiconductor device theory. It is used in the analysis of a number of devices, including rectifiers, Zener diodes, solar cells, photodiodes, light-emitting diodes, semiconductor lasers, bipolar transistors, controlled rectifiers, and avalanche transit-time device. In this chapter *p-n* junction theory is developed for a general abrupt *p-n* heterojunction. The equations resulting from this analysis can easily be reduced to the simpler equations of *p-n* homojunctions. In many cases, they can also be applied to the analysis of metal-semiconductor and metal-oxide semiconductor contacts, which are discussed in Chapter 10.

One of the most active research areas in the field of semiconductors is in the development of heterostructure technology. This is because with heterostructures, nonuniformities in both doping and band structure can be exploited for device applications. Ideally, this additional flexibility of altering band structure and other physical properties could be used to improve or achieve almost any device function [H. Kroemer, *Jpn. J. Appl. Phys. 20,* 9 (1980)]. An important practical constraint, however, is the necessity of selecting materials that have small differences in lattice constant at the required fabrication and operating temperatures. A good lattice match is required to minimize interface states and strain fields in the device. Table 9.1 shows a number of semiconductor systems in which a small lattice mismatch can be achieved.

Figure 9.1 shows the energy band diagrams for one of these closely lattice-matched heterostructure systems: GaAs-AlAs. These diagrams were

TABLE 9.1   Some Semiconductor Systems with Small Lattice Mismatch

Material	Lattice Constant (Å)	Expansion Coefficient (10^{-6}°C)	Energy Gap (eV)	Electron Affinity (eV)
Si	5.451	2.3	1.11	4.01
GaP	5.431	5.3	2.25	4.3
ZnGeP$_2$	5.465		2.0	3.58
Ge	5.6461	6.1	0.663	4.13
GaAs	5.6535	6.0	1.435	4.07
AlAs	5.6605	5.2	2.16	3.5–2.62
Al$_{0.2}$Ga$_{0.8}$As	5.6549	5.8	1.70	3.96
Ga$_{0.51}$In$_{0.49}$P	5.6535			
ZnSe	5.667	7.0	2.67	4.09
ZnSnP$_2$	5.651		1.62	4.25
InP	5.8688	4.5	1.27	4.35
Ga$_{0.48}$In$_{0.52}$As	5.8688		0.75	
CdS			2.42	4.87
CdSnP$_2$	5.900		1.14	4.41
AlSb	6.136	3.7	1.6	3.65
GaSb	6.095	6.9	0.68	4.06
InAs	6.058	4.5	0.36	4.9
ZnTe	6.103	8.2	2.26	3.5
CdSnAs$_2$	6.093	8.2	0.26	4.9
InSb	6.479	4.9	0.17	4.59
CdTe	6.477		1.44	4.28
PbTe	6.52		0.29	

constructed with the energy gap and electron affinity values given in Table 9.1 and (8.113), (8.115), and (8.116). An abrupt interface was assumed, and any effects due to interface states have been ignored. In part (a) the GaAs is p-type and the AlAs is n-type, while in part (b) the AlAs is p-type and the GaAs is n-type. When the carrier mean free paths are less than the widths of the space-charge regions, the effects of the band structure on the transport of charge are different in these two cases. For example, in part (a) the minority electron flow from the p-side to the n-side is retarded by the notch due to the conduction band discontinuity, while the minority hole flow from the n-side to the p-side is aided by the valence band continuity. In part (b), however, the flow of minority electrons from the p-side to the n-side is enhanced by the conduction band discontinuity, while the flow of minority holes from the n-side to the p-side is reduced by the valence band notch. These and other such effects can profitably be used in the design of heterostructure devices.

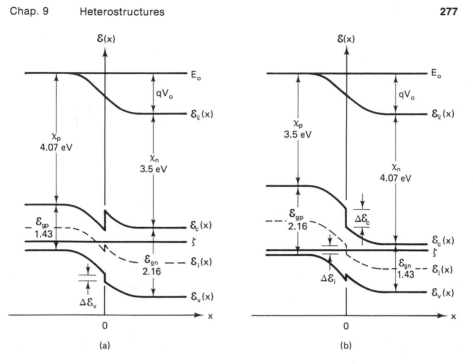

Figure 9.1 Energy band diagrams for ideal abrupt GaAs-AlAs *p-n* heterojunctions. In (a) the GaAs is *p*-type. In (b) the AlAs is *p*-type.

In Fig. 9.1(a) the conduction band discontinuity is assumed to be given by

$$\Delta \mathscr{E}_c = \chi_p - \chi_n \qquad (9.1)$$

which for this heterojunction is 0.57 eV. This is referred to as the *electron affinity rule* [R. L. Anderson, *Solid-State Electron. 5,* 341 (1962)]. (Although there is considerable controversy concerning the applicability of this *rule* to heterostructures, it seems to work fairly well.) The same value for $\Delta \mathscr{E}_c$ is obtained in Fig. 9.1(b), except that here

$$\Delta \mathscr{E}_c = \chi_n - \chi_p \qquad (9.2)$$

That is, $\Delta \mathscr{E}_c$ must have a positive value. The valence band discontinuity can then be obtained from

$$\Delta \mathscr{E}_g = \Delta \mathscr{E}_c + \Delta \mathscr{E}_v \qquad (9.3)$$

For the heterojunction in Fig. 9.1, $\Delta \mathscr{E}_g$ is 0.73 eV, so $\Delta \mathscr{E}_v$ is 0.16 eV. The built-in potential, V_0, can also be determined. Referring to the figure, we see that

$$qV_0 = \chi_p + (\mathscr{E}_{cp} - \zeta) - \chi_n - (\mathscr{E}_{cn} - \zeta) \qquad (9.4)$$

From (8.142),

$$\mathscr{E}_{cp} - \zeta = kT \ln \frac{N_{cp}}{n_{p0}} \tag{9.5}$$

$$-(\mathscr{E}_{cn} - \zeta) = kT \ln \frac{n_{n0}}{N_{cn}} \tag{9.6}$$

so

$$qV_0 = \Delta\mathscr{E}_c + kT \ln \frac{n_{n0}N_{cp}}{n_{p0}N_{cn}} \tag{9.7}$$

In a similar manner, an expression for the built-in potential with reference to the valence band is

$$qV_0 = -\Delta\mathscr{E}_v + kT \ln \frac{p_{p0}N_{vn}}{p_{n0}N_{vp}} \tag{9.8}$$

With (8.142), (8.143), and (9.3), equations (9.7) and (9.8) can be shown to be identical.

When voltage division between the p- and n-regions and the extent of the space-charge region is known, these and (9.1), (9.3), and (9.7) or (9.8) completely determine the equilibrium properties of the heterojunction. To determine the first two quantities, however, it is necessary to examine the interfacial region in some detail.

9.1 SPACE-CHARGE REGION

The analysis of the space-charge region will be based on the quantities defined in Fig. 9.2. First examine the boundary conditions. If there is no dipole layer at the interface, the potential is continuous or

$$\psi_1(0) = \psi_2(0) = \psi(0) \tag{9.9}$$

Assuming no interface states, the electric flux density D must also be continuous,

$$\epsilon_1 \frac{d\psi_1(0)}{dx} = \epsilon_2 \frac{d\psi_2(0)}{dx} \tag{9.10}$$

Far away from the interface the material on both sides is uniform, so we have

$$\frac{d\psi_1(-\infty)}{dx} = \frac{d\psi_2(\infty)}{dx} = 0 \tag{9.11}$$

With these equations, solutions to Poisson's equation can be obtained on both sides of the interface.

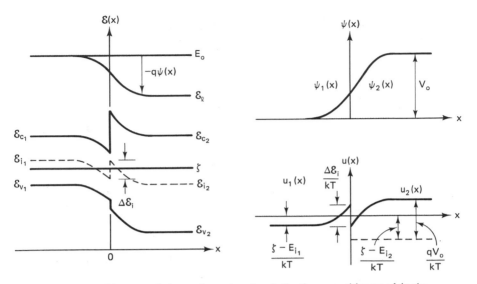

Figure 9.2 Diagram of abrupt heterojunction indicating quantities used in the analysis.

First consider region 1 on the left side of the diagrams in Fig. 9.2. Assuming that all the donors and acceptors are ionized, Poisson's equation is

$$\frac{d^2\psi_1(x)}{dx^2} = \frac{q}{\epsilon_1}[n_1(x) - p_1(x) + N_{a1} - N_{d1}] \tag{9.12}$$

From (8.112), (8.119), and (8.120), this becomes

$$\frac{d^2\psi_1(x)}{dx^2} = \frac{2qn_{i1}}{\epsilon_1}\left[\sinh\frac{\zeta - E_{i1} + q\psi_1(x)}{kT} - \frac{N_{d1} - N_{a1}}{2n_{i1}}\right] \tag{9.13}$$

Far from the interface there is space-charge neutrality, and

$$\frac{N_{d1} - N_{a1}}{2n_{i1}} = \sinh\frac{\zeta - E_{i1} + q\psi_1(-\infty)}{kT} \tag{9.14}$$

Using the dimensionless variable,

$$u(x) \equiv \frac{\zeta - E_i + q\psi(x)}{kT} \tag{9.15}$$

and (9.14), equation (9.13) becomes

$$\lambda_{i1}^2\frac{d^2u_1(x)}{dx^2} = \sinh u_1(x) - \sinh u_1(-\infty) \tag{9.16}$$

where we have defined an *intrinsic Debye* length,

$$\lambda_i^2 \equiv \frac{kT\epsilon}{2q^2 n_i} \qquad (9.17)$$

Equation (9.16) can be solved for the dimensionless electric field,

$$e_1(x) \equiv -\lambda_{i1} \frac{du_1(x)}{dx} = \frac{q\lambda_{i1}}{kT} E \qquad (9.18)$$

by noting that

$$\lambda_{i1}^2 \frac{d^2 u_1(x)}{dx^2} = \lambda_{i1} \frac{d}{dx}\left[\lambda_{i1} \frac{du_1(x)}{dx}\right]$$

$$= -\lambda_{i1} \frac{de_1(x)}{dx} = \frac{de_1(x)}{du_1(x)}\left[-\lambda_{i1} \frac{du_1(x)}{dx}\right]$$

$$= e_1(x) \frac{de_1(x)}{du_1(x)} \qquad (9.19)$$

Using (9.19) in (9.16) and separating variables gives us

$$\int_{-\infty}^{x} e_1(x)\, de_1(x) = \int_{-\infty}^{x} [\sinh u_1(x) - \sinh u_1(-\infty)]\, du_1(x) \qquad (9.20)$$

Performing the integration and using boundary conditions (9.11), equation (9.20) gives

$$e_1(x) = \pm\sqrt{2}\,\{\cosh u_1(x) - \cosh u_1(-\infty)$$

$$- [u_1(x) - u_1(-\infty)]\sinh u_1(-\infty)\}^{1/2} \qquad (9.21)$$

After taking the square root in this analysis, the minus sign must be used since the field is in the negative x direction. With (9.18), equation (9.21) can be integrated to obtain an implicit expression for $u_1(x)$,

$$+\frac{x}{\lambda_{i_1}} = \frac{1}{\sqrt{2}}\int_{-x}^{0} \frac{du_1(x)}{\{\cosh u_1(x) - \cosh u_1(-\infty) - [u_1(x) - u_1(-\infty)]\sinh u_1(-\infty)\}^{1/2}}$$

$$(9.22)$$

Region 2 on the right side of the diagrams in Fig. 9.2 can be treated in the same way, to obtain

$$e_2(x) = \pm\sqrt{2}\,\{\cosh u_2(x) - \cosh u_2(\infty) - [u_2(x) - u_2(\infty)]\sinh u_2(\infty)\}^{1/2}$$

$$(9.23)$$

and

$$-\frac{x}{\lambda_{i2}} = \frac{1}{\sqrt{2}} \int_0^x \frac{du_2(x)}{\{\cosh u_2(x) - \cosh u_2(\infty) - [u_2(x) - u_2(\infty)]\sinh u_2(\infty)\}^{1/2}}$$

$$(9.24)$$

An analytic solution to (9.22) and (9.24) can be obtained for intrinsic material.

9.1.1 Intrinsic Material

Although the intrinsic case is of little interest in the analysis of devices at their operating temperature, it is of importance in the evaluation of problems that can occur during device fabrication. That is, during high-temperature fabrication processes, the material is often intrinsic. Problems can occur under these conditions due to the drift of ionized impurities in the internal fields [C. M. Wolfe and K. H. Nichols, *Appl. Phys. Lett. 31*, 356 (1977)]. As indicated in Fig. 9.3, the analysis is simplified considerably when both sides of the heterostructure are intrinsic. From (9.15) for region 1 on the left,

$$u_1(x) = \frac{q\psi_1(x)}{kT} \tag{9.25}$$

Setting the potential reference at minus infinity, $u_1(-\infty) = 0$, (9.21) for the electric field becomes

$$e_1(x) = -\sqrt{2}\,[\cosh u_1(x) - 1]^{1/2} \tag{9.26}$$

or

$$E_1(x) = \frac{-2kT}{q\lambda_{i1}}\sinh\frac{q\psi_1(x)}{2kT} \tag{9.27}$$

Equation (9.22) can now be integrated to obtain

$$-\frac{x}{\lambda_{i1}} = \ln\frac{\tanh[u_1(0)/4]}{\tanh[u_1(x)/4]} \tag{9.28}$$

or

$$\psi_1(x) = \frac{4kT}{q}\tanh^{-1}\left[\exp\left(\frac{+x}{\lambda_{i1}}\right)\tanh\left[\frac{q\psi(0)}{4kT}\right]\right] \tag{9.29}$$

For region 2 equation (9.15) gives

$$u_2(x) = \frac{q\psi_2(x)}{kT} \tag{9.30}$$

but here $u_2(\infty) = qV_0/kT$. In this situation (9.23) and (9.24) are difficult to

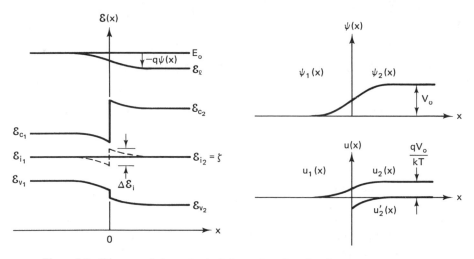

Figure 9.3 Diagram of abrupt intrinsic heterojunction showing quantities used in the analysis.

evaluate. However, if the substitution

$$u_2'(x) = \frac{q}{kT}[\psi_2(x) - V_0] \tag{9.31}$$

is made, the equations have the same form as for region 1. The electric field and the potential are then

$$E_2(x) = \frac{2kT}{q\lambda_{i2}} \sinh\left\{\frac{q[\psi_2(x) - V_0]}{2kT}\right\} \tag{9.32}$$

and

$$\psi_2(x) = V_0 + \frac{4kT}{q} \tanh^{-1}\left\{\exp\left(\frac{-x}{\lambda_{i2}}\right) \tanh\frac{q[\psi(0) - V_0]}{4kT}\right\} \tag{9.33}$$

Since V_0 can be obtained from (9.7) or (9.8), all that remains is to determine $\psi(0)$ and the analysis is complete.

We can determine $\psi(0)$ from boundary condition (9.10) for the continuity of the electric flux density. Evaluating (9.27) and (9.32) at $x = 0$,

$$\psi(0) = \frac{2kT}{q} \tanh^{-1}\left[\frac{\alpha \sinh\left(\frac{qV_0}{2kT}\right)}{1 + \alpha \cosh\left(\frac{qV_0}{2kT}\right)}\right] \tag{9.34}$$

where

$$\alpha \equiv \frac{\lambda_{i1}\epsilon_2}{\lambda_{i2}\epsilon_1} \tag{9.35}$$

With these equations an analytic expression for the capacitance per unit area of the space-charge region can be obtained [A. Chatterjee and A. H. Marshak, *Solid-State Electron.* 24, 1111 (1981)] as

$$C_i = \left| \frac{dQ_i}{dV_0} \right| \tag{9.36}$$

$$C_i = \frac{1}{\lambda_{i1}} \frac{\alpha(\alpha + \cosh u_0)}{1 + \alpha^2 + 2\alpha \cosh u_0} \cosh \left(\tanh^{-1} \frac{\alpha \sinh u_0}{1 + \alpha \cosh u_0} \right) \tag{9.37}$$

where

$$u_0 \equiv \frac{qV_0}{2kT} \tag{9.38}$$

9.1.2 Extrinsic Material

For extrinsic material it is apparently not possible to integrate (9.22) and (9.24) to find the potential analytically. Because of this it is customary to introduce what is called the *depletion approximation*. The basic assumption in the depletion approximation is that, to first order, there are no electrons or holes in the space-charge region. The extent of the space-charge region is defined as that region over which the potential varies spatially. When there is no spatial variation in potential, the electric field is zero. Poisson's equation is then solved for the potential under these assumptions, and with this potential the spatial variations of the electron and hole concentrations can be determined. This constitutes a first-order solution in the depletion approximation. If desired, this first-order solution for the electron and hole concentrations can be used, iteratively, in Poisson's equation to obtain a second-order solution for the potential, and so on. For many purposes, the first-order solution is reasonably good. Thus our analysis of extrinsic heterojunctions will be restricted to the depletion approximation.

The spatial variations of the energy bands, electrostatic potential, and electric field for an abrupt extrinsic heterojunction in the depletion approximation are shown in Fig. 9.4. The distances $-x_{p0}$ and x_{n0} define the extent of the equilibrium space-charge region in the p- and n-type material, respectively. In the depletion approximation, these are the distances from the interface ($x = 0$), by definition, at which the electric field is zero. At $x = 0$ there is a discontinuity in the electric field and, therefore, in the gradient of the potential due to the change in dielectric constant. The total built-in potential, V_0, due to charge transfer, consists of V_{p0} across the p-side of the space-charge region and V_{n0} across the n-side.

Analysis of the space-charge region proceeds as before with some modifications. The boundary conditions are a continuous potential at the interface,

$$\psi_p(0) = \psi_n(0) = \psi(0) \tag{9.39}$$

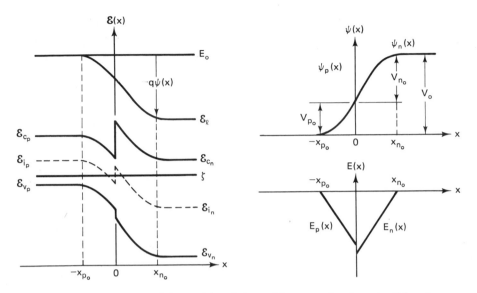

Figure 9.4 Diagrams of the spatial variations of the energy bands, potential, and electric field for an abrupt extrinsic heterojunction in the depletion approximation.

a continuous electric flux density,

$$\epsilon_p \frac{d\psi_p(0)}{dx} = \epsilon_n \frac{d\psi_n(0)}{dx} \tag{9.40}$$

and zero electric field at the edges of the depletion region,

$$\frac{d\psi_p(-x_{po})}{dx} = \frac{d\psi_n(x_{no})}{dx} = 0 \tag{9.41}$$

analogous to (9.9), (9.10), and (9.11). Since the depletion approximation assumes to first order no electrons or holes in the space-charge region, Poisson's equation becomes

$$\frac{d^2\psi_p(x)}{dx^2} = \frac{q}{\epsilon_p} N_a \tag{9.42}$$

on the p-side, and

$$\frac{d^2\psi_n(x)}{dx^2} = \frac{-q}{\epsilon_n} N_d \tag{9.43}$$

on the n-side. For simplicity of notation, in (9.42) and (9.43) we have used N_a to indicate the net acceptor concentration ($N_{ap} - N_{dp}$) on the p-side and N_d to indicate the net donor concentration ($N_{dn} - N_{an}$) on the n-side. Notice that neither N_a nor N_d are spatially dependent because of the abrupt junction assumption.

Integrating (9.42) and (9.43) once and using boundary condition (9.41), we find expressions for the electric field,

$$-\frac{d\psi_p(x)}{dx} = E_p(x) = -\frac{q}{\epsilon_p} N_a(x_{p0} + x) \tag{9.44}$$

and

$$-\frac{d\psi_n(x)}{dx} = E_n(x) = -\frac{q}{\epsilon_n} N_d(x_{n0} - x) \tag{9.45}$$

Evaluating these equations at $x = 0$ with boundary condition (9.40), we also obtain

$$N_a x_{p0} = N_d x_{n0} \tag{9.46}$$

This simply indicates that the net negative space charge on the p-side is equal to the net positive space charge on the n-side, and is a consequence of the assumption of no interface states.

Setting the potential reference on the p-side as indicated in Fig. 9.4, (9.44) and (9.45) can be integrated to obtain

$$\psi_p(x) = \frac{qN_a}{2\epsilon_p} (x_{p0} + x)^2 \tag{9.47}$$

and

$$\psi_n(x) = V_0 - \frac{qN_d}{2\epsilon_n} (x_{n0} - x)^2 \tag{9.48}$$

To determine how the built-in potential is divided between the p and n regions, (9.47) and (9.48) are evaluated at the interface with boundary condition (9.39). In this manner

$$V_{p0} = \psi(0) - 0 = \frac{qN_a}{2\epsilon_p} x_{p0}^2 \tag{9.49}$$

and

$$V_{n0} = V_0 - \psi(0) = \frac{qN_d}{2\epsilon_n} x_{n0}^2 \tag{9.50}$$

Using (9.46), we also have

$$\frac{V_{p0}}{V_{n0}} = \frac{\epsilon_n N_d}{\epsilon_p N_a} \tag{9.51}$$

The extent of the space-charge region on either side of the heterojunction can be obtained in terms of the total built-in potential from

$$V_0 = V_{p0} + V_{n0} \tag{9.52}$$

Inserting (9.49) and (9.50) into (9.52) and then using (9.46), we find that

$$x_{p0}^2 = \frac{2\epsilon_p\epsilon_n N_d V_0}{qN_a(\epsilon_p N_a + \epsilon_n N_d)} \tag{9.53}$$

and

$$x_{n0}^2 = \frac{2\epsilon_p\epsilon_n N_a V_0}{qN_d(\epsilon_p N_a + \epsilon_n N_d)} \tag{9.54}$$

For the entire equilibrium space-charge region,

$$W_0 \equiv x_{p0} + x_{n0} \tag{9.55}$$

(9.53) and (9.54) give

$$W_0^2 = \frac{2\epsilon_p\epsilon_n(N_a + N_d)^2 V_0}{qN_a N_d(\epsilon_p N_u + \epsilon_n N_d)} \tag{9.56}$$

Once V_0 has been determined from (9.7) or (9.8), the equations above completely characterize the equilibrium space-charge region in the depletion approximation.

When an external voltage, V, is applied to the heterojunction, the space-charge region is no longer in equilibrium and a net current flows through the sample. Under these conditions the total potential across the space-charge region indicated in Fig. 9.4 is decreased or increased, depending on the polarity of the applied voltage. When the n-side is biased negative with respect to the p-side, the total potential across the space-charge region decreases, which lowers the space charge by reducing the space-charge width, W. This is referred to as *forward bias*. On the other hand, when the n-side is biased positive with respect to the p-side, the total potential increases, which raises the space charge by broadening the space width. This is called *reverse bias*.

For small net current values, the equilibrium description of the space-charge region can easily be modified to account for these effects. Since most of the applied voltage, V, appears across the space-charge region, (9.53) to (9.56) become

$$x_p^2 \simeq \frac{2\epsilon_p\epsilon_n N_d(V_0 - V)}{qN_a(\epsilon_p N_a + \epsilon_n N_d)} \tag{9.57}$$

$$x_n^2 \simeq \frac{2\epsilon_p\epsilon_n N_a(V_0 - V)}{qN_d(\epsilon_p N_a + \epsilon_n N_d)} \tag{9.58}$$

$$W = x_p + x_n \tag{9.59}$$

and

$$W^2 \simeq \frac{2\epsilon_p\epsilon_n(N_a + N_d)^2(V_0 - V)}{qN_a N_d(\epsilon_p N_a + \epsilon_n N_d)} \tag{9.60}$$

In these equations we have used the convention that an applied voltage in the forward direction is $+V$, while an applied voltage in the reverse direction is $-V$. It should be pointed out that these equations are only approximate, because of the voltage drop across the space-charge-neutral p and n regions due to current flow. For relatively larger net current values, the voltage across the space-charge region can be substantially lower than V.

The capacitance per unit area of the space-charge region can be obtained from (9.36). In the depletion approximation,

$$Q_i = qN_d x_n = -qN_a x_p \qquad (9.61)$$

Taking the derivative of this charge per unit area with respect to voltage and using (9.60), we find that

$$C_i = \frac{\epsilon_p \epsilon_n (N_a + N_d)}{(\epsilon_p N_a + \epsilon_n N_d) W} \qquad (9.62)$$

9.1.3 Interface States

In reality, most heterostructures have interface states, due to lattice constant differences or other effects. When these interface states cannot be ignored, the analyses above must be modified to account for their effects. Several possible situations can occur. If a net positive or negative charge exists at the interface, the electrostatic potential is continuous but the electric flux density is not. For an abrupt p-n heterojunction, (9.40) would become

$$\epsilon_p \frac{d\psi_p(0)}{dx} \pm qN_s = \epsilon_n \frac{d\psi_n(0)}{dx} \qquad (9.63)$$

where N_s is the number of interface states per unit area and the $\pm$ signs indicate the interface charge. The analyses can then proceed as before with this boundary condition.

It is also possible to have a dipole layer at the interface directed in the positive or negative x direction. In the presence of a dipole layer, the electrostatic potential is not continuous and the energy band diagram must be modified as indicated in Fig. 9.5. For an abrupt p-n heterojunction, (9.39) would be

$$\psi_p(0) = \psi_n(0) \pm \Delta\psi(0) \qquad (9.64)$$

where $\Delta\psi(0)$ is the discontinuity induced by the dipole and the $\pm$ signs indicate the direction of the dipole. Notice, also, that the electron affinity rule must be modified to account for the discontinuity in potential,

$$\Delta\mathscr{E}_c = \chi_p - \chi_n \pm q\,\Delta\psi(0) \qquad (9.65)$$

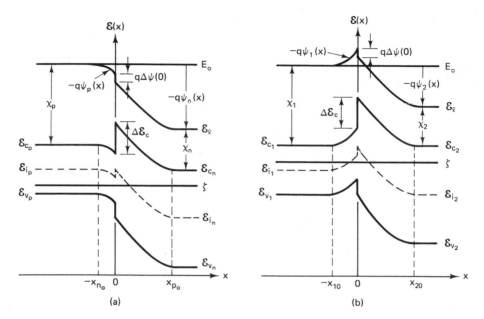

Figure 9.5 Energy band diagrams for (a) an abrupt *p-n* heterojunction and (b) an abrupt *n-n* heterojunction with an interfacial dipole layer in the minus *x* direction. Notice that the electron affinity rule is not valid.

9.2 CURRENT FLOW

Let us now consider the problem of current flow in an abrupt *p-n* heterojunction. We will assume a one-dimensional structure, as indicated in Fig. 9.6, with surface recombination velocities S_p on the *p* surface at $-W_p$ and S_n on the *n* surface at W_n. $-x_p$ and x_n are the edges of the space-charge region in the depletion approximation on the *p* and *n* sides, respectively. We will also assume a time-independent direct current with no excess electron–hole pair generation due to incident radiation or impact ionization.

Under these conditions the continuity equations (8.57) and (8.58) reduce to

$$\frac{dJ_p(x)}{dx} = -q[R(x) - G_t(x)] \tag{9.66}$$

$$\frac{dJ_n(x)}{dx} = +q[R(x) - G_t(x)] \tag{9.67}$$

Notice that the divergences or rates of change with respect to distance of the electron and hole current densities are equal but opposite in direction. That is, if we add (9.66) and (9.67), we find that

$$\frac{dJ_p(x)}{dx} + \frac{dJ_n(x)}{dx} = 0 \tag{9.68}$$

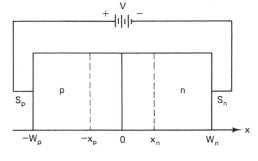

Figure 9.6 An abrupt *p-n* heterojunction with surface recombination velocities S_p at $-W_p$ and S_n at W_n. The applied voltage V is shown in the forward-bias direction.

Integrating over distance, we then have

$$J_p(x) + J_n(x) = J \qquad (9.69)$$

where the integration constant J is the total current density.

Considering (9.68) and (9.69) it is apparent that a diagram for the spatial dependence of the current density as indicated in Fig. 9.7 can be constructed. Here $J_p(x)$ is positive in the positive $J(x)$ direction, while $J_n(x)$ is positive in the negative $J(x)$ direction, producing a constant current density J throughout the heterojunction. The discontinuity in $J_p(x)$ and $J_n(x)$ at $x = 0$ is due to the abrupt change in material parameters.

From (9.69) or Fig. 9.7 it is obvious that we also have

$$J = J_p(-W_p) + J_n(-W_p)$$

or   (9.70)

$$J = J_p(W_n) + J_n(W_n)$$

Subtracting $J_n(-W_p) + J_p(W_n)$ from both sides of these equations gives us

$$J - J_n(-W_p) - J_p(W_n) = - \int_{-W_p}^{W_n} dJ_p(x)$$

or   (9.71)

$$J - J_n(-W_p) - J_p(W_n) = \int_{-W_p}^{W_n} dJ_n(x)$$

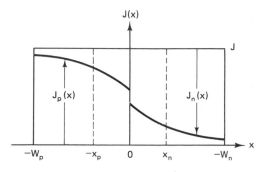

Figure 9.7 Spatial variations of the electron and hole current densities in an abrupt *p-n* heterojunction.

Using (9.66) and (9.67) in (9.71), we obtain one equation for the total current density,

$$J = J_n(-W_p) + J_p(W_n) + q \int_{-W_p}^{W_n} [R(x) - G_t(x)] \, dx \qquad (9.72)$$

The first term, $J_n(-W_p)$, in (9.72) is a contribution to the total current due to the recombination of minority injected electrons on the p surface. The second term, $J_p(W_n)$, is due to the recombination of holes on the n surface. The third term represents the current component due to net recombination in the entire bulk of the heterojunction, including the space-charge region. If this third term is separated into three components, corresponding to net recombination in the p region, the space-charge region, and the n region, and (8.50) and (8.51) for surface recombination currents are used, the total current density can be represented by

$$\frac{J}{q} = \left[S_p \, \delta n_p(-W_p) + \int_{-W_p}^{-x_p} \frac{\delta n_p(x)}{\tau_{np}} \, dx \right]$$

$$+ \left[S_n \, \delta p_n(W_n) + \int_{x_n}^{W_n} \frac{\delta p_n(x)}{\tau_{pn}} \, dx \right] + \int_{-x_p}^{x_n} [R(x) - G_t(x)] \, dx \quad (9.73)$$

In (9.73) we have separated the total current density into three components. The first term in square brackets is the component due to net recombination of injected excess minority electrons on the p-side, including the p surface, while the second term in square brackets is due to recombination of excess holes on the n-side, including the n surface. Although these two terms are due to recombination of excess minority carriers, they are referred to as *diffusion currents*. The reason for this nomenclature will be apparent when we evaluate $\delta n_p(x)$ and $\delta p_n(x)$. The last term in (9.73) is due to net recombination of injected excess electrons and holes in the space-charge region. It is called the *recombination–generation current*.

Notice that with expressions for the excess minority carriers, the surface recombination currents can be determined when S_p and S_n are known. If it is possible to obtain good ohmic contacts on the p and n surfaces, a good assumption is $S_p = S_n = \infty$. This gives $\delta n_p(-W_p) = \delta p_n(W_n) = 0$ and the surface recombination currents are undetermined in (9.73). The surface recombination currents, however, can still be obtained from the gradients of the excess minority concentrations at the surfaces.

An alternative method for determining the total current density is shown in Fig. 9.8. Instead of evaluating $\delta n_p(x)$ and $\delta p_n(x)$ and using (9.73), we can use expressions for $\delta n_p(x)$ and $\delta p_n(x)$ to determine the minority currents at the edges of the space-charge regions. Notice that $J_n(-x_p)$ includes the current due to recombination of excess minority electrons in both the bulk and surface of the p region, while $J_p(x_n)$ includes both components in the n region. When these diffusion currents are added to the recombi-

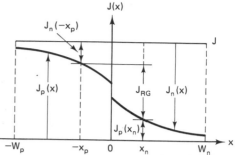

Figure 9.8 Method for determining the total current density: $J = J_n(-x_p) + J_{RG} + J_p(x_n)$.

nation–generation current, J_{RG}, the total current density is obtained,

$$J = J_n(-x_p) + J_{RG} + J_p(x_n) \tag{9.74}$$

We will use the simpler approach.

9.2.1 Diffusion Currents

First, consider the minority hole current in the n region (outside the space-charge region). From (8.141),

$$J_p(x) = \mu_{pn} p_n(x) \frac{d}{dx} [\zeta_p(x)] \tag{9.75}$$

where $\zeta_p(x)$ is the quasi-Fermi energy for holes in the n region given by

$$\zeta_p(x) = E_{in} - q\psi(x) - kT \ln \frac{p_n(x)}{n_{in}} \tag{9.76}$$

Inserting (9.76) into (9.75) and using the Einstein relationship gives us

$$J_p(x) = q\mu_{pn} p_n(x) E(x) - q D_{pn} \frac{d}{dx} [p_n(x)] \tag{9.77}$$

In the depletion approximation we assume that the electric field is zero outside the space-charge region, so that the hole current is simply

$$J_p(x) = -q D_{pn} \frac{d}{dx} [p_n(x)] \tag{9.78}$$

For high currents a significant voltage drop can occur across the n region (and across the p region). Under these conditions (9.78) will be a reasonable approximation when the field across the n region is much less than [see (8.81)]

$$\left(\frac{D_{pn}}{\mu_{pn}^2 \tau_{pn}} \right)^{1/2}$$

Finally, since we are assuming uniform doping in the n region, the hole current density is

$$J_p(x) = -qD_{pn} \frac{d}{dx} [\delta p_n(x)] \tag{9.79}$$

The continuity equation for holes in the n region from (9.66) is

$$\frac{dJ_p(x)}{dx} = -q \frac{\delta p_n(x)}{\tau_{pn}} \tag{9.80}$$

Combining (9.79) and (9.80), the differential equation governing the distribution of holes is

$$\frac{d^2}{dx^2} [\delta p_n(x)] = \frac{\delta p_n(x)}{L_{pn}^2} \tag{9.81}$$

where

$$L_{pn}^2 \equiv D_{pn}\tau_{pn} \tag{9.82}$$

This equation has the solution

$$\delta p_n(x) = A \exp\left(-\frac{x}{L_{pn}}\right) + B \exp\left(\frac{+x}{L_{pn}}\right) \tag{9.83}$$

where A and B are determined from the boundary conditions at x_n and W_n.

The boundary condition at W_n is easily determined when an infinite surface recombination velocity can be assumed. With this assumption $\delta p_n(W_n) = 0$ and (9.83) becomes

$$\delta p_n(x) = \delta p_n(x_n) \frac{\sinh [(W_n - x)/L_{pn}]}{\sinh [(W_n - x_n)/L_{pn}]} \tag{9.84}$$

Using this expression in (9.79) the spatial variation of the hole current density in the n region is

$$J_p(x) = \frac{qD_{pn}}{L_{pn}} \delta p_n(x_n) \frac{\cosh [(W_n - x)/L_{pn}]}{\sinh [(W_n - x_n)/L_{pn}]} \tag{9.85}$$

Notice in (9.84) and (9.85) that the distribution of excess holes and hole current density is completely determined by the relative width of the n region compared to the hole diffusion length. For example, the hole current due to surface recombination (for $S_n = \infty$) is

$$J_p(W_n) = \frac{qD_{pn}}{L_{pn}} \delta p(x_n) \operatorname{csch} \frac{W_n - x_n}{L_{pn}} \tag{9.86}$$

while the hole current at the edge of the space-charge region is

$$J_p(x_n) = \frac{qD_{pn}}{L_{pn}} \delta p(x_n) \coth \frac{W_n - x_n}{L_{pn}} \tag{9.87}$$

It is now only necessary to determine the boundary condition for (9.83) at x_n: that is, we have to determine $\delta p(x_n)$.

A similar analysis for the excess minority electrons in the p region gives

$$\delta n_p(x) = \delta n_p(-x_p) \frac{\sinh [(W_p + x)/L_{np}]}{\sinh [(W_p - x_p)/L_{np}]} \tag{9.88}$$

and

$$J_n(x) = \frac{qD_{np}}{L_{np}} \delta n_p(-x_p) \frac{\cosh [(W_p + x)/L_{np}]}{\sinh [(W_p - x_p)/L_{np}]} \tag{9.89}$$

In this case we have to determine $\delta n_p(-x_p)$.

There are several ways to obtain analytic expressions for $\delta p_n(x_n)$ and $\delta n_p(-x_p)$: the excess minority carrier concentrations injected across the space-charge region. The most common method (at least for homojunctions) is to use (8.149) or (8.150) for the electron or hole current densities in terms of the quasi-Fermi energy gradients. The assumption is made that $d(\zeta_n)/dx$ or $d(\zeta_p)/dx$ is small across the space-charge region: that is, J_n or J_p is approximately zero. Equation (8.146) or (8.147) is then used to obtain ζ_n or ζ_p in terms of n or p, and the result is integrated over the space-charge region. It is then assumed that $\zeta_n - \zeta_p = qV$, and V_0 is eliminated by (9.7) or (9.8) to obtain the desired result.

A much easier way, which uses equivalent assumptions and gives the same result, is to relate the spatial variations of the concentrations to the change in electrostatic potential near equilibrium. For the hole concentrations, (8.134) and (8.132) give

$$p_n(x) = p_{n0} \exp \left[\frac{qV_0 - q\psi(x)}{kT} \right] \tag{9.90}$$

and

$$p_p(x) = p_{p0} \exp \left[\frac{-q\psi(x)}{kT} \right] \tag{9.91}$$

Evaluating these equations at the edges of the space-charge region and taking the ratio of the two gives us

$$\frac{p_n(x_n)}{p_p(-x_p)} = \frac{p_{n0}}{p_{p0}} \exp \left[\frac{q\psi(-x_p) - q\psi(x_n) + qV_0}{kT} \right] \tag{9.92}$$

The assumptions used in the analysis are illustrated schematically in the forward-bias energy band diagram of Fig. 9.9(a). Notice that with these

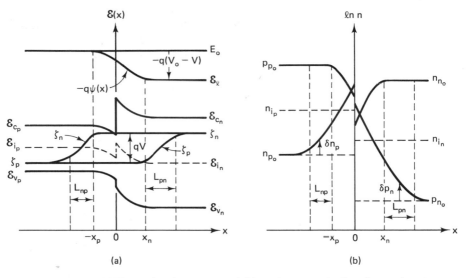

Figure 9.9 (a) Energy band structure and (b) carrier concentrations for an abrupt *p-n* heterojunction under forward-bias conditions.

assumptions $\psi(-x_p) = 0$ and $\psi(x_n) = V_0 - V$. Using these results in (9.92), we have

$$\frac{p_{n0} + \delta p_n(x_n)}{p_{p0} + \delta p_p(-x_p)} = \frac{p_{n0}}{p_{p0}} \exp\left(\frac{qV}{kT}\right) \tag{9.93}$$

From (8.131) and (8.133) and the potential indicated above,

$$\frac{n_{p0} + \delta n_p(-x_p)}{n_{n0} + \delta n_n(x_n)} = \frac{n_{p0}}{n_{n0}} \exp\left(\frac{qV}{kT}\right) \tag{9.94}$$

Because of space-charge neutrality in the *p* and *n* regions, we also have

$$\delta p_n(x_n) = \delta n_n(x_n) \tag{9.95}$$
$$\delta n_p(-x_p) = \delta p_p(-x_p)$$

Equation (9.95) reduces (9.93) and (9.94) to a set of coupled equations,

$$\frac{\delta p_n(x_n) + p_{n0}}{\delta n_p(-x_p) + p_{p0}} = \frac{p_{n0}}{p_{p0}} \exp\left(\frac{qV}{kT}\right) \tag{9.96}$$
$$\frac{\delta n_p(-x_p) + n_{p0}}{\delta p_n(x_n) + n_{n0}} = \frac{n_{p0}}{n_{n0}} \exp\left(\frac{qV}{kT}\right)$$

which can be solved for the desired quantities. After some manipulation we find that

$$\delta p_n(x_n) = \frac{p_{n0}}{\Delta}\left[\exp\left(\frac{qV}{kT}\right) - 1\right]\left[\left(\frac{n_{ip}}{N_a}\right)^2 \exp\left(\frac{qV}{kT}\right) + 1\right] \tag{9.97}$$

and

$$\delta n_p(-x_p) = \frac{n_{p0}}{\Delta}\left[\exp\left(\frac{qV}{kT}\right) - 1\right]\left[\left(\frac{n_{in}}{N_d}\right)^2 \exp\left(\frac{qV}{kT}\right) + 1\right] \quad (9.98)$$

where

$$\Delta \equiv 1 - \left(\frac{n_{ip}n_{in}}{N_aN_d}\right)^2 \exp\left(\frac{2qV}{kT}\right) \quad (9.99)$$

The denominator Δ can also be put in the form

$$\Delta = 1 - \exp\left[\frac{2q}{kT}\left(V - V_0 - \frac{\Delta\mathscr{E}_i}{q}\right)\right] \quad (9.100)$$

which goes to zero as V approaches $V_0 + \Delta\mathscr{E}_i/q$. Thus $V_0 + \Delta\mathscr{E}_i/q$ is the limiting voltage across the space-charge region. The current is, of course, limited by the resistance of the p and n regions.

Equations (9.97) and (9.98) can now be used in (9.84), (9.85) and (9.88), (9.89), respectively, to characterize the voltage dependence of the excess carriers and diffusion currents. Notice, however, that (9.97) and (9.98) have terms of the form $(n_{ip}/N_a)^2 \exp(qV/kT)$, which are small when V is small. Under this so-called *low-level injection* condition, (9.97) and (9.98) reduce to

$$\delta p_n(x_n) = p_{n0}\left[\exp\left(\frac{qV}{kT}\right) - 1\right] \quad (9.101)$$

$$\delta n_p(-x_p) = n_{p0}\left[\exp\left(\frac{qV}{kT}\right) - 1\right] \quad (9.102)$$

The resulting diffusion currents are then

$$J_p(x_n) = \frac{qD_{pn}p_{n0}}{L_{pn}}\coth\frac{W_n - x_n}{L_{pn}}\left[\exp\left(\frac{qV}{kT}\right) - 1\right] \quad (9.103)$$

$$J_n(-x_p) = \frac{qD_{np}n_{p0}}{L_{np}}\coth\frac{W_p - x_p}{L_{np}}\left[\exp\left(\frac{qV}{kT}\right) - 1\right] \quad (9.104)$$

which have the form of the ideal diode equation.

Although (9.103) and (9.104) (and the other equations in this section) were derived assuming forward bias, there is nothing in the analysis to preclude their use in reverse bias. For (9.103) and (9.104), in forward bias a positive V results in an exponential rise in current which easily dominates the equations. On the other hand, a negative V, corresponding to reverse bias, reduces the exponential term from unity to produce diffusion currents in the opposite direction. These currents quickly saturate at small values determined by the material parameters and the geometry of the device.

Also under low-level injection conditions the excess minority carrier distributions are given by

$$\delta p_n(x) = p_{n0} \frac{\sinh\left[(W_n - x)/L_{pn}\right]}{\sinh\left[(W_p - x_p)/L_{pn}\right]} \left[\exp\left(\frac{qV}{kT}\right) - 1\right] \qquad (9.105)$$

and

$$\delta n_p(x) = n_{p0} \frac{\sinh\left[(W_p + x)/L_{np}\right]}{\sinh\left[(W_p - x_p)/L_{np}\right]} \left[\exp\left(\frac{qV}{kT}\right) - 1\right] \qquad (9.106)$$

These spatial variations are indicated in Fig. 9.9(b) for a given forward bias. Notice in (9.105) and (9.106) that in reverse bias the excess carrier distributions become negative. This is because under these conditions, carriers are extracted and there is thus a deficit of carriers.

9.2.2 Recombination–Generation Current

We saw from (9.73) that the recombination–generation in the space-charge region was given by

$$J_{RG} = q \int_{-x_p}^{x_n} [R(x) - G_t(x)]\, dx \qquad (9.107)$$

For *direct* recombination in a heterostructure,

$$R(x) - G_t(x) = r(x)[n(x)p(x) - n_i^2(x)] \qquad (9.108)$$

In the case of an abrupt *p-n* heterojunction, (9.107) and (9.108) indicate that

$$J_{RG} = q \int_{-x_p}^{0} r_p[n_p(x)p_p(x) - n_{ip}^2]\, dx + q \int_{0}^{x_n} r_n[n_n(x)p_n(x) - n_{in}^2]\, dx$$

$$(9.109)$$

where from (8.148),

$$n(x)p(x) = n_i^2 \exp\left[\frac{\zeta_n(x) - \zeta_p(x)}{kT}\right] \qquad (9.110)$$

Assuming, as indicated in Fig. 9.9(a), that

$$\zeta_n(x) - \zeta_p(x) = qV \qquad (9.111)$$

across the space-charge region, then

$$J_{RG} = q(r_p n_{ip}^2 x_p + r_n n_{in}^2 x_n)\left[\exp\left(\frac{qV}{kT}\right) - 1\right] \qquad (9.112)$$

Using (9.57) to (9.60) in (9.112), we have for the recombination–generation current,

$$J_{\text{RG}} = q \left[\frac{r_p n_{ip}^2 N_d + r_n n_{in}^2 N_a}{N_d + N_a} \right] \left[\exp\left(\frac{qV}{kT}\right) - 1 \right] W \qquad (9.113)$$

It can be seen that the recombination–generation current depends on the volume of the space-charge region through its dependence on W. Also, since W varies with $V^{1/2}$, J_{RG} does not have the ideal diode form. The voltage dependence for *indirect* recombination is also less than ideal because of an $\exp(qV/kT)$ dependence in the denominator (see the problems). In the forward-bias direction, J_{RG} is dominated by its exponential voltage dependence and is primarily a recombination current. In reverse bias, J_{RG} is dominated by the $V^{1/2}$ dependence of W and is primarily a generation current.

Because of the complicated dependence of J_{RG} at any bias and $J_p(x_n)$ and $J_n(x_p)$ at high forward bias on voltage, it is often useful to represent the total current density in the form

$$J(V) = J_0(V) \left[\exp\left(\frac{qV}{kT}\right) - 1 \right] \qquad (9.114)$$

or

$$J(V) = J_0 \left[\exp\left(\frac{qV}{\eta kT}\right) - 1 \right] \qquad (9.115)$$

where $J_0(V)$ or $\eta(V)$ are determined empirically.

9.3 IMPACT IONIZATION

When a reverse bias is applied to a p-n heterojunction, the electric field across the space-charge region can become very large. Under these conditions carriers in the space-charge region can gain sufficient energy from the field to create secondary electron–hole pairs when they collide with the lattice. This process is called *impact* ionization. The primary carriers that initiate this process are the minority carriers within a diffusion length of either side of the space-charge region and the thermally or optically generated carriers within the space-charge region. The secondary electron–hole pairs that the primary carriers create are separated and accelerated by the field so that they can, in turn, create electron–hole pairs, and so on. When, on the average, each electron–hole pair creates an electron–hole pair, the reverse-bias current increases abruptly to a value limited by the resistance of the circuit. This is *avalanche breakdown,* and the reverse-bias voltage at which this occurs is the *breakdown voltage.*

9.3.1 Threshold Energy

To determine the energy a carrier must have to initiate impact ionization, consider an electron with energy $\mathscr{E}_e$ and velocity $\mathbf{v}_i$ just prior to a collision. We will assume parabolic, isotropic, nondegenerate energy bands. The initial energy of this primary electron, referenced to the bottom of the conduction band, is

$$\mathscr{E}_e = \tfrac{1}{2}m_e\mathbf{v}_i^2 \qquad (9.116)$$

and its momentum is

$$\mathbf{p}_e = m_e\mathbf{v}_i \qquad (9.117)$$

Assume that, after the collision, the primary electron has a final velocity $\mathbf{v}_f$ and it has created a secondary electron with velocity $\mathbf{v}_e$ and a secondary hole with velocity $\mathbf{v}_h$. Conservation of energy and momentum requires that

$$\mathscr{E}_e = \tfrac{1}{2}m_e\mathbf{v}_f^2 + \tfrac{1}{2}m_e\mathbf{v}_e^2 + \tfrac{1}{2}m_h\mathbf{v}_h^2 + \mathscr{E}_g \qquad (9.118)$$

and

$$\mathbf{p}_e = m_e\mathbf{v}_f + m_e\mathbf{v}_e + m_h\mathbf{v}_h \qquad (9.119)$$

We want to find the minimum initial electron energy (threshold energy), $\mathscr{E}_{et}$, which will produce impact ionization.

First, consider momentum conservation during the collision process. Figure 9.10 illustrates schematically two ways in which momentum can be conserved. In Fig. 9.10(a) momentum is conserved, but since energy depends on the magnitude of velocity, energy values are high. To conserve momentum at lowest energy during a collision, all the momentum vectors must be collinear, as indicated in Fig. 9.10(b). Thus, to determine the threshold energy, we can treat the momentum vectors as scalar.

The problem, then, is to find the minimum energy for an electron to

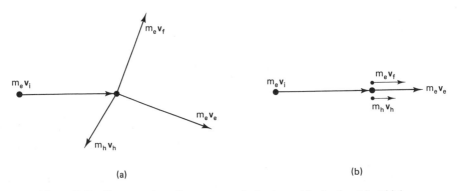

(a)   (b)

Figure 9.10 Conservation of momentum during impact ionization (a) at high energy and (b) at low energy.

create an electron–hole pair subject to the constraint that the magnitude of the momentum is conserved. The easiest way of treating this problem is by the method of Lagrange's undetermined multiplier, as discussed in Chapter 4. In this case we want to find an extrema (minimum) of $\mathscr{E}_e$ subject to the constraint that $|\mathbf{p}_e| = p_e$ remain constant. The threshold energy is then determined from

$$d\mathscr{E}_e(v_f, v_e, v_h) + \alpha\, dp_e(v_f, v_e, v_h) = 0 \qquad (9.120)$$

which is independent of the undetermined multiplier, α.

Taking the first partial derivative with respect to v_f,

$$\frac{\partial \mathscr{E}_e}{\partial v_f} + \alpha\, \frac{\partial p_e}{\partial v_f} = m_e v_f + \alpha m_e = 0 \qquad (9.121)$$

or

$$\alpha = -v_f \qquad (9.122)$$

The partial derivatives with respect to v_e and v_h give similar results and we find that the minimum energy occurs when

$$v_f = v_e = v_h \qquad (9.123)$$

Using this equation in (9.119), we have

$$v_f = \frac{m_e}{2m_e + m_h} v_i \qquad (9.124)$$

Substituting (9.124) into (9.118) and using (9.116), the threshold energy for electrons is

$$\mathscr{E}_{et} = \frac{2m_e + m_h}{m_e + m_h} \mathscr{E}_g \qquad (9.125)$$

A similar analysis for holes (see the problems) results in the threshold energy

$$\mathscr{E}_{ht} = \left[1 + \frac{m_l(1 - \Delta/\mathscr{E}_g)}{2m_h + m_e - m_l}\right] \mathscr{E}_g \qquad (9.126)$$

where Δ is the spin-orbit splitting energy, and m_h and m_l are the heavy and light hole masses, respectively.

Figure 9.11(a) and (b) show schematically the electron and hole transitions corresponding to the threshold energies given by (9.125) and (9.126), respectively. Notice that at threshold a primary electron creates a heavy hole in (a), while impact ionization is initiated by a light hole in the spin-orbit split band in (b). It can also be seen from the equations that the threshold energies for either electron- or hole-initiated impact ionization are constrained between $\mathscr{E}_g$ and $2\mathscr{E}_g$.

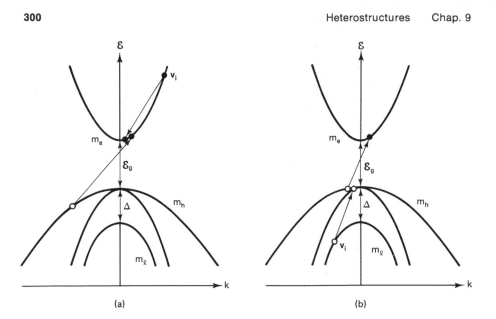

Figure 9.11 Electron and hole transitions corresponding to the threshold energy for impact ionization initiated by (a) electrons and (b) holes.

9.3.2 Multiplication and Breakdown

Let us now consider the effect of impact ionization on the reverse steady-state current in a one-dimensional *p-n* heterojunction. From (8.57) and (8.58) we obtain the continuity equations for this problem,

$$\frac{d}{dx} J_n(x) = qG'(x)$$

$$-\frac{d}{dx} J_p(x) = qG'(x)$$

(9.127)

In (9.127) we have changed the sign convention so that the reverse current is positive; neglected the recombination term because the generated carriers are quickly separated by the large field; and included in $G'(x)$ optical, thermal, and impact ionization generation of electron–hole pairs.

To take into account impact ionization, we define an ionization coefficient for electrons, $\alpha(x)$, and an ionization coefficient for holes, $\beta(x)$. These ionization coefficients are usually determined empirically and are strongly dependent on the electric field (see problems), which in general varies spatially. They represent the probability that an electron or hole, traveling a unit distance, will create an electron–hole pair. Thus the number of electrons and holes created per unit volume in unit time by an electron current $J_n(x)$

and a hole current $J_p(x)$ is

$$\frac{1}{q} [\alpha(x)J_n(x) + \beta(x)J_p(x)]$$

Using this concept in (9.127), we can separate $G'(x)$ into components produced by impact ionization and by optical and thermal generation, $G(x)$.

This gives us a set of coupled differential equations for the electron and hole current densities,

$$\frac{d}{dx} J_n(x) = \alpha(x)J_n(x) + \beta(x)J_p(x) + qG(x)$$

$$-\frac{d}{dx} J_p(x) = \alpha(x)J_n(x) + \beta(x)J_p(x) + qG(x)$$

(9.128)

Notice that when J_n and J_p in these equations have units of amperes cm^{-2}, α and β are in cm^{-1} and G is in cm^{-3} s^{-1}. In the space-charge region of a heterojunction, it is understood that not only can J_n and J_p be discontinuous at the interface, but also α, β, and G. Equations (9.128) can be decoupled with (9.69) to obtain

$$\frac{d}{dx} J_n(x) = \alpha(x)J_n(x) + \beta(x)[J - J_n(x)] + qG(x)$$

$$-\frac{d}{dx} J_p(x) = \alpha(x)[J - J_p(x)] + \beta(x)J_p(x) + qG(x)$$

(9.129)

where the total current density, J, is the only quantity that is independent of x.

Equations (9.129) can be simplified by defining a difference in ionization coefficients,

$$\gamma(x) \equiv \alpha(x) - \beta(x)$$

(9.130)

with the results,

$$\frac{d}{dx} J_n - \gamma J_n = \beta J + qG$$

(9.131)

$$\frac{d}{dx} J_p - \gamma J_p = \alpha J - qG$$

(9.132)

Except for J, the dependence of the quantities on x is implicit. These equations can be solved by using the integrating factor,

$$\exp\left(-\int_{-x_p}^{x} \gamma \, d\xi \right)$$

Multiplying both sides of (9.131) by this quantity, we can see that the left-hand side becomes an exact differential,

$$\frac{d}{dx}\left[J_n \exp\left(-\int_{-x_p}^{x} \gamma \, d\xi \right) \right] = \exp\left(-\int_{-x_p}^{x} \gamma \, d\xi \right)(\beta J + qG) \quad (9.133)$$

Because we are interested in the minority electron current that enters the space-charge region at $-x_p$, we integrate (9.133) from $-x_p$ to x and obtain

$$J_n \exp\left(-\int_{-x_p}^{x} \gamma \, d\xi \right) - J_n(-x_p) = \int_{-x_p}^{x} \exp\left(-\int_{-x_p}^{\eta} \gamma \, d\xi \right)(\beta J + qG) \, d\eta$$

$$(9.134)$$

The same integrating factor can be used for (9.132). The boundary conditions are different, however, since we are interested in the hole current at x_n. Integrating from x to x_n gives us

$$J_p(x_n) \exp\left(-\int_{-x_p}^{x_n} \gamma \, d\xi \right) - J_p \exp\left(-\int_{-x_p}^{x} \gamma \, d\xi \right)$$

$$= -\int_{x}^{x_n} \exp\left(-\int_{-x_p}^{\eta} \gamma \, d\xi \right)(\alpha J + qG) \, d\eta \quad (9.135)$$

To simplify notation, let

$$\phi(x) \equiv \exp\left(-\int_{-x_p}^{x} \gamma \, d\xi \right) \quad (9.136)$$

Equations (9.134) and (9.135) now become

$$\phi(x)J_n(x) = J_n(-x_p) + \int_{-x_p}^{x} \phi(\eta)(\beta J + qG) \, d\eta \quad (9.137)$$

$$\phi(x)J_p(x) = \phi(x_n)J_p(x_n) + \int_{x}^{x_n} \phi(\eta)(\alpha J + qG) \, d\eta \quad (9.138)$$

Adding (9.137) and (9.138) and using (9.69) and (9.130), we obtain

$$\phi(x)J = J_n(-x_p) + \phi(x_n)J_p(x_n)$$

$$+ J\left[\int_{-x_p}^{x_n} \phi(\eta)\alpha \, d\eta - \int_{-x_p}^{x} \phi(\eta)\gamma \, d\eta \right] + q\int_{-x_p}^{x_n} G \, d\eta \quad (9.139)$$

From (9.136) we see that the integrand in the term

$$-\int_{-x_p}^{x} \phi(\eta)\gamma \, d\eta = \phi(x) - 1 \quad (9.140)$$

is an exact differential. Using this in (9.139) and solving for J, we find that

$$J = \frac{J_n(-x_p) + \phi(x_n)J_p(x_n) + q \int_{-x_p}^{x_n} \phi(\eta)G \, d\eta}{1 - \int_{-x_p}^{x_n} \phi(\eta)\alpha \, d\eta} \tag{9.141}$$

Notice that (9.141) has the form

$$J = M_n J_n(-x_p) + M_p J_p(x_n) + M_G J_G \tag{9.142}$$

where M_n, M_p, and M_G are multiplication factors. That is, each component of current in the space-charge region is multiplied differently depending on the material parameters in each region. Avalanche breakdown occurs when the denominator in (9.141) goes to zero. Under these conditions the current is limited by the resistance of the circuit, including the resistance of the ohmic parts of the heterojunction. When the ionization coefficients are known, the breakdown voltage can be obtained from the condition

$$\int_{-x_p}^{x_n} \phi(\eta)\alpha \, d\eta = 1 \tag{9.143}$$

9.4 TUNNELING

Another phenomenon that can affect the current–voltage characteristics of a heterostructure is tunneling. Electrons can tunnel through notches in the conduction band and holes can tunnel through notches in the valence band. For heavily doped structures the space-charge region can be small enough to allow appreciable tunneling of electrons from the conduction band into empty states in the valence band. In forward-biased p-n heterojunctions this can produce the negative differential resistance of the Esaki or tunnel diode [L. Esaki, *Phys. Rev. 109*, 603 (1958)], while in reverse bias it can lead to Zener breakdown [C. Zener, *Proc. R. Soc. London A145*, 523 (1934)].

To examine the problem of tunneling, consider a one-dimensional energy barrier $U(x)$ to an electron with energy $\mathscr{E}$. In the effective mass approximation Schrödinger's equation (3.31) is

$$\left[\frac{-\hbar^2}{2m^*}\frac{d^2}{dx^2} + U(x)\right]A(x) = \mathscr{E}A(x) \tag{9.144}$$

Assume a solution for the wavefunction of the form

$$A(x) = \exp[\alpha(x)] \tag{9.145}$$

Substituting (9.145) into (9.144), we obtain

$$\frac{\hbar^2}{2m^*}\left[\frac{d^2\alpha}{dx^2} + \left(\frac{d\alpha}{dx}\right)^2\right] = U(x) - \mathscr{E} \tag{9.146}$$

For slowly varying α it can be assumed that

$$\frac{d^2\alpha}{dx^2} \ll \left(\frac{d\alpha}{dx}\right)^2$$

and (9.146) becomes

$$\frac{d\alpha}{dx} \simeq \pm \left\{\frac{2m^*}{\hbar^2}[U(x) - \mathscr{E}]\right\}^{1/2} \qquad (9.147)$$

This is the WKB approximation [G. Wentzel, *Z. Phys. 38,* 518 (1926); H. A. Kramers, *Z. Phys. 39,* 828 (1926); L. Brillouin, *C. R. Acad. Sci. 183,* 24 (1926)]. Equation (9.147) is easily solved to obtain

$$A(\mathscr{E}) = \exp\left(\pm \int_x \left\{\frac{2m^*}{\hbar^2}[U(\xi) - \mathscr{E}]\right\}^{1/2} d\xi\right) \qquad (9.148)$$

where the integral is taken over the region where $U(x) > \mathscr{E}$. The tunneling probability is given by

$$P(\mathscr{E}) = A(\mathscr{E})A^*(\mathscr{E}),$$

$$= \exp\left(\pm 2 \int_x \left\{\frac{2m^*}{\hbar^2}[U(\xi) - \mathscr{E}]\right\}^{1/2} d\xi\right) \qquad (9.149)$$

To illustrate the use of (9.149), consider the triangular energy barrier shown in Fig. 9.12. In this case

$$U(x) - \mathscr{E} = \Delta\mathscr{E}\left(1 - \frac{x}{x_0}\right) \qquad (9.150)$$

and the tunneling probability is

$$P(\mathscr{E}) = \exp\left(-2 \int_0^{x_0} \left[\frac{2m^*}{\hbar^2}\Delta\mathscr{E}\left(1 - \frac{x}{x_0}\right)\right]^{1/2} dx\right) \qquad (9.151)$$

The minus sign is used in the positive x direction to obtain an exponentially

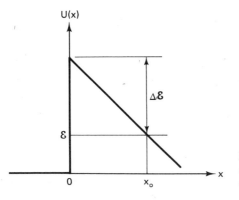

Figure 9.12 Triangular barrier of height $\Delta\mathscr{E}$ and width x_0 to an electron with energy $\mathscr{E}$.

decaying wavefunction. Performing the integration yields

$$P(\mathscr{E}) = \exp\left[-\frac{4}{3}\left(\frac{2m^*}{\hbar^2}\Delta\mathscr{E}\right)^{1/2} x_0 \right] \qquad (9.152)$$

The width of the triangular barrier can be eliminated from (9.152) with

$$x_0 = \frac{\Delta\mathscr{E}}{qE}$$

where E is the constant electric field, to give

$$P(\mathscr{E}) = \exp\left[\frac{-4(2m^*)^{1/2}\,\Delta\mathscr{E}^{3/2}}{3q\hbar E} \right] \qquad (9.153)$$

To determine the tunneling current through an arbitrary barrier, it is necessary to integrate the tunneling probability over the energy distribution of initial and final states. The current in the forward direction through the barrier is proportional to the number of filled initial states, the number of empty final states, and the tunneling probability. That is,

$$J_f = C \int_{\mathscr{E}} N_i(\mathscr{E})f_i(\mathscr{E})N_f(\mathscr{E})[1 - f_f(\mathscr{E})]P(\mathscr{E})\,d\mathscr{E} \qquad (9.154)$$

where C is a proportionality constant. Conversely, the current in the reverse direction is

$$J_r = C \int_{\mathscr{E}} N_f(\mathscr{E})f_f(\mathscr{E})N_i(\mathscr{E})[1 - f_i(\mathscr{E})]P(\mathscr{E})\,d\mathscr{E} \qquad (9.155)$$

The net tunneling current is then the difference between (9.154) and (9.155), or

$$J = C \int_{\mathscr{E}} N_i(\mathscr{E})N_f(\mathscr{E})[f_i(\mathscr{E}) - f_f(\mathscr{E})]P(\mathscr{E})\,d\mathscr{E} \qquad (9.156)$$

In this equation the integration is performed over the energy range where tunneling can occur.

9.5 CAPACITANCE

In Section 9.1 we obtained expressions for the capacitance C_i of the space-charge region in intrinsic (9.37) and extrinsic (9.62) heterostructures. This capacitance, in the extrinsic case, is due to changes in uncovered ionized impurities at the edges of the space-charge region with changes in applied bias. When current flows, there are also capacitive components due to changes in excess charge carriers with applied bias. This is because the change of charge with voltage produces a capacitance.

For a *p-n* heterojunction the total excess charge per unit area can be represented by

$$\delta Q = q \int_{-W_p}^{W_n} \delta p(x) \, dx$$

or   (9.157)

$$-\delta Q = q \int_{-W_p}^{W_n} \delta n(x) \, dx$$

The total excess charge in (9.157) can be broken up into components for the *p*-region, the space-charge region, and the *n*-region:

$$\delta Q = q \int_{-W_p}^{-x_p} \delta p_p(x) \, dx + q \int_{-x_p}^{x_n} \delta p(x) \, dx + q \int_{x_n}^{W_n} \delta p_n(x) \, dx$$

or   (9.158)

$$-\delta Q = q \int_{-W_p}^{-x_p} \delta n_p(x) \, dx + q \int_{-x_p}^{x_n} \delta n(x) \, dx + q \int_{x_n}^{W_n} \delta n_n(x) \, dx$$

For times longer than the dielectric relaxation times of the majority carriers,

$$\int_{-W_p}^{-x_p} \delta p_p(x) \, dx = \int_{-W_p}^{-x_p} \delta n_p(x) \, dx$$

$$\int_{x_n}^{W_n} \delta n_n(x) \, dx = \int_{x_n}^{W_n} \delta p_n(x) \, dx$$

                                                                             (9.159)

Therefore,

$$\int_{-x_p}^{x_n} \delta p(x) \, dx = \int_{-x_p}^{x_n} \delta n(x) \, dx \qquad\qquad (9.160)$$

The total excess charge in the heterojunction can thus be separated into the excess minority charge in the *p*- and *n*-regions plus the excess charge in the space-charge region,

$$\delta Q = \delta Q_p + \delta Q_n + \delta Q_s \qquad\qquad (9.161)$$

Let us consider first the excess charge in the *p*-region,

$$-\delta Q_p = q \int_{-W_p}^{-x_p} \delta n_p(x) \, dx \qquad\qquad (9.162)$$

Using (9.106) in (9.162) and performing the integration, we have

$$-\delta Q_p = qL_{np}n_{p0} \frac{\cosh \left[(W_p - x_p)/L_{np} \right] - 1}{\sinh \left[(W_p - x_p)/L_{np} \right]} \left[\exp \left(\frac{qV}{kT} \right) - 1 \right] \qquad (9.163)$$

The capacitance can now be obtained from

$$C_p = \left| \frac{d}{dV} (\delta Q_p) \right| \tag{9.164}$$

Notice, however, from (9.57) that x_p is proportional to the square root of V. Since the change in x_p with V will be small in comparison to the exponential V dependence,

$$C_p \simeq \frac{q^2 L_{np} n_{p0}}{kT} \left(\coth \frac{W_p - x_p}{L_{np}} - \operatorname{csch} \frac{W_p - x_p}{L_{np}} \right) \exp \left(\frac{qV}{kT} \right) \tag{9.165}$$

In a similar manner, the capacitance of the excess charge in the n-region is,

$$C_n \simeq \frac{q^2 L_{pn} P_{n0}}{kT} \left(\coth \frac{W_n - x_n}{L_{pn}} - \operatorname{csch} \frac{W_n - x_n}{L_{pn}} \right) \exp \left(\frac{qV}{kT} \right) \tag{9.166}$$

Equations (9.165) and (9.166) were derived for low-level injection. C_p and C_n are referred to as *diffusion capacitances* corresponding to the diffusion currents of (9.103) and (9.104).

Let us examine the excess charge in the space-charge region,

$$\delta Q_s = q \int_{-x_p}^{x_n} \delta p(x) \, dx \tag{9.167}$$

Because of the discontinuity at the heterojunction, we divide this excess charge into two components,

$$\delta Q_s = q \int_{-x_p}^{0} \delta p_p(x) \, dx + q \int_{0}^{x_n} \delta p_n(x) \, dx \tag{9.168}$$

Although in the depletion approximation it is assumed that, to first order, there are no holes (or electrons) in the space-charge region, the spatial dependence of the hole (or electron) concentration can be obtained on the first iteration. Using the nonequilibrium forms of (9.47) in (9.91) and (9.48) in (9.90), we obtain

$$p_p(x) = p_{p0} \exp \left[- \left(\frac{x_p + x}{2\lambda_p} \right)^2 \right] \tag{9.169}$$

and

$$p_n(x) = p_{n0} \exp \left[+ \left(\frac{x_n - x}{2\lambda_n} \right)^2 \right] \exp \left(\frac{qV}{kT} \right) \tag{9.170}$$

respectively. In these equations λ_p and λ_n are the extrinsic Debye lengths given by

$$\lambda_p^2 \equiv \frac{kT\epsilon_p}{2q^2 N_a} \tag{9.171}$$

and

$$\lambda_n^2 \equiv \frac{kT\epsilon_n}{2q^2 N_d} \qquad (9.172)$$

Equation (9.168) can also be expressed as

$$\delta Q_s = q \int_{-x_p}^{0} [p_p(x) - p_{p0}] \, dx + q \int_{0}^{x_n} [p_n(x) - p_{n0}] \, dx \qquad (9.173)$$

In this form there are four components to the excess charge in the space-charge region. The last term, involving p_{n0}, is clearly small in comparison to the first three, so it will be neglected. From (9.170) we see that the third term in (9.173) can be significant at high forward bias. We will assume low-level injection and neglect this term also. This leaves us with the first two terms, which can be put in the form

$$\delta Q_s \simeq -q N_a x_p + q \int_{-x_p}^{0} p_p(x) \, dx \qquad (9.174)$$

Now we can see that the first term in (9.174) is the charge due to the ionized acceptors, while the second term is due to the holes injected into the space-charge region. The capacitance due to the ionized acceptors (and donors) is given by

$$C_i = \frac{\epsilon_p \epsilon_n (N_a + N_d)}{(\epsilon_p N_a + \epsilon_n N_d) W} \qquad (9.175)$$

It remains to evaluate the last term, which we will call δQ_c.

Inserting (9.169) in (9.174) and changing variables, we have

$$\delta Q_c \simeq 2 q p_{p0} \lambda_p \int_{0}^{x_p/2\lambda_p} \exp(-\xi^2) \, d\xi \qquad (9.176)$$

Differentiating with respect to voltage, the capacitance of the space-charge region due to free carriers is

$$C_c \simeq q p_{p0} \exp\left[-\left(\frac{x_p}{2\lambda_p} \right)^2 \right] \left| \frac{dx_p}{dV} \right| \qquad (9.177)$$

Using (9.57) and (9.62), this becomes

$$C_c \simeq C_i \exp\left[\frac{-q\epsilon_n N_d (V_0 - V)}{kT(\epsilon_n N_d + \epsilon_p N_a)} \right] \qquad (9.178)$$

From (9.178) we see that the space-charge capacitance is dominated by the ionized impurity contribution for low-level injection.

To determine the response of a heterojunction to an applied voltage that varies with time, the capacitive components given by (9.62), (9.165), (9.166), and (9.178) should be considered. Notice that these components all operate in parallel since they are produced by the same bias.

PROBLEMS

9.1. Find the capacitance per unit area of an abrupt intrinsic heterojunction [A. Chatterjee and A. H. Marshak, *Solid-State Electron. 24*, 1111 (1981)].

9.2. Plot and compare the electric field and excess carrier distributions in the equilibrium space-charge region of an abrupt *p-n* heterojunction in the depletion approximation and in the exact case.

9.3. An abrupt heterostructure is formed between GaAs with a lattice constant of 5.654 Å and Ge with a lattice constant of 5.658 Å. Determine the number of dangling bonds per square centimeter for {100}, {110}, and {111} interfaces.

9.4. Show that the heterojunction theory proposed by M. J. Adams and A. Nussbaum [*Solid-State Electron. 22*, 783 (1979)] is not self-consistent, and therefore must be wrong.

9.5. Show that integrating the continuity equations gives the same expressions for the diffusion currents $J_{pn}(x)$ and $J_{np}(x)$ as were obtained from the current equations. Is the current due to recombination or diffusion? Explain.

9.6. For indirect recombination where $\mathcal{E}_t(x) = \mathcal{E}_i(x)$ show for a *p-n* homojunction that

$$J_{RG} \leq \frac{qn_i W}{2\sqrt{\tau_{n0}\tau_{p0}}\, e(qV/2kT) + \tau_{p0} + \tau_{n0}} \left(e^{\frac{qV}{kT}} - 1 \right)$$

[*Hint:* First show that the maximum value of an integral occurs for $\mathcal{E}_i(x) = [(\zeta_n + \zeta_p)/2] + (kT/2) \ln (\tau_{p0}/\tau_{n0})$.]

9.7. For isotropic, parabolic, nondegenerate energy bands, show that the threshold energy for impact ionization initiated by a hole is given by

$$\mathcal{E}_{ht} = \left[1 + \frac{m_l(1 - \Delta/\mathcal{E}_g)}{2m_h + m_e - m_l} \right] \mathcal{E}_g$$

where Δ is the spin-orbit splitting energy.

9.8. The ionization coefficients for electrons and holes in GaAs are

$$\alpha(x) = 2 \times 10^6 \exp \left[\frac{-2 \times 10^6}{E(x)} \right], \qquad \beta(x) = 1 \times 10^5 \exp \left[\frac{-5 \times 10^5}{E(x)} \right] \text{cm}^{-1}$$

for $E(x)$ in V/cm. The dielectric constant is 12.5. Determine the breakdown voltage at 300 K for an abrupt *p-n* junction with $N_a = 10^{18}$ cm^{-3}, $N_d = 10^{15}$ cm^{-3}.

9.9. In intrinsic material show that the potential due to band bending at the surface can be approximated by

$$\psi(x) \simeq \psi(0) - \frac{2kT}{q} \frac{x}{\lambda_i} \sinh \frac{q\psi(0)}{2kT} \qquad \text{for} \quad \frac{x}{\lambda_i} \text{ small}$$

9.10. In GaAs with $N_d - N_a = 1 \times 10^{15}$ cm^{-3}, $n_i = 1 \times 10^7$ cm^{-3}, $\epsilon_r = 12.5$, and $T = 300$ K, plot and compare the number of surface states required to give values of built-in potential $0 \leq V_0 \leq 1.4$ eV in the depletion approximation and in the exact case.

9.11. A p-n junction of 1-cm^2 area with a current–voltage characteristic

$$I = I_s \left[\exp \left(\frac{qV}{\beta kT} \right) - 1 \right]$$

has light incident through the p-region of which 3 μm is undepleted. The other parameters of the device are $\alpha = 10^4$ cm^{-1}, $R = 0.32$, $L_n = 2$ μm, $L_p = 4$ μm, and $I_s = 10$ nA. To a good approximation the width of the depletion region remains constant at 5 μm.

(a) If the open-circuit voltage is 0.31 V and the short-circuit current is 4 μA at 300 K, what is the ideality factor β?

(b) What is the incident photon flux?

9.12. Consider an abrupt n^+p junction such that in the forward direction all the current is due to the injection of electrons into the p-region. That is, just on the p-side of the junction, $J = J_n$ and $J_p = 0$. Use the current equations to obtain expressions for the low-level ($\delta n_p \ll N_a$) and high-level ($\delta n_p \gg N_a$) injected current density. Assume that $\delta p_p = \delta n_p \gg n_{p0}$.

9.13. A p-n junction has a capacitance of 10 pF at zero bias, a capacitance of 8.736 pF at a reverse bias of 0.5 V, and a capacitance of 7.937 pF at a reverse bias of 1.0 V.

(a) Is this a linear or an abrupt junction?

(b) What is the built-in voltage?

9.14. Draw the equilibrium energy band diagram and electric field distribution for a p-i-n. homojunction with an intrinsic region of width t. Assume that $N_a = N_d \gg n_i$.

(a) Determine the total built-in voltage for $N_a = N_d = 10^{19}$ cm^{-3}, $n_i = 10^7$ cm^{-3}, and $t = 2 \times 10^{-4}$ cm.

(b) Show that a good approximation to the reverse-bias breakdown condition is given by $(\alpha - \beta)t = \ln(\alpha/\beta)$.

9.15. (a) For an n-n heterojunction with interfacial states, devise a capacitance–voltage measurement from which the two built-in voltages can be determined.

(b) Find $\Delta \mathscr{E}_c$, $\Delta \mathscr{E}_v$, and V_0 for the p-InAs, n-GaSb heterostructure indicated below. Neglect interface states.

	a_0 (Å)	$\mathscr{E}_g$ (eV)	χ (eV)	$\dfrac{m_e^*}{m}$	$\dfrac{m_h^*}{m}$	Doping (cm^{-3})
p-InAs	6.058	0.36	4.90	0.028	0.33	1×10^{18}
n-GaSb	6.095	0.68	4.06	0.045	0.39	1×10^{17}

9.16. Consider a heterojunction formed from p-type GaSb with an energy gap of 0.68 eV and an electron affinity of 4.06 eV and n-type InAs with an energy gap of 0.36 eV and an electron affinity of 4.90 eV.

(a) Draw the equilibrium energy band diagram and indicate how this heterojunction differs from the usual p-n junction.

(b) Neglecting interface states, what are the discontinuities in the conduction

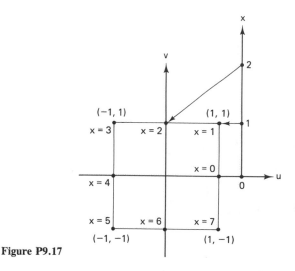

Figure P9.17

and valence bands? How are these different from those of the usual heterojunction?

9.17. By wrapping the line $0 \rightarrow x$ around a unit square we can define a new set of special functions, $v = \text{set } x$ and $u = \text{corset } x$ (abbreviated $u = \text{cor } x$). Consider an n-type heterostructure of width $0 \le x \le 4$ with electron concentration $n = 1$, temperature $kT = 1$, and force-free energy bands given by $\chi(x) = 2 + \text{cor } x$, $E_g(x) = 1 + \text{set } x$, and $N_c(x) = 1 + \text{set } x$ (Fig. P9.17). Obtain expressions for and plot to *scale* all the parameters in the force-free and equilibrium energy band diagram using E_0 as the zero of energy and $\psi(0) = 0$.

9.18. For the heterostructure in Problem 9.17, obtain an expression for and plot to *scale* the electron current for the force-free case at the beginning of the charge transfer process toward equilibrium. Assume that $\mu_n = 1$.

Surface Structures

In Chapter 9 we discussed the properties of heterostructures and their application to abrupt *p-n* junctions. To utilize these devices in electronic circuits, ohmic contacts must be fabricated on their surfaces. There are also a number of semiconductor devices that use surface properties and their interactions with metals and insulators as the basis for operation. These include Schottky devices (metal–semiconductor diodes), MIS capacitors (metal–insulator–semiconductor diodes), and a variety of transistors that depend on these devices for the controlling electrodes. Since the formalism for treating surfaces, ohmic contacts, Schottky barriers, and MIS capacitors is the same, we examine them together as "surface structures."

10.1 SURFACE REGIONS

In Section 3.7 a quantum mechanical plausibility argument indicated that the unsaturated bonds at a surface can result in allowed acceptor-like states within the energy gap. For a semiconductor with 10^{22} atoms/cm³, this could produce a total of

$$(10^{22} \text{ cm}^{-3})^{2/3} \simeq 5 \times 10^{14} \text{ cm}^{-2}$$

intrinsic surface states, many of which would be in the energy gap. In a practical situation these acceptor-like surface states can attract ambient impurities by chemisorption or physical adsorption and form compounds with metals and insulators deposited on the surface. The simplest possible consequences of these effects for the region of a semiconductor near the surface are summarized in Fig. 10.1.

This figure shows the energy band diagrams for an n-type semiconductor near the surface under (a) depletion, (b) inversion, and (c) accumulation. In Fig. 10.1(a) a relatively low concentration of acceptor-like surface states, N_S^- (cm^{-2}), captures electrons from the bulk donor atoms near the surface, producing a positive space-charge region, depleted of electrons, near the surface. The relationship among the concentrations in the depletion approximation is

$$N_S^- \simeq \int_0^{W_0} (N_d^+ - n)\, dx \qquad (10.1)$$

In Fig. 10.1(b) a higher concentration of acceptor-like surface states produces a sufficiently high barrier so that there is an excess of holes over electrons near the surface of this n-type semiconductor. (Notice that $\mathscr{E}_i$ is above ζ near the surface.) Here

$$N_S^- \simeq \int_0^{W_0} (N_d^+ + p)\, dx \qquad (10.2)$$

and the surface states have induced a p-n junction near the surface. In Fig. 10.1(c) donor-like surface states attract and accumulate excess electrons into the surface region, such that

$$N_S^+ \simeq \int_0^{W_0} (n - N_d^+)\, dx \qquad (10.3)$$

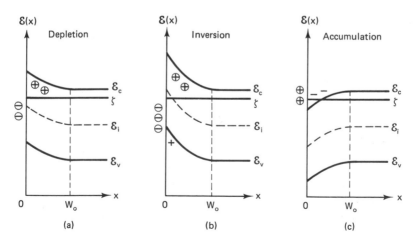

Figure 10.1   Depending on the concentration and sign of surface states the region near the surface may be in (a) depletion, (b) inversion, or (c) accumulation.

10.1.1 Surfaces Charges and Dipoles

This model of the surface shows only the concentration and charge (occupancy) of the surface states and implicitly assumes that the surface structure is a simple termination of the bulk crystal structure. Low-energy electron diffraction measurements, however, indicate that the structure of most surfaces deviates markedly from that of the underlying material. These changes in the positions of the surface atoms are driven by free-energy considerations, and the process is referred to as *surface reconstruction* [W. A. Harrison, *Electronic Structure and the Properties of Solids* (San Francisco: W. H. Freeman, 1980)]. Although there has been a substantial amount of work on reconstructed semiconductor surfaces, in many instances the surface structures and the reconstruction processes are not well understood. This is because the microscopic details vary with crystal orientation, adsorbed or chemisorbed impurities, preparation conditions, and the properties of the material itself.

To avoid these complications, we will take a simpler macroscopic viewpoint, which provides some qualitative insight into surface properties. This viewpoint has the following features or assumptions:

1. The surface states can be electrically neutral, positive, or negative. This provides a means for obtaining "Fermi-level pinning" in either *n*-type or *p*-type material. We discuss this in more detail in Section 10.2.1.

2. In the absence of charge transfer with the bulk, the electrochemical potential of the surface is located near the middle of the forbidden energy gap.

3. Surface reconstruction can produce a surface dipole that adds to (or increases) the electron affinity. The magnitude of this potential discontinuity is expected to change with surface orientation.

The formation of surface dipoles is indicated schematically in Fig. 10.2. Assuming a column IV semiconductor, in Fig. 10.2(a) each surface atom contributes three electrons to the covalent bonds with its three nearest-neighbor atoms, leaving one electron in a dangling bond. In this case each surface atom is neutral and the surface has the ideal or bulk structure. A more energetically favorable condition will occur, however, when some of the dangling bonds pair up. This is indicated in Fig. 10.2(b). Here, the surface atoms with the pair bonds will be locally negative, while those with no dangling bond will be locally positive. For an atom with a pair bond the ground state becomes degenerate and the atom itself must deform to remove the degeneracy [H. A. Jahn and E. Teller, *Proc. R. Soc. London A161,* 220 (1937)]. A back-bonding effect can also distort the position of an atom with no dangling bonds [W. A. Harrison, *Surf. Sci. 55,* 1 (1976)]. The net result

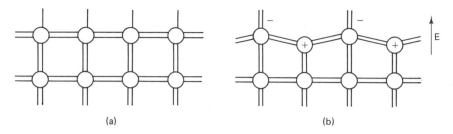

Figure 10.2 Schematic illustration of surface dipole formation. In (a) each neutral surface atom has one dangling bond. In (b) the dangling bonds pair and produce a shift in atomic positions.

is a spatial separation of positive and negative charge which produces an electric field normal to the surface. This dipolar field acting over atomic scale dimensions appears macroscopically as a discontinuity in electrostatic potential at the surface, $\Delta\psi(0)$.

Although the reconstruction of real surfaces is more complicated and often more pronounced than indicated in Fig. 10.2, a simple analysis suggests the possible magnitude of the effect. If half of the N_s surface states per unit area are positively charged and half are negatively charged by this process, Poisson's equation indicates an electric field,

$$E' = \frac{qN_s}{2\epsilon} \tag{10.4}$$

For $N_s = 10^{15}$ cm^{-2} and $\epsilon = 10^{-12}$ F/cm, $E = 0.8 \times 10^8$ V/cm. Assuming a charge separation of 1 Å, the discontinuity in potential, $\Delta\psi'(0) = 0.8$ eV, which is of the same magnitude as the energy gap in semiconductors.

The effects of such dipoles on the energy band structure near the surface, before we allow charge transfer between the surface and the bulk, are shown schematically in Fig. 10.3(a). In this diagram we show states below the surface electrochemical potential, ζ_s, as being negative (pair bonds), states near ζ_s as being neutral (single bonds), and states above ζ_s as being positive (no bonds). The dipole field E' is in a direction to retard the motion of electrons from the surface so that the resulting potential energy discontinuity, $q\,\Delta\psi'(0)$, adds to the electron affinity due to polarization (image force), χ_p. As indicated in the figure, $q\,\Delta\psi'(0)$ can be regarded either as the energy difference between the infinite or force-free vacuum level, E_0, and the local vacuum level, $\mathcal{E}_l$, or as a dipole contribution, χ_d', to the total electron affinity,

$$\chi = \chi_d' + \chi_p \tag{10.5}$$

We use the latter interpretation (see Section 8.4.2).

Figure 10.3(b) shows the results of allowing electron transfer from the n-type material into the surface states to achieve thermal equilibrium. As expected, this electron transfer uncovers donors, producing a space-charge

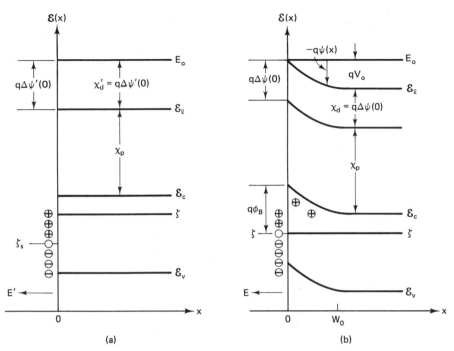

Figure 10.3 Effects of surface dipoles on energy band structure (a) before and (b) after charge transfer between the surface and the bulk material.

region in the bulk material near the surface. Notice, however, that this electron transfer also decreases the concentration of surface dipoles and thus the potential discontinuity or dipole contribution to the electron affinity at the surface. That is, χ_d is less than χ_d'. Thus the net result of surface reconstruction and charge transfer is a *dipole layer and a charge sheet at the surface.* For N_s total surface states, N_s^0 neutral states, N_s^- negative states, and N_p dipoles,

$$N_s = N_s^0 + N_s^- + 2N_p \tag{10.6}$$

10.1.2 Space-Charge Region

The analysis of the space-charge region in the semiconductor produced by the surface states is a straightforward extension of the heterostructure analysis in Section 9.1. Here, however, we set the reference potential at the surface ($x = 0$), as indicated in Fig. 10.3(b). That is, we treat the dipole potential discontinuity as part of the electron affinity and set

$$\psi(0) = 0 \tag{10.7}$$

For intrinsic material, (10.7) and (9.32) and (9.33) for the electric field and electrostatic potential, respectively, become

$$E(x) = -\frac{2kT}{q\lambda_i} \sinh \frac{q[V_0 - \psi(x)]}{2kT} \tag{10.8}$$

and

$$\psi(x) = V_0 - \frac{4kT}{q} \tanh^{-1}\left[\exp\left(\frac{-x}{\lambda_i}\right) \tanh \frac{qV_0}{4kT}\right] \tag{10.9}$$

From (9.35) and (9.37) the capacitance of the space-charge region is

$$C_i = \frac{\epsilon}{\lambda_i} \cosh \frac{qV_0}{2kT} \tag{10.10}$$

The intrinsic Debye length in (10.8), (10.9), and (10.10) is

$$\lambda_i^2 \equiv \frac{kT\epsilon}{2q^2 n_i} \tag{9.17}$$

A relationship between the built-in potential, V_0, and the concentration of negatively charged surface states, N_s^-, on an *ideal* (not reconstructed) surface can be obtained rather easily. Using Poisson's equation, (10.8) for $E(0)$, and (9.17) for λ_i, we have

$$V_0 = \frac{2kT}{q} \sinh^{-1} \frac{N_s^-}{4n_i\lambda_i} \tag{10.11}$$

The relationship for a reconstructed surface is not so easy to obtain.

For extrinsic material it is necessary to use the depletion approximation. Equations (9.45), (9.48), and (9.62) from the heterostructure analysis easily reduce to

$$E(x) = -\frac{qN_d}{\epsilon}(W_0 - x) \tag{10.12}$$

$$\psi(x) = V_0 - \frac{qN_d}{2\epsilon}(W_0 - x)^2 \tag{10.13}$$

and

$$C = \frac{\epsilon}{W_0} \tag{10.14}$$

Using (9.54) or (9.56), the space-charge width in equilibrium is

$$W_0^2 = \frac{2\epsilon V_0}{qN_d} \tag{10.15}$$

The relationship between V_0 and N_s^- for an ideal surface in the depletion approximation is

$$V_0 = \frac{qN_s^2}{2\epsilon N_d} \qquad (10.16)$$

which is a rather unphysical result. We note in passing, however, that an exact relationship between these two quantities can be obtained for extrinsic material.

10.2 METAL–SEMICONDUCTOR CONTACTS

A metal–semiconductor contact can be rectifying or nonrectifying depending on the properties of the metal, the semiconductor, and the semiconductor surface and the preparation method. By rectifying we mean a contact that has a larger barrier to charge carrier transport in one direction than in the opposite direction as evidenced by an asymmetric, nonlinear current–voltage (I–V) characteristic. Although to purists there are several subtype barriers, in common usage these contacts are called *Schottky diodes*. By nonrectifying we mean a contact that has no *effective* barrier to charge carrier transport in either direction, as evidenced by a symmetric, linear I–V characteristic.

As will be shown, metal–semiconductor contacts with no barrier are difficult to come by, and preparation techniques are usually developed to reduce the effectiveness of the barrier. Typically, these techniques may produce thin barriers that carriers can tunnel through, short barriers that carriers can be emitted over, or high recombination rate surfaces that will not support excess carrier injection (see Section 8.3.3). Although purists may object, these contacts are usually referred to as *ohmic*.

10.2.1 Energy Bands

If there were no surface states, the properties of the contact would be determined by the energy difference between the work function of the metal, Φ_m, and the work function of the semiconductor, Φ_s. Figure 10.4 shows the energy band diagrams of a metal–semiconductor pair with Φ_m greater than Φ_s (a) before and (b) after contact is made and equilibrium is established. In (a) all the bands are force-free and the chemical potential of the semiconductor, μ, is higher in energy than the chemical potential of the metal, μ_m. Thus, to achieve thermal equilibrium, on contact electrons must be transferred from the n-type semiconductor to the metal.

The result is shown in Fig. 10.4(b), where we have left a small distance δ between the two materials and have assumed that contact is made in some direction other than that shown. The electron transfer produces space-charge

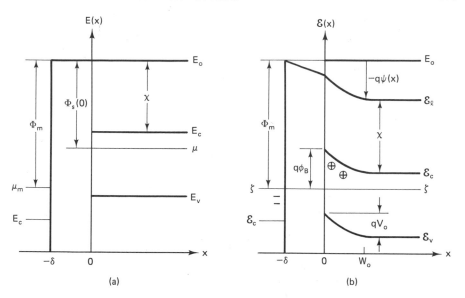

Figure 10.4 Energy band diagrams of a metal–semiconductor pair with $\Phi_m >$ $\Phi_s(0)$ (a) before and (b) after contact is made, assuming no surface states.

regions in the metal and in the semiconductor. In the metal the negative space-charge region is located very near the surface and is of little consequence due to the large equilibrium electron concentration. In the semiconductor, however, the positive space-charge region due to depletion of electrons extends over a relatively long distance (W_0 in the depletion approximation) and has a potential barrier, V_0, to electron flow from the semiconductor to the metal. There is also a potential barrier, Φ_B, to electron flow from the metal to the semiconductor which is larger than V_0. This metal–semiconductor contact is a Schottky diode. Notice, also, in Fig. 10.4(b) that the electrostatic potential from the semiconductor to the metal is shared by the space-charge region and the space between the semiconductor and the metal. If there is no charge in this space, $\psi(x)$ is linear in x and the electric field is constant. When we let δ approach zero, more electrons are transferred from the semiconductor to the metal, the space-charge region gets larger, and less of the electrostatic potential appears across the space between the two materials. For δ equal to zero, all of the potential is across the space-charge region.

Figure 10.5 shows the energy band diagrams of a metal–semiconductor pair with Φ_m less than Φ_s (a) before and (b) after contact. In this case the chemical potential of the metal, μ_m, is higher in energy than the chemical potential of the semiconductor, μ. Thus electrons must be transferred from the metal to the semiconductor to achieve equilibrium. This produces a positive space charge in the metal and a negative space-charge region, due

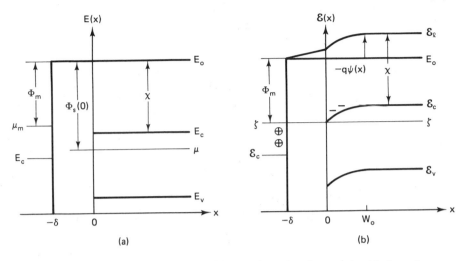

Figure 10.5 Energy band diagrams of a metal–semiconductor pair with $\Phi_m < \Phi_s$ (a) before and (b) after contact is made, assuming no surface states.

to an accumulation of electrons, in the semiconductor. Notice that this accumulation layer is not a barrier to electron flow from semiconductor to metal and is essentially not a barrier to electron flow from metal to semiconductor. This, therefore, is an ohmic contact.

The energy band diagrams for metal–semiconductor contacts with surface states are more complicated than those shown in Figs. 10.4 and 10.5. The properties of virtually all real contacts, however, are dominated by these states, whether they exist before, during, or after the deposition of metal on the surface.

Figure 10.6 indicates the energy band diagrams for a metal–semiconductor pair with Φ_m greater than Φ_s (a) before and (b) after contact. In (a) we assume that the semiconductor surface has the same dipole layer and negative charge sheet as the surface in Fig. 10.3(b). Since the electrochemical potential, ζ, of the semiconductor is higher in energy than the chemical potential, μ_m, of the metal, electrons must flow from the semiconductor and its surface to the metal to achieve equilibrium. As indicated in (b), this increases the width of the space-charge region, empties some of the surface states, and *changes* the dipole contribution to the electron affinity by $\Delta\chi_d$. This change, $\Delta\chi_d$, can be positive or negative, depending on the number of electrons transferred from the surface states. That is, for small values of $\Phi_m - \Phi_s$ it will be positive, while for large values it can be negative.

For large concentrations of occupied surface states essentially all the electrons required to achieve equilibrium can be obtained from the surface alone, with little transfer from the space-charge region. Under these con-

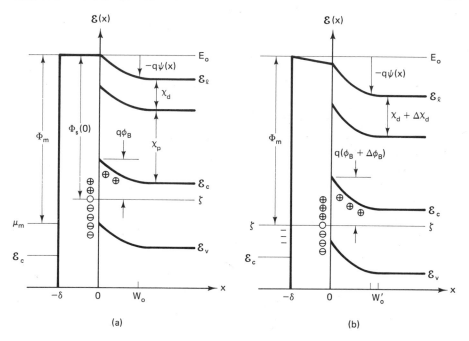

Figure 10.6   Energy band diagrams of a metal–semiconductor pair with $\Phi_m > \Phi_s$ and surface states (a) before and (b) after thermal equilibrium.

ditions the changes described above in the semiconductor can be small, perhaps unmeasurable. This is referred to as "Fermi-level pinning." Also, under these conditions, when we let the separation δ approach zero, the potential across the region between the metal and the semiconductor becomes a potential discontinuity at the surface.

Energy band diagrams for Φ_m less than Φ_s and including surface states are shown in Fig. 10.7. Here the transfer of electrons from the metal to the semiconductor and its surface to achieve equilibrium decreases the space-charge region, fills some of the surface states, decreases the barriers to electron flow, and *decreases* the dipole contribution to electron affinity, $\Delta\chi_d$. For large occupied surface state concentrations all of these changes will be small.

Figures 10.6 and 10.7 indicate that metal–semiconductor contacts with surface states are Schottky diodes independent of the relative magnitude of the metal work function. That is, the surface states produce the Schottky barriers and (unless the metal creates the states) the metal does little to change the barriers. Although we looked only at n-type semiconductor-metal contacts, the same surface states will also produce Schottky barriers on p-type material.

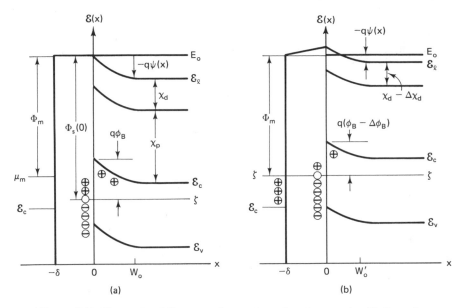

Figure 10.7 Energy band diagrams of a metal–semiconductor pair with $\Phi_m < \Phi_s$ and surface states (a) before and (b) after thermal equilibrium.

10.2.2 Charge Flow

Beginning with the original drift-diffusion [W. Schottky, *Naturwissen-schaften 26,* 843 (1938)] and thermionic emission [H. A. Bethe, *MIT Radiat. Lab. Rep.* 43-12 (1942)] models, there have been many treatments of charge conduction in Schottky barrier diodes. Many of these treatments attempt to combine the physics of drift-diffusion and thermionic emission to obtain quantitative agreement with experimental current–voltage (I–V) character-istics. In one sense these two models are mutually exclusive. The drift-diffusion model assumes that the space-charge region of the diode is much longer than the mean free path of the carriers, so that the carriers undergo many collisions as they travel through the space-charge region. The ther-mionic emission model, however, implicitly assumes that the mean free path of the carriers is longer than the space-charge region of the diode, so that few collisions occur. In this section we develop Schottky's original drift-diffusion model.

Figure 10.8 shows the energy band diagram and spatial dependence of the electron concentration for a forward-biased Schottky barrier diode. In the band diagram we have included any surface dipoles in the electron af-finity, χ. The dashed lines in this figure indicate several parameters under equilibrium conditions. Before the bias is applied, the large electron con-centration gradient from the metal to the semiconductor surface is balanced by the electric field of the barrier, ϕ_B, while the concentration gradient from

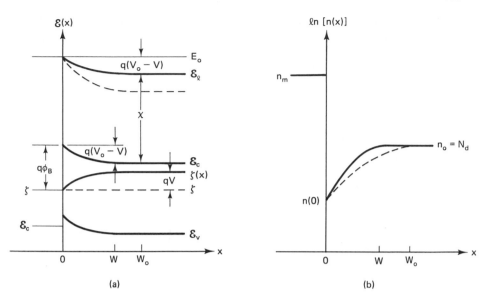

Figure 10.8 Energy band diagram and electron concentration for a forward-biased Schottky barrier diode. The dashed lines indicate equilibrium conditions.

the semiconductor to the surface is balanced by the space-charge region electric field of the built-in voltage V_0. That is, there is no net transport of electrons, and the diode is in equilibrium.

When a forward bias, V, is applied, the drift-diffusion balance in the metal is not affected. In the semiconductor, however, the electrochemical potential, $\zeta(x)$, increases by V producing a decrease in the barrier to $(V_0 - V)$ and an increase in the concentration gradient. Under these conditions there is a net diffusion of electrons from the semiconductor to the metal (or net *electric* current from the metal to the semiconductor). If we assume that there is no inversion layer near the surface, the current is essentially all electron current and the Schottky diode is a *majority* carrier device.

Using (8.140), the current density is

$$J = \mu_n n(x) \frac{d}{dx} [\zeta(x)] \tag{10.17}$$

where from (8.144),

$$d\zeta(x) = kT \frac{dn(x)}{n(x)} - q \, d\psi(x) \tag{10.18}$$

Inserting (10.18) into (10.17) and rearranging gives us

$$J \, dx = kT\mu_n \left\{ dn(x) + n(x)d \left[\frac{-q\psi(x)}{kT} \right] \right\} \tag{10.19}$$

The right-hand side of (10.19) can be made an exact differential with an integrating factor, so that

$$J \exp\left[\frac{-q\psi(x)}{kT}\right] dx = kT\mu_n d\left\{n(x) \exp\left[\frac{-q\psi(x)}{kT}\right]\right\} \qquad (10.20)$$

Integrating (10.20) over the space-charge region, we obtain

$$J \int_0^W \exp\left[\frac{-q\psi(x)}{kT}\right] dx$$

$$= kT\mu_n \left\{n(W) \exp\left[\frac{-q\psi(W)}{kT}\right] - n(0) \exp\left[\frac{-q\psi(0)}{kT}\right]\right\} \qquad (10.21)$$

The boundary values in (10.21) are relatively easy to obtain. Referring to Fig. 10.8, the metal is grounded as our reference, so

$$\psi(0) = 0 \qquad (10.22)$$

and

$$\psi(W) = V_0 - V \qquad (10.23)$$

In equilibrium, we have, from (8.121),

$$n(0) = N_d \exp\left(\frac{-qV_0}{kT}\right) \qquad (10.24)$$

Assuming "Fermi-level pinning" at the surface, this boundary condition should also be valid with an applied bias. The last boundary value is

$$n(W) = N_d \qquad (10.25)$$

Using (10.22) to (10.25) in (10.21) gives us

$$J \int_0^W \exp\left[\frac{-q\psi(x)}{kT}\right] = kT\mu_n N_d \exp\left(\frac{-qV_0}{kT}\right)\left[\exp\left(\frac{qV}{kT}\right) - 1\right] \qquad (10.26)$$

The left-hand side of (10.26) can be evaluated in the following manner. From the equilibrium equations (10.13) and (10.15), we obtain the non-equilibrium equation

$$\psi(x) = (V_0 - V)\left(\frac{2x}{W} - \frac{x^2}{W^2}\right) \qquad (10.27)$$

by replacing V_0 with $(V_0 - V)$ and W_0 with W. Figure 10.9(a) shows the contribution of each term in (10.27) to the potential and Fig. 10.9(b) indicates a typical spatial dependence for $\exp[-q\psi(x)/kT]$. From these figures it is clear that the major contribution to the integral in (10.26) occurs for small values of x where the potential can be approximated by its linear term. With this approximation,

$$\int_0^W \exp\left[\frac{-2q(V_0 - V)x}{kTW}\right]$$

$$= \frac{kTW}{2q(V_0 - V)}\left\{1 - \exp\left[\frac{-2q(V_0 - V)}{kT}\right]\right\} \tag{10.28}$$

$$\approx \frac{kTW}{2q(V_0 - V)} \tag{10.29}$$

Equations (10.26) and (10.29) result in the current–voltage characteristic

$$J = \frac{q^2\mu_n N_d^2}{\epsilon}\exp\left(\frac{-qV_0}{kT}\right)\left[\exp\left(\frac{qV}{kT}\right) - 1\right]W \tag{10.30}$$

where

$$W = \left[\frac{2\epsilon(V_0 - V)}{qN_d}\right]^{1/2} \tag{10.31}$$

Notice that this drift-diffusion model indicates nonideal behavior for a Schottky barrier diode. In fact, the voltage dependence is the same as was obtained in (9.113) for recombination–generation current in an abrupt p-n heterojunction. We therefore expect the reverse-bias current to increase with voltage until a reverse breakdown condition is obtained. Under these conditions (9.141), with straightforward modifications for the Schottky barrier space-charge region, can be used for impact-ionization-induced breakdown, while (9.156) can be used for tunneling.

10.2.3 Barrier Lowering Effects

One of the more significant modifications of the Schottky barrier energy band model is the barrier lowering effect of the image force. As indicated in Fig. 10.10, an electron in the vicinity of a metal surface exhibits a Coulomb

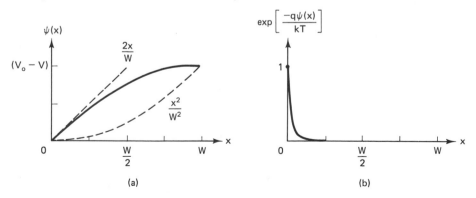

Figure 10.9   Diagram indicating (a) the spatial dependence of $\psi(x)$ and the contribution of each term in (10.27), and (b) a typical spatial dependence of $\exp\left[-q\psi(x)/kT\right]$.

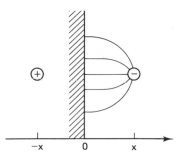

Figure 10.10 Illustration of the image force induced on an electron at x near a metal surface at $x = 0$. The force is the same as that of a positive charge at $-x$.

attractive force which is equivalent to that between a negative and a positive charge at equal and opposite distances from the surface. This *image force* is given by

$$F_i(x) = \frac{-q^2}{4\pi\epsilon(2x)^2} \tag{10.32}$$

$$F_i(x) = \frac{-q^2}{16\pi\epsilon x^2} \tag{10.33}$$

where x is the distance from the surface. The electrostatic energy associated with this image force is determined by the work that would be required to move the electron from the point x to the infinite or force-free vacuum level E_0. Thus this energy is a component of the local vacuum level,

$$\mathcal{E}_{li}(x) = \int_x^\infty F_i(\xi)\, d\xi \tag{10.34}$$

Inserting (10.33) into (10.34), we have

$$\mathcal{E}_{li}(x) = \frac{-q^2}{16\pi\epsilon x} \tag{10.35}$$

Before we included the image force, the local vacuum level had only one component due to the space charge,

$$\mathcal{E}_{ls}(x) = -q\psi_s(x) \tag{10.36}$$

In the depletion approximation, (10.27), the electrostatic potential due to space charge is

$$\psi_s(x) = \frac{qN_d}{\epsilon}\left(Wx - \frac{x^2}{2}\right) \tag{10.37}$$

and from (10.36),

$$\mathcal{E}_{ls}(x) = \frac{-q^2 N_d}{\epsilon}\left(Wx - \frac{x^2}{2}\right) \tag{10.38}$$

The local vacuum level is now obtained by adding the image force component

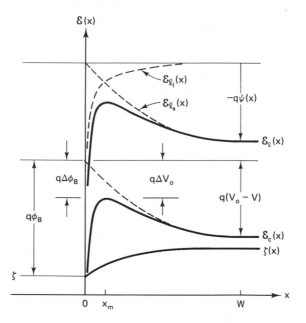

Figure 10.11   Energy band diagram for Schottky barrier diode under forward bias, illustrating image force barrier lowering.

(10.35) to the space-charge component (10.38), so that

$$\mathscr{E}_l(x) = \frac{-q^2 N_d}{\epsilon}\left(Wx - \frac{x^2}{2} + \frac{1}{16\pi N_d x}\right) \qquad (10.39)$$

The result is shown in Fig. 10.11. Since according to (8.114) all the energy levels follow the local vacuum level, the barrier in the conduction band is lowered by the image force.

Notice that the maximum in the barrier has moved from $x = 0$ to some point x_m. From (10.39) we obtain

$$x_m^2 = \frac{1}{16\pi N_d(W - x_m)} \qquad (10.40)$$

Since x_m is typically an order of magnitude or more smaller than W,

$$x_m \simeq \left(\frac{1}{16\pi N_d W}\right)^{1/2} \qquad (10.41)$$

Using (10.41) in (10.39), the barrier is reduced by

$$\Delta\phi_B \simeq \frac{q}{\epsilon}\left(\frac{WN_d}{4\pi}\right)^{1/2} \qquad (10.42)$$

10.3 METAL–INSULATOR–SEMICONDUCTOR CONTACTS

In contrast to metal–semiconductor contacts, there is essentially no conduction current in forward or reverse bias with metal–insulator–semiconductor (MIS) contacts. This makes them very attractive for high-impedance-
controlling electrodes in field-effect transistors. The major problem in the
fabrication of MIS contacts, in general, however, is the minimization of
semiconductor surface states. As was discussed above, high concentrations
of surface states lead to "Fermi-level pinning," which makes it difficult (if
not impossible) to vary the properties of the semiconductor near the surface.
In fact, the single process that makes silicon technology superior to all other
semiconductor technologies is the ability to fabricate thermally grown SiO_2–
Si interfaces with very low concentrations of interface states. Although it
has been common practice to use heavily doped polycrystalline silicon as
the "metal" in this technology, the demands of larger-scale and higher-speed
integrated circuits probably require the use of real metals. We will, therefore,
use Al–SiO_2–Si as our model MIS contact in the following discussion.

10.3.1 Energy Bands

The energy band diagrams for an Al–SiO_2–Si MIS structure (a) before
and (b) after equilibrium are shown in Fig. 10.12. In drawing these diagrams
we assumed zero interface and insulator states, and grounded the bulk semiconductor as reference. In (a), observe that the work function of the metal,
$\Phi_m = 4.1$ eV, while the work function of the semiconductor, $\Phi_s = 5.0$ eV.
In (b), to achieve equilibrium, electrons have transferred from the metal to
the p-type semiconductor (through some other contact), ionizing acceptors
in the semiconductor. The difference in work functions

$$\Phi_s - \Phi_m = 0.9 \text{ eV} \tag{10.43}$$

produces a potential across the insulation of 0.6 V and a potential across
the semiconductor depletion region of 0.3 V. Notice in Fig. 10.12 that a
potential of 0.9 V applied to the metal will remove the depletion region from
the semiconductor and change the band diagram of (b) into that of (a). This
is known as the *flat-band condition*, and the applied voltage that will produce
this condition is the *flat-band voltage*. With no interface or insulator charge,

$$V_{fb} = - \frac{\Phi_s - \Phi_m}{q} \tag{10.44}$$

in this case a negative number (-0.9 V).

When a potential more negative than the flat-band voltage is applied
to the metal, excess holes are induced near the surface of the otherwise
uniform p-type material and an accumulation layer is obtained. As shown
in Fig. 10.13(a), this accumulation of holes simply produces a p^+ surface

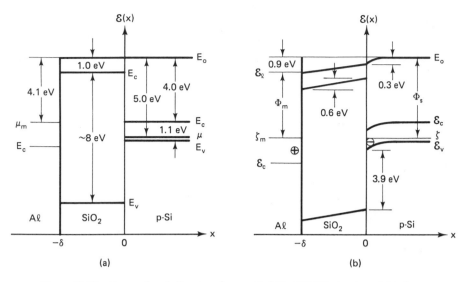

Figure 10.12   Energy band diagrams for an Al–SiO$_2$–Si MIS structure (a) before and (b) after contact is made. The diagrams are constructed approximately to scale with 10^{15} cm^{-3} doped p-Si.

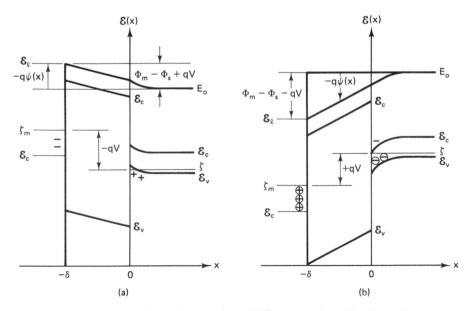

Figure 10.13   Energy band diagrams for an MIS structure. In (a) $V < V_{fb}$ produces an accumulation layer. In (b) $V_t < V$ produces a strong inversion layer.

which is electrically conducting to the p substrate. Notice that the electro-chemical potential in the semiconductor is level in this diagram, since there is no steady-state charge flow and the applied potential energy qV is the difference between ζ_m and ζ.

When a positive potential is applied to the metal, it ionizes more ac-ceptors than are ionized in equilibrium, and the width of the depletion region increases. As the applied potential becomes more and more positive, mi-nority electrons begin to accumulate near the surface and an inversion layer is created. This is shown in Fig. 10.13(b). At this point the width of the depletion region no longer increases and the applied potential induces more and more electrons in the inversion region.

The potential at which *strong* inversion is obtained is called the *thresh-old voltage, V_t*. Notice that under this condition, the n-region induced near the surface is electrically isolated from the p substrate by the space-charge region of the ionized acceptors. This is the basis of operation for *n-channel* MIS transistors. In these transistors the threshold voltage is the minimum gate voltage under which electron conduction from source to drain occurs.

10.3.2 Threshold Voltage

The threshold voltage, V_t, can be determined by reference to Fig. 10.14. In (a) the structure is in equilibrium and the built-in potential produced by the work function difference is divided between the insulator, V_{i0}, and the semiconductor space-charge region, V_{s0}. That is,

$$\Phi_s - \Phi_m = q(V_{i0} + V_{s0}) \tag{10.45}$$

Assuming no insulator charges, the electric field in the insulator, E_i, is uni-form and

$$V_{i0} = E_i\delta \tag{10.46}$$

If there are no charged states at the insulator–semiconductor interface, the electric flux density is continuous and

$$\epsilon_i E_i = \epsilon_s E_s(0) \tag{10.47}$$

where $E_s(0)$ is the electric field in the semiconductor at the interface. Using (10.47) in (10.46), we obtain

$$V_{i0} = \frac{\epsilon_s E_s(0)}{C_i} \tag{10.48}$$

where

$$C_i = \frac{\epsilon_i}{\delta} \tag{10.49}$$

is the capacitance per unit area of the insulator.

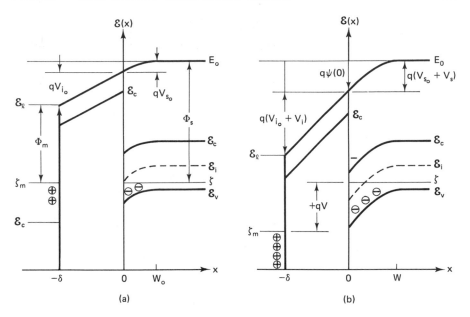

Figure 10.14   Energy bands for an MIS structure (a) in equilibrium and (b) with a positive voltage, V, applied to the metal assuming no interface or insulator charges.

The relationship between the electric field in the semiconductor and the net equilibrium positive charge per unit area, $Q_s(W_0)$, is obtained by integrating Poisson's equation over the space-charge region. In this way we find

$$\epsilon_s E_s(0) = -Q_s(W_0) \tag{10.50}$$

Inserting (10.50) into (10.48) yields

$$V_{i0} = -\frac{Q_s(W_0)}{C_i} \tag{10.51}$$

For the p-type semiconductor shown in Fig. 10.14, we find from (10.15) that

$$Q_s(W_0) = -(2q\epsilon_s N_a V_{s0})^{1/2} \tag{10.52}$$

so that the built-in voltage across the insulator, V_{i0}, is positive. The built-in voltage across the semiconductor space-charge region, V_{s0}, using (10.45) and (10.52), can be determined from

$$V_{s0} = \frac{\Phi_s - \Phi_m}{q} + \frac{Q_s(W_0)}{C_i} \tag{10.53}$$

or for p-type material,

$$V_{s0} = \frac{\Phi_s - \Phi_m}{q} - \frac{(2q\epsilon_s N_a V_{s0})^{1/2}}{C_i} \tag{10.54}$$

When we apply a positive voltage, V, to the metal, Fig. 10.14(b) indicates that it is divided between the insulator, V_i, and the space-charge region, V_s. That is,

$$V = V_i + V_s \tag{10.55}$$

Now, instead of (10.46), we have

$$V_{i0} + V_i = E_i\delta \tag{10.56}$$

and using the same procedure as in equilibrium,

$$V_{i0} + V_i = -\frac{Q_s(W)}{C_i} \tag{10.57}$$

Now, however, for p-type material,

$$Q_s(W) = -[2q\epsilon_s N_a\psi(0)]^{1/2} - q\int_0^W n(x)\,dx \tag{10.58}$$

where Fig. 10.14(b) shows that

$$\psi(0) = V_{s0} + V_s \tag{10.59}$$

For applied voltages smaller than the voltage required to obtain strong inversion ($V < V_t$), the second term on the right-hand side of (10.58) will be much smaller than the first term and can be neglected.

The condition for strong inversion is somewhat arbitrarily chosen as the point where the minority carrier concentration at the interface is equal to the majority carrier concentration in the bulk. In our case this condition is

$$n(0) = p(W) = N_a \tag{10.60}$$

Since the mass-action law is valid when there are no injected excess carriers, (8.130) gives us

$$n(x) = \frac{n_i^2}{N_a} \exp\left[\frac{q\psi(x)}{kT}\right] \tag{10.61}$$

With (10.60) and (10.61) we find that the interfacial potential corresponding to the onset of strong inversion for p-type material is

$$\psi_t(0) = \frac{2kT}{q} \ln\left(\frac{N_a}{n_i}\right) \tag{10.62}$$

Finally, substituting (10.57), (10.59), and (10.45) into (10.55), the threshold voltage is

$$V_t = \psi_t(0) - \frac{Q_s(W_t)}{C_i} - \frac{\Phi_s - \Phi_m}{q} \tag{10.63}$$

For *p*-type material,

$$V_t = \psi_t(0) + \frac{[2q\epsilon_s N_a \psi_t(0)]^{1/2}}{C_i} - \frac{\Phi_s - \Phi_m}{q} \qquad (10.64)$$

where $\psi_t(0)$ is given by (10.62). Referring to Fig. 10.14, we can see that the applied voltage required to achieve strong inversion is equal to the potential required at the interface plus the potential dropped across the insulator minus the potential that was already across both in equilibrium. This is the English equivalent of (10.64).

10.3.3 Insulator Charges

When the insulator has a relatively large net positive charge, it is possible for the surface of *p*-type material to be in strong inversion under *equilibrium* conditions. Such a situation is shown in Fig. 10.15 for an arbitrary net positive charge per unit area,

$$Q_i = q \int_0^\delta N_i(x) \, dx \qquad (10.65)$$

For notation convenience, we have shifted the origin of the x axis to the metal–insulator interface. As before, the built-in potential energy is equal to the work function difference, so (10.45) is also valid in this equilibrium situation. The relationship between the field and charge in the insulator is

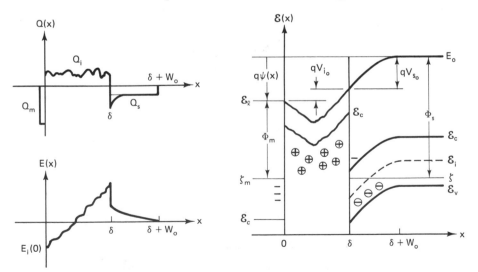

Figure 10.15 Diagrams showing the spatial dependence of the charge per unit area, the electric field, and the energy bands of an MIS structure when the insulator has an arbitrary net positive charge distribution.

obtained by integrating Poisson's equation, which gives

$$E_i(x) - E_i(0) = \frac{q}{\epsilon_i} \int_0^x N_i(\xi)\, d\xi \tag{10.66}$$

Integrating once more, the potential in the insulator is

$$-\psi(x) + \psi(0) = E_i(0)x + \frac{q}{\epsilon_i} \int_0^x \int_0^\xi N_i(\eta)\, d\eta\, d\xi \tag{10.67}$$

With (10.67) we can formally evaluate the built-in potential across the insulator,

$$V_{i0} = \psi(0) - \psi(\delta) \tag{10.68}$$

$$V_{i0} = E_i(0)\delta + \frac{q}{\epsilon_i} \int_0^\delta \int_0^x N_i(\xi)\, d\xi\, dx \tag{10.69}$$

Since the net positive charge must equal the net negative charge,

$$Q_i = -[Q_m + Q_s(W_0)] \tag{10.70}$$

where Q_i is given by (10.65) and $Q_s(W_0)$ by (10.52) when the material is p-type and the induced electron concentration is small. Assuming no charged states at the metal–insulator interface,

$$Q_m = \epsilon_i E_i(0) \tag{10.71}$$

since both are negative. Using (10.65) in (10.70), (10.70) in (10.71), and (10.71) in (10.69), we obtain

$$V_{i0} = \frac{-Q_s(W_0)}{C_i} - \frac{q}{C_i} \int_0^\delta N_i(x)\, dx + \frac{q}{C_i \delta} \int_0^\delta \int_0^x N_i(\xi)\, d\xi\, dx \tag{10.72}$$

$$V_{i0} = \frac{-Q_s(W_0)}{C_i} - \frac{q}{C_i \delta} \int_0^\delta \int_x^\delta N_i(\xi)\, d\xi\, dx \tag{10.73}$$

Integrating by parts, the built-in voltage across the insulator in equilibrium is

$$V_{i0} = \frac{-Q_s(W_0)}{C_i} - \frac{q}{C_i \delta} \int_0^\delta x N_i(x)\, dx \tag{10.74}$$

For p-type material the first term on the right-hand side of (10.74) is positive, so the insulator charges of the second term reduce the built-in potential across the insulator, as indicated in Fig. 10.15.

The built-in voltage of the semiconductor space-charge region, from (10.45) and (10.74), is

$$V_{s0} = \frac{\Phi_s - \Phi_m}{q} + \frac{Q_s(W_0)}{C_i} + \frac{q}{C_i \delta} \int_0^\delta x N_i(x)\, dx \tag{10.75}$$

In comparison to (10.53), we see that the positive insulator charge increases the space-charge built-in voltage. Although it is now impossible to obtain a flat-band condition in the insulator, the semiconductor bands will be flat when a voltage

$$V_{fb} = -\frac{\Phi_s - \Phi_m}{q} - \frac{q}{C_i\delta} \int_0^\delta xN_i(x)\, dx \qquad (10.76)$$

is applied to the metal. Compared to (10.44), we see that a larger negative bias is required to obtain flat-band conditions when the insulator has positive charge.

There is also an equivalent shift in the threshold voltage. In comparison with (10.63), the positive insulator charge decreases the threshold voltage to

$$V_t = \psi_t(0) - \frac{Q_s(W_t)}{C_i} - \frac{\Phi_s - \Phi_m}{q} - \frac{q}{C_i\delta} \int_0^\delta xN_i(x)\, dx \qquad (10.77)$$

where $\psi_t(0)$ is given by (10.62).

10.3.4 Capacitance

The flat-band and threshold voltages for an MIS structure can be determined from capacitance–voltage (C–V) measurements. The capacitance of an MIS structure has two components: the capacitance per unit area of the insulator,

$$C_i = \frac{\epsilon_i}{\delta} \qquad (10.49)$$

and the capacitance per unit area of the semiconductor space charge region, C_s. For p-type material in the depletion approximation,

$$C_s = \frac{\epsilon_s}{W} \qquad (10.78)$$

where

$$W = \left[\frac{2\epsilon_s(V_{s0} + V_s)}{qN_a}\right]^{1/2} \qquad (10.79)$$

Since the voltage applied to the MIS structure divides between the two regions, the two components are in series.

A typical C–V characteristic for a p-type MIS capacitor is shown in Fig. 10.16. With large negative bias on the metal, the semiconductor surface has an accumulation of majority holes [see Fig. 10.13(a)] and only the capacitance of the insulator, C_i, is observed. Since C_i is not voltage dependent, the measured capacitance, C, remains constant with smaller negative bias

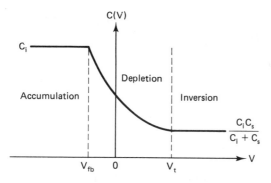

Figure 10.16 Capacitance–voltage variation at high measurement frequency for an MIS structure.

until the flat band condition is obtained. For lower negative bias, a depletion region begins to form in the semiconductor. The two components, C_i and C_s, then start to operate in series, lowering C. As the bias across the depletion region increases, W in (10.79) increases until inversion occurs. Under strong inversion conditions, additional positive bias on the metal acts only to increase the concentration of minority electrons near the interface. Beyond this point W and, therefore, C remain constant.

Figure 10.16 indicates how V_{fb} and V_t can be determined from the behavior of a C–V characteristic. Notice also that with (10.76) and (10.77), some information regarding the relative importance of charges and work function differences can be obtained from these measurements.

PROBLEMS

10.1. Gallium arsenide at 77 K has a bandgap of ≈ 1.5 eV and at a doping of $N_d \approx 10^{17}$ atoms/cm³ is degenerate. Surface states and ions pin the Fermi level at the surface at midgap. Sketch the energy band diagram (equilibrium) of this sample from the surface into the material and *determine* how far an electric field extends into the material. Where is the field maximum, and in which direction is it oriented?

10.2. In some Schottky barrier diodes a significant component of the reverse current is due to the tunneling of electrons from the metal to the semiconductor. As indicated in Fig. P10.2, at the Fermi level these electrons must tunnel through a barrier of height $q\phi_B$ and thickness t, whose shape is determined by $\mathscr{E}(x)$. If an electron at the Fermi level in the metal absorbs a photon of energy $\hbar\omega$, both the barrier height and thickness, t, are substantially reduced. This is referred to as photon-assisted tunneling. Obtain expressions for both the barrier height and thickness, t, in terms of $\hbar\omega$ and the parameters shown in the diagram.

10.3. It is found experimentally that a Schottky barrier is formed when Al is evaporated on *n*-type GaAs with a donor concentration of 5×10^{15} cm⁻³. Al has a work function of 4.36 V and GaAs has an electron affinity of 4.07 V. De-

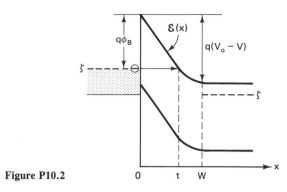

Figure P10.2

termine the built-in voltage V_0, the depletion layer width W_0, and the capacitance at zero bias assuming:

(a) No surface states

(b) That the Fermi level at the semiconductor surface is pinned at $\frac{1}{3}\mathscr{E}_g$ above the valence band by surface states.

10.4. p-Type polycrystalline Si is often used as the "metal" gate in an MOS capacitor. Assume that the gate of such a device is uniformly doped with N_a acceptors, the oxide has a capacitance C_{ox}, and the substrate is uniformly doped with N_d donors. If the total capacitance, C, of the device is measured, voltage division among the three regions of the MOS capacitor can be determined. Obtain the three equations for voltage division in terms of the parameters above.

Appendix A

Semiconductor Properties (300 K)

Material	Crystal Structure	Lattice Constant (Å)	Energy Gap (eV)	Electron Effective Mass	Hole Effective Mass	Static Dielectric Constant	Refractive Index	Electron Mobility (cm²/V·s)	Hole Mobility (cm²/V·s)
C	Dia								
Si	Dia	5.4310	1.11X	m_l 0.98, m_t 0.19	0.16, 0.5	11.7	3.44	1,350	480
Ge	Dia	5.6461	0.67L	m_l 1.58, m_t 0.08	0.04, 0.3	16.3	4.00	3,900	1,900
α-Sn	Dia	6.4892	0.08Γ	0.02				2,000	1,000
α-SiC	Wur	a 3.0865, c 15.117	2.8			10.2	2.65	500	
AlN	Wur	a 3.111, c 4.978	5.9i						
AlP	Sph	5.4625	2.43X	0.13		9.8	3.0	80	
AlAs	Sph	5.6605	2.16X	0.5	m_l 0.49, m_h 1.06	12.0		1,000	180
AlSb	Sph	6.1355	1.52X	0.11	0.39	11	3.4	200	300
GaN	Wur	a 3.189, c 5.185	3.4Γ	0.2	0.8	12	2.4	300	
GaP	Sph	5.4506	2.26X	0.13	m_l 0.67	10	3.37	300	150
GaAs	Sph	5.6535	1.43Γ	0.067	m_l 0.12, m_h 0.5	12.5	3.4	8,500	400
GaSb	Sph	6.0954	0.72Γ	0.045	0.39	15	3.9	5,000	1,000
InN	Wur	a 3.533, c 5.693	2.4Γ						
InP	Sph	5.8688	1.35Γ	0.07	0.40	12.1	3.37	4,000	600
InAs	Sph	6.0584	0.36Γ	0.028	0.33	12.5	3.42	22,600	200
InSb	Sph	6.4788	0.18Γ	0.013	0.18	18	3.75	100,000	1,700
ZnS	Sph	5.4109	3.6Γ	0.39		8.3	2.4	110	
ZnSe	Sph	5.6686	2.58Γ	0.17		8.1	2.89	600	
ZnTe	Sph	6.1037	2.25Γ	0.15		9.7	3.56		
CdS	Wur	a, c	2.42Γ	0.20	0.7	8.9	2.5	250	

		a	c	E_g	m^*/m_0		m^*/m_0					
CdSe	Wur	6.4816		1.73Γ	0.13		0.4	10.6		650		CdSe
CdTe	Sph	5.852		1.50Γ	0.11		0.35	10.9	2.75	1,050	100	CdTe
HgS	Sph	6.084		2.0Γ						50		HgS
HgSe	Sph	6.4616		−0.15Γ	0.045			25		18,500	100	HgSe
HgTe	Sph			−0.15Γ	0.029		0.3	20	3.7	22,000		HgTe
GeS	Orth			1.8							70	GeS
GeSe	Orth			1.16							100	GeSe
GeTe	NaCl	5.986									100	GeTe
SnS	Orth			1.08							90	SnS
SnSe	Orth			0.9							110	SnSe
SnTe	NaCl	6.325		0.18L					3.7		400	SnTe
PbS	NaCl	5.936		0.37L	0.1 (m_l), 0.07, 0.039	m_l, m_t	0.1, 0.06, 0.03	170		500	600	PbS
PbSe	NaCl	6.147		0.26L	0.24, 0.02	m_l, m_t	0.3	250		1,800	930	PbSe
PbTe	NaCl	6.45		0.29L	0.07	m_l, m_h	0.02	412		1,400	1,100	PbTe
ZnSiP$_2$	Chal	a 5.400 c 10.441		2.96Γ						100		ZnSiP$_2$
ZnSiAs$_2$	Chal	a 5.606 c 10.890		2.12Γ			0.07				140	ZnSiAs$_2$
ZnGeP$_2$	Chal	a 5.465 c 10.771		2.34Γ			0.5				20	ZnGeP$_2$
ZnGeAs$_2$	Chal	a 5.670 c 11.153		1.15Γ			0.4				23	ZnGeAs$_2$
ZnSnP$_2$	Chal	a 5.651 c 11.302		1.66Γ							55	ZnSnP$_2$
ZnSnAs$_2$	Chal	a 5.852 c 11.705		0.73Γ			0.35				200	ZnSnAs$_2$
CdSiP$_2$	Chal	a 5.678 c 10.431		2.45Γ	0.09					150		CdSiP$_2$
CdSiAs$_2$	Chal	a 5.884 c 10.882		1.55Γ							500	CdSiAs$_2$

(Continued)

	Crystal Structure	Lattice Constant (Å)	Energy Gap (eV)	Electron Effective Mass	Hole Effective Mass	Static Dielectric Constant	Refractive Index	Electron Mobility (cm²/V·s)	Hole Mobility (cm²/V·s)	
$CdGeP_2$	Chal	a 5.740 c 10.775	1.72Γ					100	25	$CdGeP_2$
$CdGeAs_2$	Chal	a 5.942 c 11.224	0.57Γ	0.035	0.3			2,500	1,500	$CdGeAs_2$
$CdSnP_2$	Chal	a 5.900 c 11.518	1.17Γ	0.036				2,000		$CdSnP_2$
$CdSnAs_2$	Chal	a 6.093 c 11.936	0.26Γ	0.015	m_l 0.020 m_h 1.0			11,000	190	$CdSnAs_2$

Appendix B

Fermi–Dirac Integrals

η	$F_{-3/2}$	F_{-1}	$F_{-1/2}$	F_0	$F_{1/2}$	F_1	$F_{3/2}$	F_2	$F_{5/2}$	F_3	$F_{7/2}$	F_4	η
−4.0	1.78 (−2)	1.799 (−2)	1.808 (−2)	1.815 (−2)	1.8199 (−2)	1.8232 (−2)	1.8256 (−2)	1.8274 (−2)	1.8287 (−2)	1.8295 (−2)	1.8301 (−2)	1.8305 (−2)	−4.0
−3.9	1.96 (−2)	1.984 (−2)	1.995 (−2)	2.003 (−2)	2.0099 (−2)	2.0140 (−2)	2.0170 (−2)	2.0191 (−2)	2.0206 (−2)	2.0216 (−2)	2.0224 (−2)	2.0229 (−2)	−3.9
−3.8	2.17 (−2)	2.188 (−2)	2.203 (−2)	2.213 (−2)	2.2195 (−2)	2.2247 (−2)	2.2283 (−2)	2.2309 (−2)	2.2327 (−2)	2.2340 (−2)	2.2349 (−2)	2.2355 (−2)	−3.8
−3.7	2.38 (−2)	2.413 (−2)	2.429 (−2)	2.442 (−2)	2.4510 (−2)	2.4572 (−2)	2.4617 (−2)	2.4648 (−2)	2.4670 (−2)	2.4686 (−2)	2.4697 (−2)	2.4705 (−2)	−3.7
−3.6	2.63 (−2)	2.660 (−2)	2.681 (−2)	2.696 (−2)	2.7063 (−2)	2.7139 (−2)	2.7191 (−2)	2.7231 (−2)	2.7259 (−2)	2.7277 (−2)	2.7291 (−2)	2.7301 (−2)	−3.6
−3.5	2.89 (−2)	2.931 (−2)	2.956 (−2)	2.975 (−2)	2.9880 (−2)	2.9972 (−2)	3.0037 (−2)	3.0084 (−2)	3.0118 (−2)	3.0141 (−2)	3.0158 (−2)	3.0169 (−2)	−3.5
−3.4	3.18 (−2)	3.230 (−2)	3.260 (−2)	3.283 (−2)	3.2986 (−2)	3.3099 (−2)	3.3179 (−2)	3.3235 (−2)	3.3276 (−2)	3.3304 (−2)	3.3325 (−2)	3.3339 (−2)	−3.4
−3.3	3.50 (−2)	3.557 (−2)	3.595 (−2)	3.625 (−2)	3.6412 (−2)	3.6549 (−2)	3.6645 (−2)	3.6715 (−2)	3.6764 (−2)	3.6799 (−2)	3.6824 (−2)	3.6841 (−2)	−3.3
−3.2	3.85 (−2)	3.917 (−2)	3.962 (−2)	3.995 (−2)	4.0187 (−2)	4.0354 (−2)	4.0473 (−2)	4.0557 (−2)	4.0617 (−2)	4.0659 (−2)	4.0690 (−2)	4.0711 (−2)	−3.2
−3.1	4.23 (−2)	4.311 (−2)	4.367 (−2)	4.407 (−2)	4.4349 (−2)	4.4552 (−2)	4.4696 (−2)	4.4800 (−2)	4.4872 (−2)	4.4924 (−2)	4.4961 (−2)	4.4986 (−2)	−3.1
−3.0	4.65 (−2)	4.743 (−2)	4.810 (−2)	4.858 (−2)	4.8933 (−2)	4.9181 (−2)	4.9356 (−2)	4.9482 (−2)	4.9571 (−2)	4.9634 (−2)	4.9679 (−2)	4.9710 (−2)	−3.0
−2.9	5.10 (−2)	5.215 (−2)	5.298 (−2)	5.356 (−2)	5.3984 (−2)	5.4284 (−2)	5.4498 (−2)	5.4651 (−2)	5.4759 (−2)	5.4836 (−2)	5.4891 (−2)	5.4929 (−2)	−2.9
−2.8	5.60 (−2)	5.732 (−2)	5.831 (−2)	5.904 (−2)	5.9545 (−2)	5.9910 (−2)	6.0170 (−2)	6.0356 (−2)	6.0488 (−2)	6.0582 (−2)	6.0649 (−2)	6.0695 (−2)	−2.8
−2.7	6.13 (−2)	6.297 (−2)	6.417 (−2)	6.504 (−2)	6.5665 (−2)	6.6109 (−2)	6.6425 (−2)	6.6652 (−2)	6.6813 (−2)	6.6927 (−2)	6.7009 (−2)	6.7066 (−2)	−2.7
−2.6	6.71 (−2)	6.914 (−2)	7.059 (−2)	7.164 (−2)	7.2398 (−2)	7.2938 (−2)	7.3323 (−2)	7.3599 (−2)	7.3795 (−2)	7.3934 (−2)	7.4033 (−2)	7.4103 (−2)	−2.6
−2.5	7.35 (−2)	7.586 (−2)	7.762 (−2)	7.889 (−2)	7.9804 (−2)	8.0459 (−2)	8.0927 (−2)	8.1263 (−2)	8.1501 (−2)	8.1671 (−2)	8.1791 (−2)	8.1877 (−2)	−2.5
−2.4	8.02 (−2)	8.317 (−2)	8.529 (−2)	8.684 (−2)	8.7944 (−2)	8.8740 (−2)	8.9309 (−2)	8.9716 (−2)	9.0006 (−2)	9.0213 (−2)	9.0360 (−2)	9.0464 (−2)	−2.4
−2.3	8.76 (−2)	9.112 (−2)	9.369 (−2)	9.555 (−2)	9.6887 (−2)	9.7852 (−2)	9.8544 (−2)	9.9038 (−2)	9.9391 (−2)	9.9643 (−2)	9.9822 (−2)	9.9949 (−2)	−2.3
−2.2	9.55 (−2)	9.975 (−2)	1.0284 (−1)	1.051 (−1)	1.0671 (−1)	1.0788 (−1)	1.0872 (−1)	1.0932 (−1)	1.0975 (−1)	1.1005 (−1)	1.1027 (−1)	1.1042 (−1)	−2.2
−2.1	1.040 (−1)	1.091 (−1)	1.1280 (−1)	1.155 (−1)	1.1748 (−1)	1.1890 (−1)	1.1992 (−1)	1.2065 (−1)	1.2117 (−1)	1.2154 (−1)	1.2181 (−1)	1.2200 (−1)	−2.1
−2.0	1.132 (−1)	1.192 (−1)	1.2366 (−1)	1.269 (−1)	1.2930 (−1)	1.3101 (−1)	1.3225 (−1)	1.3313 (−1)	1.3377 (−1)	1.3422 (−1)	1.3454 (−1)	1.3477 (−1)	−2.0
−1.9	1.229 (−1)	1.301 (−1)	1.3546 (−1)	1.394 (−1)	1.4225 (−1)	1.4432 (−1)	1.4581 (−1)	1.4689 (−1)	1.4766 (−1)	1.4821 (−1)	1.4860 (−1)	1.4888 (−1)	−1.9
−1.8	1.331 (−1)	1.419 (−1)	1.4826 (−1)	1.530 (−1)	1.5642 (−1)	1.5893 (−1)	1.6074 (−1)	1.6204 (−1)	1.6297 (−1)	1.6364 (−1)	1.6412 (−1)	1.6446 (−1)	−1.8
−1.7	1.442 (−1)	1.545 (−1)	1.6213 (−1)	1.678 (−1)	1.7193 (−1)	1.7496 (−1)	1.7714 (−1)	1.7872 (−1)	1.7986 (−1)	1.8067 (−1)	1.8125 (−1)	1.8166 (−1)	−1.7
−1.6	1.558 (−1)	1.680 (−1)	1.7712 (−1)	1.839 (−1)	1.8889 (−1)	1.9253 (−1)	1.9517 (−1)	1.9708 (−1)	1.9846 (−1)	1.9944 (−1)	2.0015 (−1)	2.0066 (−1)	−1.6
−1.5	1.680 (−1)	1.824 (−1)	1.9330 (−1)	2.014 (−1)	2.0740 (−1)	2.1178 (−1)	2.1497 (−1)	2.1728 (−1)	2.1895 (−1)	2.2015 (−1)	2.2099 (−1)	2.2162 (−1)	−1.5
−1.4	1.808 (−1)	1.978 (−1)	2.1074 (−1)	2.204 (−1)	2.2759 (−1)	2.3286 (−1)	2.3671 (−1)	2.3950 (−1)	2.4152 (−1)	2.4297 (−1)	2.4401 (−1)	2.4476 (−1)	−1.4
−1.3	1.941 (−1)	2.142 (−1)	2.2948 (−1)	2.410 (−1)	2.4959 (−1)	2.5592 (−1)	2.6055 (−1)	2.6392 (−1)	2.6636 (−1)	2.6812 (−1)	2.6938 (−1)	2.7029 (−1)	−1.3
−1.2	2.080 (−1)	2.315 (−1)	2.4958 (−1)	2.633 (−1)	2.7353 (−1)	2.8112 (−1)	2.8669 (−1)	2.9075 (−1)	2.9370 (−1)	2.9583 (−1)	2.9736 (−1)	2.9846 (−1)	−1.2
−1.1	2.222 (−1)	2.497 (−1)	2.7108 (−1)	2.873 (−1)	2.9955 (−1)	3.0863 (−1)	3.1533 (−1)	3.2022 (−1)	3.2378 (−1)	3.2636 (−1)	3.2822 (−1)	3.2955 (−1)	−1.1
−1.0	2.367 (−1)	2.689 (−1)	2.9402 (−1)	3.133 (−1)	3.2780 (−1)	3.3865 (−1)	3.4667 (−1)	3.5256 (−1)	3.5686 (−1)	3.5997 (−1)	3.6222 (−1)	3.6384 (−1)	−1.0
−0.9	2.517 (−1)	2.891 (−1)	3.1845 (−1)	3.412 (−1)	3.5841 (−1)	3.7135 (−1)	3.8096 (−1)	3.8804 (−1)	3.9321 (−1)	3.9698 (−1)	3.9970 (−1)	4.0166 (−1)	−0.9
−0.8	2.667 (−1)	3.100 (−1)	3.4438 (−1)	3.711 (−1)	3.9154 (−1)	4.0695 (−1)	4.1844 (−1)	4.2693 (−1)	4.3316 (−1)	4.3770 (−1)	4.4098 (−1)	4.4336 (−1)	−0.8
−0.7	2.820 (−1)	3.318 (−1)	3.7181 (−1)	4.032 (−1)	4.2733 (−1)	4.4564 (−1)	4.5936 (−1)	4.6953 (−1)	4.7702 (−1)	4.8249 (−1)	4.8646 (−1)	4.8933 (−1)	−0.7
−0.6	2.971 (−1)	3.543 (−1)	4.0077 (−1)	4.375 (−1)	4.6595 (−1)	4.8766 (−1)	5.0400 (−1)	5.1617 (−1)	5.2515 (−1)	5.3174 (−1)	5.3653 (−1)	5.4000 (−1)	−0.6
−0.5	3.121 (−1)	3.775 (−1)	4.3123 (−1)	4.741 (−1)	5.0754 (−1)	5.3322 (−1)	5.5265 (−1)	5.6718 (−1)	5.7795 (−1)	5.8587 (−1)	5.9164 (−1)	5.9584 (−1)	−0.5
−0.4	3.268 (−1)	4.013 (−1)	4.6318 (−1)	5.130 (−1)	5.5224 (−1)	5.8255 (−1)	6.0561 (−1)	6.2294 (−1)	6.3583 (−1)	6.4533 (−1)	6.5229 (−1)	6.5736 (−1)	−0.4
−0.3	3.410 (−1)	4.256 (−1)	4.9657 (−1)	5.544 (−1)	6.0022 (−1)	6.3590 (−1)	6.6321 (−1)	6.8382 (−1)	6.9923 (−1)	7.1063 (−1)	7.1899 (−1)	7.2510 (−1)	−0.3

−0.2	3.548 (−1)	4.502 (−1)	5.3137 (−1)	5.981 (−1)	6.5161 (−1)	6.9350 (−1)	7.2577 (−1)	7.5026 (−1)	7.6863 (−1)	7.8228 (−1)	7.9234 (−1)	7.9969 (−1)	−0.2
−0.1	3.677 (−1)	4.750 (−1)	5.6750 (−1)	6.444 (−1)	7.0654 (−1)	7.5561 (−1)	7.9365 (−1)	8.2267 (−1)	8.4455 (−1)	8.6088 (−1)	8.7294 (−1)	8.8179 (−1)	−0.1
0.0	3.800 (−1)	5.000 (−1)	6.0490 (−1)	6.932 (−1)	7.6515 (−1)	8.2247 (−1)	8.6720 (−1)	9.0154 (−1)	9.2755 (−1)	9.4703 (−1)	9.6148 (−1)	9.7212 (−1)	0.0
0.1	3.915 (−1)	5.250 (−1)	6.4348 (−1)	7.444 (−1)	8.2756 (−1)	8.9430 (−1)	9.4680 (−1)	9.8730 (−1)	1.0182 (0)	1.0414 (0)	1.0587 (0)	1.0715 (0)	0.1
0.2	4.019 (−1)	5.498 (−1)	6.8317 (−1)	7.981 (−1)	8.9388 (−1)	9.7150 (−1)	1.0328 (0)	1.0806 (0)	1.1171 (0)	1.1448 (0)	1.1654 (0)	1.1807 (0)	0.2
0.3	4.114 (−1)	5.744 (−1)	7.2384 (−1)	8.544 (−1)	9.6422 (−1)	1.0541 (0)	1.1257 (0)	1.1818 (0)	1.2250 (0)	1.2578 (0)	1.2824 (0)	1.3008 (0)	0.3
0.4	4.196 (−1)	5.987 (−1)	7.6540 (−1)	9.130 (−1)	1.0387 (0)	1.1424 (0)	1.2258 (0)	1.2916 (0)	1.3425 (0)	1.3814 (0)	1.4107 (0)	1.4326 (0)	0.4
0.5	4.269 (−1)	6.225 (−1)	8.0774 (−1)	9.7410 (−1)	1.1173 (0)	1.2367 (0)	1.3336 (0)	1.4105 (0)	1.4704 (0)	1.5164 (0)	1.5513 (0)	1.5774 (0)	0.5
0.6	4.328 (−1)	6.457 (−1)	8.5074 (−1)	1.0375 (0)	1.2003 (0)	1.3373 (0)	1.4494 (0)	1.5391 (0)	1.6095 (0)	1.6638 (0)	1.7052 (0)	1.7363 (0)	0.6
0.7	4.378 (−1)	6.682 (−1)	8.9429 (−1)	1.1032 (0)	1.2875 (0)	1.4443 (0)	1.5738 (0)	1.6782 (0)	1.7606 (0)	1.8246 (0)	1.8736 (0)	1.9106 (0)	0.7
0.8	4.415 (−1)	6.900 (−1)	9.3826 (−1)	1.1711 (0)	1.3791 (0)	1.5580 (0)	1.7071 (0)	1.8282 (0)	1.9246 (0)	1.9998 (0)	2.0577 (0)	2.1017 (0)	0.8
0.9	4.441 (−1)	7.110 (−1)	9.8255 (−1)	1.2412 (0)	1.4752 (0)	1.6786 (0)	1.8497 (0)	1.9900 (0)	2.1023 (0)	2.1906 (0)	2.2589 (0)	2.3111 (0)	0.9
1.0	4.457 (−1)	7.311 (−1)	1.0271 (0)	1.3133 (0)	1.5756 (0)	1.8063 (0)	2.0023 (0)	2.1642 (0)	2.2948 (0)	2.3982 (0)	2.4787 (0)	2.5404 (0)	1.0
1.1	4.463 (−1)	7.503 (−1)	1.0717 (0)	1.3873 (0)	1.6806 (0)	1.9413 (0)	2.1650 (0)	2.3515 (0)	2.5031 (0)	2.6239 (0)	2.7184 (0)	2.7913 (0)	1.1
1.2	4.459 (−1)	7.685 (−1)	1.1163 (0)	1.4633 (0)	1.7900 (0)	2.0838 (0)	2.3385 (0)	2.5527 (0)	2.7282 (0)	2.8690 (0)	2.9799 (0)	3.0658 (0)	1.2
1.3	4.447 (−1)	7.858 (−1)	1.1608 (0)	1.5410 (0)	1.9038 (0)	2.2340 (0)	2.5232 (0)	2.7685 (0)	2.9712 (0)	3.1349 (0)	3.2647 (0)	3.3658 (0)	1.3
1.4	4.427 (−1)	8.022 (−1)	1.2052 (0)	1.6204 (0)	2.0221 (0)	2.3921 (0)	2.7194 (0)	2.9997 (0)	3.2332 (0)	3.4232 (0)	3.5747 (0)	3.6936 (0)	1.4
1.5	4.398 (−1)	8.176 (−1)	1.2493 (0)	1.7014 (0)	2.1449 (0)	2.5582 (0)	2.9278 (0)	3.2472 (0)	3.5155 (0)	3.7354 (0)	3.9120 (0)	4.0513 (0)	1.5
1.6	4.365 (−1)	8.320 (−1)	1.2931 (0)	1.7839 (0)	2.2720 (0)	2.7324 (0)	3.1486 (0)	3.5116 (0)	3.8192 (0)	4.0732 (0)	4.2786 (0)	4.4415 (0)	1.6
1.7	4.325 (−1)	8.455 (−1)	1.3366 (0)	1.8678 (0)	2.4035 (0)	2.9150 (0)	3.3823 (0)	3.7939 (0)	4.1456 (0)	4.4383 (0)	4.6766 (0)	4.8668 (0)	1.7
1.8	4.281 (−1)	8.582 (−1)	1.3796 (0)	1.9530 (0)	2.5393 (0)	3.1060 (0)	3.6294 (0)	4.0949 (0)	4.4961 (0)	4.8326 (0)	5.1085 (0)	5.3301 (0)	1.8
1.9	4.233 (−1)	8.699 (−1)	1.4222 (0)	2.0394 (0)	2.6794 (0)	3.3056 (0)	3.8903 (0)	4.4154 (0)	4.8719 (0)	5.2580 (0)	5.5767 (0)	5.8344 (0)	1.9
2.0	4.182 (−1)	8.808 (−1)	1.4643 (0)	2.1269 (0)	2.8237 (0)	3.5139 (0)	4.1654 (0)	4.7563 (0)	5.2746 (0)	5.7164 (0)	6.0838 (0)	6.3828 (0)	2.0
2.1	4.126 (−1)	8.909 (−1)	1.5058 (0)	2.2155 (0)	2.9722 (0)	3.7310 (0)	4.4552 (0)	5.1185 (0)	5.7055 (0)	6.2099 (0)	6.6325 (0)	6.9788 (0)	2.1
2.2	4.070 (−1)	9.002 (−1)	1.5468 (0)	2.3051 (0)	3.1249 (0)	3.9571 (0)	4.7600 (0)	5.5028 (0)	6.1662 (0)	6.7408 (0)	7.2258 (0)	7.6261 (0)	2.2
2.3	4.013 (−1)	9.089 (−1)	1.5872 (0)	2.3956 (0)	3.2816 (0)	4.1921 (0)	5.0803 (0)	5.9102 (0)	6.6580 (0)	7.3113 (0)	7.8668 (0)	8.3283 (0)	2.3
2.4	3.954 (−1)	9.168 (−1)	1.6271 (0)	2.4868 (0)	3.4423 (0)	4.4362 (0)	5.4164 (0)	6.3416 (0)	7.1827 (0)	7.9237 (0)	8.5585 (0)	9.0897 (0)	2.4
2.5	3.893 (−1)	9.241 (−1)	1.6663 (0)	2.5789 (0)	3.6070 (0)	4.6895 (0)	5.7689 (0)	6.7978 (0)	7.7419 (0)	8.5804 (0)	9.3044 (0)	9.9145 (0)	2.5
2.6	3.833 (−1)	9.309 (−1)	1.7049 (0)	2.6716 (0)	3.7755 (0)	4.9520 (0)	6.1380 (0)	7.2798 (0)	8.3371 (0)	9.2841 (0)	1.0108 (+1)	1.0807 (+1)	2.6
2.7	3.772 (−1)	9.370 (−1)	1.7430 (0)	2.7650 (0)	3.9480 (0)	5.2238 (0)	6.5241 (0)	7.7885 (0)	8.9700 (0)	1.0037 (+1)	1.0973 (+1)	1.1773 (+1)	2.7
2.8	3.712 (−1)	9.427 (−1)	1.7804 (0)	2.8590 (0)	4.1241 (0)	5.5050 (0)	6.9277 (0)	8.3249 (0)	9.6425 (0)	1.0843 (+1)	1.1903 (+1)	1.2817 (+1)	2.8
2.9	3.654 (−1)	9.478 (−1)	1.8172 (0)	2.9536 (0)	4.3040 (0)	5.7957 (0)	7.3491 (0)	8.8898 (0)	1.0356 (+1)	1.1703 (+1)	1.2903 (+1)	1.3943 (+1)	2.9
3.0	3.595 (−1)	9.526 (−1)	1.8535 (0)	3.0486 (0)	4.4876 (0)	6.0958 (0)	7.7886 (0)	9.4843 (0)	1.1113 (+1)	1.2622 (+1)	1.3976 (+1)	1.5159 (+1)	3.0
3.1	3.537 (−1)	9.569 (−1)	1.8891 (0)	3.1441 (0)	4.6747 (0)	6.4054 (0)	8.2467 (0)	1.0109 (+1)	1.1915 (+1)	1.3601 (+1)	1.5127 (+1)	1.6470 (+1)	3.1
3.2	3.481 (−1)	9.608 (−1)	1.9242 (0)	3.2400 (0)	4.8653 (0)	6.7246 (0)	8.7237 (0)	1.0766 (+1)	1.2763 (+1)	1.4645 (+1)	1.6360 (+1)	1.7882 (+1)	3.2
3.3	3.425 (−1)	9.644 (−1)	1.9588 (0)	3.3363 (0)	5.0595 (0)	7.0534 (0)	9.2199 (0)	1.1455 (+1)	1.3660 (+1)	1.5755 (+1)	1.7681 (+1)	1.9401 (+1)	3.3
3.4	3.370 (−1)	9.677 (−1)	1.9927 (0)	3.4328 (0)	5.2571 (0)	7.3918 (0)	9.7357 (0)	1.2177 (+1)	1.4608 (+1)	1.6937 (+1)	1.9094 (+1)	2.1035 (+1)	3.4
3.5	3.319 (−1)	9.707 (−1)	2.0262 (0)	3.5298 (0)	5.4580 (0)	7.7400 (0)	1.0271 (+1)	1.2933 (+1)	1.5608 (+1)	1.8192 (+1)	2.0605 (+1)	2.2791 (+1)	3.5
3.6	3.267 (−1)	9.734 (−1)	2.0591 (0)	3.6270 (0)	5.6623 (0)	8.0978 (0)	1.0827 (+1)	1.3725 (+1)	1.6662 (+1)	1.9524 (+1)	2.2218 (+1)	2.4676 (+1)	3.6

(Continued)

η	$F_{-3/2}$	F_{-1}	$F_{-1/2}$	F_0	$F_{1/2}$	F_1	$F_{3/2}$	F_2	$F_{5/2}$	F_3	$F_{7/2}$	F_4	η
3.7	3.216 (−1)	9.759 (−1)	2.0915 (0)	3.7244 (0)	5.8699 (0)	8.4654 (0)	1.1404 (+1)	1.4553 (+1)	1.7774 (+1)	2.0938 (+1)	2.3939 (+1)	2.6698 (+1)	3.7
3.8	3.167 (−1)	9.781 (−1)	2.1235 (0)	3.8221 (0)	6.0806 (0)	8.8427 (0)	1.2001 (+1)	1.5418 (+1)	1.8944 (+1)	2.2436 (+1)	2.5774 (+1)	2.8866 (+1)	3.8
3.9	3.120 (−1)	9.802 (−1)	2.1549 (0)	3.9200 (0)	6.2945 (0)	9.2298 (0)	1.2620 (+1)	1.6322 (+1)	2.0175 (+1)	2.4023 (+1)	2.7730 (+1)	3.1188 (+1)	3.9
4.0	3.08 (−1)	9.820 (−1)	2.1859 (0)	4.0181 (0)	6.5115 (0)	9.6267 (0)	1.3260 (+1)	1.7265 (+1)	2.1469 (+1)	2.5702 (+1)	2.9812 (+1)	3.3674 (+1)	4.0
4.2	2.99 (−1)	9.852 (−1)	2.2465 (0)	4.2149 (0)	6.9548 (0)	1.0450 (+1)	1.4607 (+1)	1.9272 (+1)	2.4254 (+1)	2.9353 (+1)	3.4379 (+1)	3.9173 (+1)	4.2
4.4	2.90 (−1)	9.879 (−1)	2.3054 (0)	4.4122 (0)	7.4100 (0)	1.1313 (+1)	1.6043 (+1)	2.1147 (+1)	2.7317 (+1)	3.3422 (+1)	3.9532 (+1)	4.5443 (+1)	4.4
4.6	2.82 (−1)	9.901 (−1)	2.3627 (0)	4.6100 (0)	7.8769 (0)	1.2215 (+1)	1.7572 (+1)	2.3799 (+1)	3.0677 (+1)	3.7944 (+1)	4.5326 (+1)	5.2572 (+1)	4.6
4.8	2.75 (−1)	9.918 (−1)	2.4186 (0)	4.8082 (0)	8.3550 (0)	1.3157 (+1)	1.9195 (+1)	2.6336 (+1)	3.4352 (+1)	4.2954 (+1)	5.1824 (+1)	6.0653 (+1)	4.8
5.0	2.69 (−1)	9.933 (−1)	2.4730 (0)	5.0067 (0)	8.8442 (0)	1.4138 (+1)	2.0914 (+1)	2.9065 (+1)	3.8361 (+1)	4.8491 (+1)	5.9089 (+1)	6.9788 (+1)	5.0
5.2	2.62 (−1)	9.945 (−1)	2.5261 (0)	5.2055 (0)	9.3441 (0)	1.5159 (+1)	2.2733 (+1)	3.1994 (+1)	4.2725 (+1)	5.4593 (+1)	6.7192 (+1)	8.0087 (+1)	5.2
5.4	2.56 (−1)	9.955 (−1)	2.5780 (0)	5.4045 (0)	9.8546 (0)	1.6220 (+1)	2.4653 (+1)	3.5131 (+1)	4.7462 (+1)	6.1302 (+1)	7.6204 (+1)	9.1666 (+1)	5.4
5.6	2.51 (−1)	9.963 (−1)	2.6288 (0)	5.6037 (0)	1.0375 (+1)	1.7321 (+1)	2.6676 (+1)	3.8485 (+1)	5.2593 (+1)	6.8660 (+1)	8.6203 (+1)	1.0465 (+2)	5.6
5.8	2.46 (−1)	9.970 (−1)	2.6784 (0)	5.8030 (0)	1.0906 (+1)	1.8462 (+1)	2.8804 (+1)	4.2062 (+1)	5.8139 (+1)	7.6711 (+1)	9.7268 (+1)	1.1918 (+2)	5.8
6.0	2.40 (−1)	9.975 (−1)	2.7272 (0)	6.0025 (0)	1.1447 (+1)	1.9643 (+1)	3.1039 (+1)	4.5872 (+1)	6.4121 (+1)	8.5500 (+1)	1.0949 (+2)	1.3539 (+2)	6.0
6.2	2.36 (−1)	9.980 (−1)	2.7748 (0)	6.2020 (0)	1.1997 (+1)	2.0863 (+1)	3.3383 (+1)	4.9922 (+1)	7.0561 (+1)	9.5076 (+1)	1.2295 (+2)	1.5343 (+2)	6.2
6.4	2.31 (−1)	9.983 (−1)	2.8216 (0)	6.4017 (0)	1.2556 (+1)	2.2123 (+1)	3.5838 (+1)	5.4220 (+1)	7.7482 (+1)	1.0549 (+2)	1.3774 (+2)	1.7347 (+2)	6.4
6.6	2.27 (−1)	9.986 (−1)	2.8677 (0)	6.6014 (0)	1.3125 (+1)	2.3424 (+1)	3.8406 (+1)	5.8774 (+1)	8.4904 (+1)	1.1678 (+2)	1.5397 (+2)	1.9568 (+2)	6.6
6.8	2.23 (−1)	9.989 (−1)	2.9128 (0)	6.8011 (0)	1.3703 (+1)	2.4764 (+1)	4.1089 (+1)	6.3592 (+1)	9.2852 (+1)	1.2901 (+2)	1.7174 (+2)	2.2025 (+2)	6.8
7.0	2.20 (−1)	9.991 (−1)	2.9573 (0)	7.0009 (0)	1.4290 (+1)	2.6144 (+1)	4.3888 (+1)	6.8682 (+1)	1.0135 (+2)	1.4224 (+2)	1.9115 (+2)	2.4735 (+2)	7.0
7.2	2.16 (−1)	9.993 (−1)	3.0009 (0)	7.2008 (0)	1.4886 (+1)	2.7564 (+1)	4.6806 (+1)	7.4052 (+1)	1.1041 (+2)	1.5650 (+2)	2.1232 (+2)	2.7721 (+2)	7.2
7.4	2.13 (−1)	9.994 (−1)	3.0439 (0)	7.4006 (0)	1.5491 (+1)	2.9024 (+1)	4.9843 (+1)	7.9711 (+1)	1.2008 (+2)	1.7188 (+2)	2.3536 (+2)	3.1003 (+2)	7.4
7.6	2.10 (−1)	9.995 (−1)	3.0862 (0)	7.6005 (0)	1.6104 (+1)	3.0524 (+1)	5.3003 (+1)	8.5665 (+1)	1.3036 (+2)	1.8841 (+2)	2.6039 (+2)	3.4604 (+2)	7.6
7.8	2.07 (−1)	9.996 (−1)	3.1280 (0)	7.8004 (0)	1.6725 (+1)	3.2065 (+1)	5.6286 (+1)	9.1923 (+1)	1.4129 (+2)	2.0616 (+2)	2.8754 (+2)	3.8547 (+2)	7.8
8.0	2.04 (−1)	9.997 (−1)	3.1691 (0)	8.0003 (0)	1.7355 (+1)	3.3645 (+1)	5.9693 (+1)	9.8493 (+1)	1.5288 (+2)	2.2520 (+2)	3.1695 (+2)	4.2859 (+2)	8.0
8.2	2.01 (−1)	9.997 (−1)	3.2097 (0)	8.2003 (0)	1.7993 (+1)	3.5265 (+1)	6.3228 (+1)	1.0538 (+2)	1.6517 (+2)	2.4558 (+2)	3.4874 (+2)	4.7564 (+2)	8.2
8.4	1.99 (−1)	9.998 (−1)	3.2497 (0)	8.4002 (0)	1.8639 (+1)	3.6925 (+1)	6.6891 (+1)	1.1260 (+2)	1.7818 (+2)	2.6737 (+2)	3.8307 (+2)	5.2691 (+2)	8.4
8.6	1.96 (−1)	9.998 (−1)	3.2892 (0)	8.6002 (0)	1.9293 (+1)	3.8625 (+1)	7.0684 (+1)	1.2016 (+2)	1.9194 (+2)	2.9064 (+2)	4.2007 (+2)	5.8269 (+2)	8.6
8.8	1.94 (−1)	9.999 (−1)	3.3282 (0)	8.8002 (0)	1.9954 (+1)	4.0365 (+1)	7.4609 (+1)	1.2805 (+2)	2.0646 (+2)	3.1546 (+2)	4.5989 (+2)	6.4327 (+2)	8.8
9.0	1.91 (−1)	9.999 (−1)	3.3667 (0)	9.0001 (0)	2.0624 (+1)	4.2145 (+1)	7.8666 (+1)	1.3631 (+2)	2.2179 (+2)	3.4189 (+2)	5.0271 (+2)	7.0898 (+2)	9.0
9.2	1.89 (−1)	9.999 (−1)	3.4048 (0)	9.2001 (0)	2.1301 (+1)	4.3965 (+1)	8.2859 (+1)	1.4492 (+2)	2.3794 (+2)	3.7001 (+2)	5.4867 (+2)	7.8014 (+2)	9.2
9.4	1.87 (−1)	9.999 (−1)	3.4425 (0)	9.4001 (0)	2.1986 (+1)	4.5825 (+1)	8.7187 (+1)	1.5389 (+2)	2.5494 (+2)	3.9988 (+2)	5.9794 (+2)	8.5710 (+2)	9.4
9.6	1.85 (−1)	9.999 (−1)	3.4796 (0)	9.6001 (0)	2.2678 (+1)	4.7725 (+1)	9.1654 (+1)	1.6325 (+2)	2.7282 (+2)	4.3159 (+2)	6.5070 (+2)	9.4022 (+2)	9.6

| 9.8 | 1.83 (−1) | 9.999 (−1) | 3.5164 (0) | 9.8001 (0) | 2.3378 (+1) | 4.9665 (+1) | 9.6259 (+1) | 1.7299 (+2) | 2.9161 (+2) | 4.6520 (+2) | 7.0713 (+2) | 1.0299 (+3) | 9.8 |
| 10.0 | 1.81 (−1) | 9.999 (−1) | 3.5528 (0) | 1.0000 (+1) | 2.4085 (+1) | 5.1645 (+1) | 1.0101 (+2) | 1.8312 (+2) | 3.1134 (+2) | 5.0081 (+2) | 7.6741 (+2) | 1.1264 (+3) | 10.0 |

Source: After J. S. Blakemore, *Semiconductor Statistics* (New York: Pergamon Press, 1962).

The Fermi-Dirac integral of order j is given by

$$F_j(\eta) = \frac{1}{j!} \int_0^\infty \frac{x^j \, dx}{1 + \exp(x - \eta)}$$ (5.59)

where the relationship among integrals is

$$\frac{d}{d\eta} F_j(\eta) = F_{j-1}(\eta)$$ (5.83)

The normalization constant, $j!$, is related to the gamma function by,

$$j! = \Gamma(j + 1) = j\Gamma(j)$$ (B.1)

If j is an odd multiple of $\frac{1}{2}$,

$$j! = \Gamma(n + \tfrac{1}{2}) = 1 \cdot 3 \cdot 5 \cdot \ldots \cdot (2n - 3)(2n - 1) \frac{\sqrt{\pi}}{2^n}$$ (B.2)

where n is an integer greater than zero. Also,

$$\Gamma(\tfrac{1}{2}) = \sqrt{\pi}$$ (B.3)

Useful approximations for the Fermi-Dirac integral of order $\frac{1}{2}$ are

$$F_{1/2}(\eta) \simeq e^\eta (1 + 0.27 e^\eta)^{-1}$$ (B.4)

for $\eta \geq 1.3$, and

$$F_{1/2}(\eta) \simeq \frac{4}{3\sqrt{\pi}} \eta^{3/2} (1 + 1.15/\eta^2)$$ (B.5)

for $\eta \leq 1$. Also, for $\eta \gg 1$,

$$\Gamma(j + 1)F_j(\eta) = \frac{\eta^{j+1}}{j + 1} + \frac{\pi^2}{6} j\eta^{j-1}$$ (B.6)

is a very good approximation.

Index